Springer-Verlag Berlin Heidelberg GmbH

Harro Träubel

New Materials Permeable to Water Vapor

With 106 Figures

 Springer

Dr. Harro Träubel

Dresdener Straße 14
D-51373 Leverkusen
Germany

Guest Professor
at Jiang Su Institute for Chemical Technology Nanjing
P.R. of China

Library of Congress Cataloging-in Publication Data

Träubel, Harro,
    New materials permeable to water vapor / Harro Träubel.
        p.   cm.
    Includes bibliographical references and index.

    ISBN 978-3-642-64206-7   ISBN 978-3-642-59978-1 (eBook)
    DOI 10.1007/978-3-642-59978-1

    1. Porous materials – Permeability. 2. Water vapor transport.
    3. Fibrous materials – Permeability. 4. Synthetic products – Permeability. I. Title.
TA418.9.P6T75 1999   620.1'16 – dc21

                                                                    99-17767 CIP

© Springer-Verlag Berlin Heidelberg 1999
Originally published by Springer-Verlag Berlin Heidelberg New York in 1999
Softcover reprint of the hardcover 1st edition 1999

The use of general descriptive names, registered names, trademarks, etc. in this publication does not imply, even in the absence of a specific statement, that such names are exempt from the relevant protective laws and regulations and therefore free for general use.

Cover design: de'blik, Berlin
Production: ProduServ GmbH Verlagsservice, Berlin
Typesetting: Fotosatz-Service Köhler GmbH, Würzburg

SPIN: 10634869   02/3020 - 5 4 3 2 1 0 - Printed on acid-free paper

Preface

During the past 40 years many patents and articles have been published which describe methods and materials for water vapor permeable materials. These materials were primarily designed to substitute leather in its look and performance. Later other industrial applications were found where microporosity or water vapor permeability could be used.

The aim of this book is to give a survey of this matter, describing in an abridged way all the publications existing up to 1996.

The terms "water vapor permeable material", "artificial leather", "synthetic leather", "leather substitute" or "man-made leather" are defined, and the special characteristics of leather are compared with its substitutes.

Then the special methods used to produce microporous and hydrophilic materials, suitable substrates, end uses in fields other than wearing purposes, testing methods, as well as patent strategies and ecological behavior are discussed.

Each chapter starts with general remarks about the the specific characteristics of its motto; then published examples belonging to the subjects of the chapter are described.

Sources of most of the literature have been the "Textilbericht" and the "Hochmolekularbericht" of Bayer AG – today no longer in existence. Additionally, Chemical Abstracts and original publications in the form of patents and articles in books and journals are used. During the past 40 years many patents and articles have been published which describe methods and materials for water vapor permeable materials. These materials were primarily designed to substitute leather in its look and performance.

One positive point of this book seems to be the literature published before 1980, because in commercial electronic data systems these publications have mostly up to now not been available.

I would like to express my thanks to the following people. W. Held, Dr. J. Pedain, Dr. M. Rolf and Dr. P Suchanek for granting me the permission to have this work published and to Dr. Holm and D. I. Reinfand for preparing the microphotos – even if it was more than 20 years ago. My thanks are also expressed to Dr. B. Zorn who helped to write the first draft and assisted in the organization of the material in a logical form.

My special thanks are expressed to Mrs. Käfinger for her patience in her attempts to improve my English and to eliminate my German language-based expressions.

May 1999   Harro Träubel

Table of Contents

Part 3
Porosity by Other Means

Part 4
Treatment of Man-Made Leather

Part 5
Chemistry, Testing Methods, Other Industrial Applications, Ecology

Part 6
Trade Names, Marketing History, Summary of Patent Applications

Part 7
Summary of Patent Applications and Practical Examples

Subject Index

Abbrevations

ABR	acrylonitrile-styrene polymer
BOD	Biological oxygen demand (Degradable organic substances by the action of micro organisms e. g. after a period of 5 d = BOD_5)
cip	Continuation in part (If the claims of a patent are not granted in their original manner, the patent may be reissued in parts)
CMC	Carboxymethyl cellulose
COD	Chemical oxygen demand (Consumption of organic material by oxidation with potassium dichromate)
DABCO	Diazabicyclooctane (A catalyst for isocyanate-hydroxyl reactions)
div.	Division (If the claims in a patent are to broad, do not correspond to the examples or are not homogeneous, parts of the patent can be separated and newly issued. This new issue is named patent application no. xx division of patent application yy which can be abandoned or reissued with claims being smaller than in the original application.)
DMAc	Dimethylacetamide (Solvent)
DMF	Dimethylformamide (Solvent)
DMSO	Dimethyl sulfoxide (Solvent)
FC	Fluorocarbon (Water- and stain-resistant agents)
H12MDA	4,4'-Diaminodicyclohexylmethane
H12MDA	4,4'-Diaminodicyclohexylmethane (MDA perhydrated)
H12MDI	4,4'-Diiisocyanatodicyclohexylmethane
H12MDI	Dicyclohexylmethane-4,4'-diisocyanate (MDI perhydrated)
HDA	Hexamethylenediamine
HDI	Hexamethylenediisocyanate
HDPE	High density polyethylene
HF	High frequency
IPDA	Isophoron diamine
IPDI	Isophoron diisocyanate
LDPE	Low density polyethylene
MDA	4,4'-Diaminodiphenylmethane
MDI	4,4'-Diisocyanatodiphenylmethane
MEK	Methyl ethyl ketone (Solvent)
MSA	Maleic acid–styrene polymer
PE	Polyether
PES	Polyester

PTFE   Polytetrafluoroethylene
PUR   Polyurethane
PVA   Polyvinyl alcohol
PVC   Polyvinyl chloride
SBR   Styrene-butadiene polymer
SEM   Scanning electronic micrograph
TDA   Toluenediamine
TDI   Toluene diisocyanate
THF   Tetrahydrofurane (Solvent)
WDD   Wasserdampfdurchlässigkeit (Water vapor permeability)

Part 1

Leather and Artificial Leather

After launching a synthetic substitute for silk and other natural products in the 1930s, chemists also wanted to develop a synthetic material able to substitute leather. This was a difficult task, but finally in the 1950s the chemical company DuPont developed CORFAM®, a substitute for leather.

CORFAM® was microporous, i.e. water vapor permeable, and was the first material which not only was optically similar to leather, like PVC or nitrocellulose coated textiles, but was also able to transport perspiration from inside to outside as long as a shoe consisted entirely of it.

Due to its special performance, leather and its properties will be discussed in the first chapters.

CHAPTER 1

Leather

Even at rest, humans lose about 30 g of water an hour through their skin. Physical activity and sport can increase this to 1000 g an hour. We often wear garments to protect ourselves against wind, rain, heat and cold. The denser the fabric, the better it protects us against the elements. However, such fabrics tend to be uncomfortable because the water vapor released by the skin as perspiration cannot escape quickly enough. Mountaineers, walkers and skiers know all about this problem. Coats and jackets made of PVC imitation leather provide excellent protection against wind and rain but are virtually synonymous with extremely poor wear comfort.

Textiles that have been given a water-repellent finish lose their water-repellent properties if they are exposed to high mechanical strain and, finally, when they are cleaned. Furthermore, fabrics that are really windproof are generally very heavy.

What options do we have if we wish to combine wear comfort with protection against the elements? The following ensure good wear comfort:

- Water-repellent fabrics
- Animal skins (leather and furs)
- Microporous garments
- Hydrophilic coatings

We can try to imitate nature by protecting ourselves with the skin of an animal or due to the fact that this skin will after a while decompose, with a modified version of an animal skin. In other words, we can wear *fur* or *leather* coats, clothes or jackets. These protect us against wind and rain and, provided the tanner did his job properly, they transport perspiration outward.

The drawback is that leather is often heavier than we would like. Moreover, it is not normally hard-wearing enough to be used in sportswear. Laundering and dry cleaning can be a problem for leather.

The next possibility is: We can give textiles a *microporous* coating, preferably using polymers with good physical properties. Such coatings allow water vapor to pass through but keep out wind and water.

However, there are also problems associated with such coatings: micropores weaken the polymer structure. Consequently, coatings are usually applied to the reverse (i.e. inner) side of fabrics, or the microporous membrane is sandwiched between two layers of textile material, a wind and waterproof outer coating and a lining. This is the only way of guaranteeing that the membrane is not destroyed by abrasion or rubbing during wear (see Chap. 27).

Another possibility is to produce a polymer that has good physical properties and is *hydrophilic*. The hydrophilic properties would ensure that the polymer can be applied as a homogenous film. Water vapor could migrate through this homogenous film by means of adsorption and desorption. Such polymers could be applied to a textile substrate as an outer coating and be used to manufacture thin and extremely lightweight garments. Unfortunately, it is not as easy as it sounds.

Hydrophilic polymers take up water, which acts as a plasticizer and weakens the resistance of the film to abrasion and rubbing. Worse, when drops of rain come into contact with the polymer, they cause localized swelling, producing unsightly blisters at the point of contact. These often remain once the rain has stopped and the water of the blisters has evaporated – a situation which could lead to complaints from consumers.

Due to the fact that leather is a kind of a model for the solutions chemistry could offer, the characteristic properties of leather are discussed first.

The starting materials for leather are animal skins and hides. Animal skins are byproducts of slaughter houses producing meat. Statistically 7 % of the weight of cattle consists of the hide. Cattle are only slaughtered for their meat – never for their hides [6].

Leather is regarded as the first material human beings produced by chemical methods. Paintings more than 20,000 years old show the use of leather (Fig. 1-1). Being a by product of meat production, hides and skins have limited availability. Hides and skins are both available in quantities depending on the worldwide demand for meat and they are sold and bought on a worldwide basis.

The price of hides and skins is established at auctions. Due to a worldwide marketing system the price level is also influenced by the exchange rate of the

Fig. 1-1. Painting of ice age humans: Stone-Age painting from the caves of Les Trois Frères by Montesquieu-Aventès (Ariège): A magician clothed in an animal skin which has presumably been tanned. The magician is playing a flute and charming animals. He would not have been able to get near these animals while he was wearing an untanned skin smelling of blood or mold. Middle Magdalenian period, approximately 20,000 years old

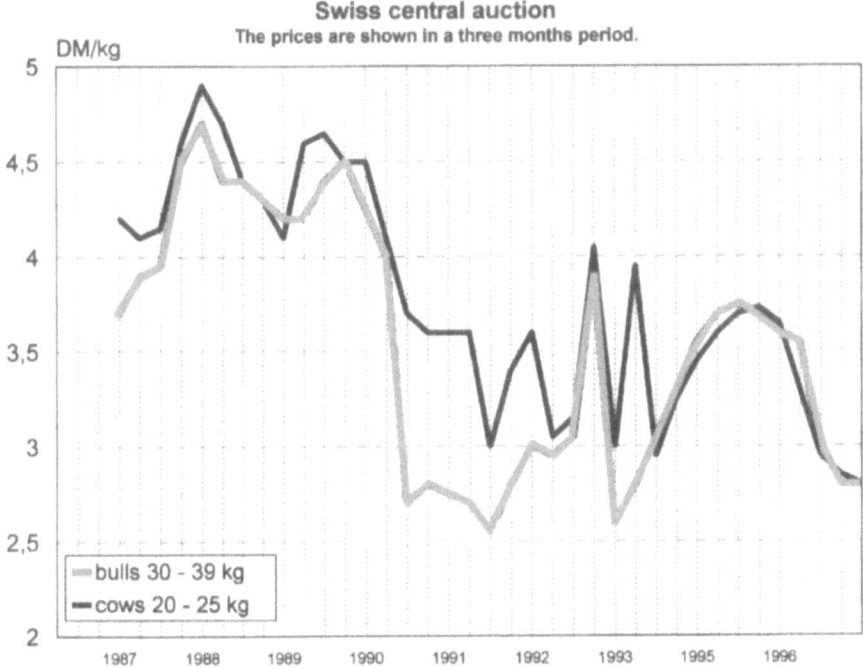

Fig. 1-2. Variation in hide prices (graph courtesy of B. Herrmann)

dollar against local currencies [2]. Their price level has an analogue movement to fluctuations of shares. Figure 1-2 shows the changes in hide prices over a ten-year period in DM.

The availability of leather differs from other natural materials such as wood, wool, natural rubber etc. These materials are not byproducts from other areas. They are produced by middle- or long-term breeding methods. Their quantity and quality is not determined by external factors.

The production, treatment after the killing of the animal and the preservation of the raw hides are decisive for the quality of the leather that can be made out of it.

The raw material for leather is produced in countries with large numbers of cattle. Countries with a large cattle production are USA, Brazil, Argentina, etc. Australia and New Zealand are big producers of sheep skins. China, India and many African countries produce large quantities of goat skins.

The availability of hides does not depend directly on the cattle living in a country. This availability depends on the killing rate of the animals, which depends on the meat production and consumption. Meat consumption is high in countries with a high standard of living like those in the USA and Western Europe. A survey of the production of hides and skins is given in Fig. 1-3.

The hides and skins are transformed into leather – most of the leather being used to produce shoes (Fig. 1-4).

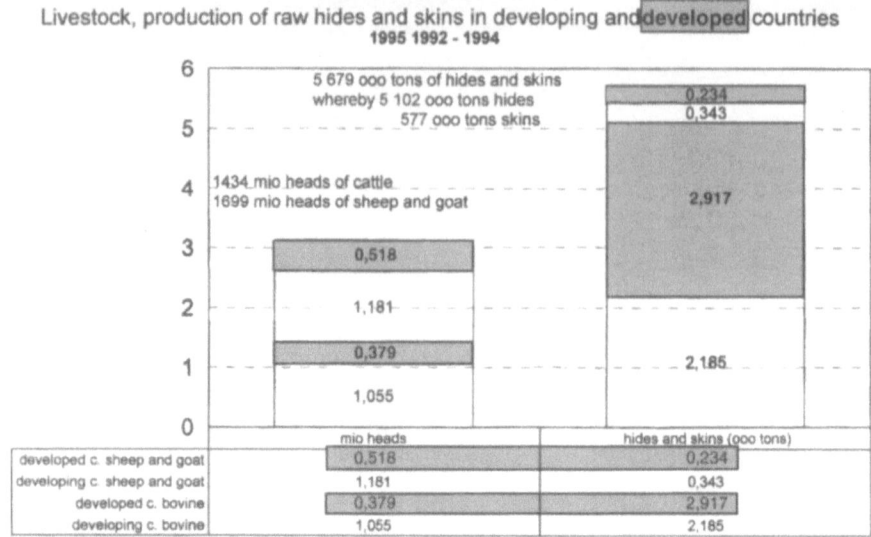

Livestock, production of raw hides and skins in developing and developed countries
1995 1992 - 1994

developed c. sheep and goat	mio heads	hides and skins (ooo tons)
developed c. sheep and goat	0,518	0,234
developing c. sheep and goat	1,181	0,343
developed c. bovine	0,379	2,917
developing c. bovine	1,055	2,185

FAO Yearbook 1996

Fig. 1-3. Livestock cattle and sheep – production of hides and skins. The numbers of cattle in developed countries is much lower than in developing countries. Due to a much higher slaughtering rate and weight of the cattle, the developed countries produce more hides than developing countries. The numbers of sheep and goats in developed countries is ca. half of the numbers the developing countries have. Skins from sheep and goats from developed countries contribute to 40 % of world production. These figures indicate more intensive breeding and slaughtering in developed countries

The worldwide trade in hides, skins and leathers is very often controlled or manipulated by government actions to protect a local leather and shoe industry [7 ,8]. These actions further influence the availability and the price of hides and leathers. A worldwide reduction in the supply of hides and skins in the 1950s also increased the efforts to produce substitutes for leather. Labor and environmental costs caused a shift in the leather and shoe production from industrial countries to developing ones (Fig. 1-5).

The production of shoes is extremely labor intensive. Starting in the 1970s the shoe industry shifted more and more towards countries with low labor costs. Therefore the leather industry moved to these countries too [9, 10] (Fig. 1-6). In addition, government action to improve the environment in industrial countries increased the costs of leather production [6]. Many tanneries either relocated or went bankrupt. Only a few tanneries with a high level of technology remained in countries like the Netherlands, Sweden, France, Switzerland and Germany.

Nevertheless, industrial countries contribute a major part to the production of hides because they have a higher consumption of shoes and leather goods and less and less production of leather and products made from it.

Goods made from leather have a touch of luxury – therefore only countries with a high average income also have a high consumption of shoes (Fig. 1-7).

	leather (Mio sq. ft.)	footwear (Mio pairs)
developed c. sheep and goat	1,455	
developing c. sheep and goat	2,442	
developed c. bovine	4,343	1,633
developing c. bovine	5,205	2,67

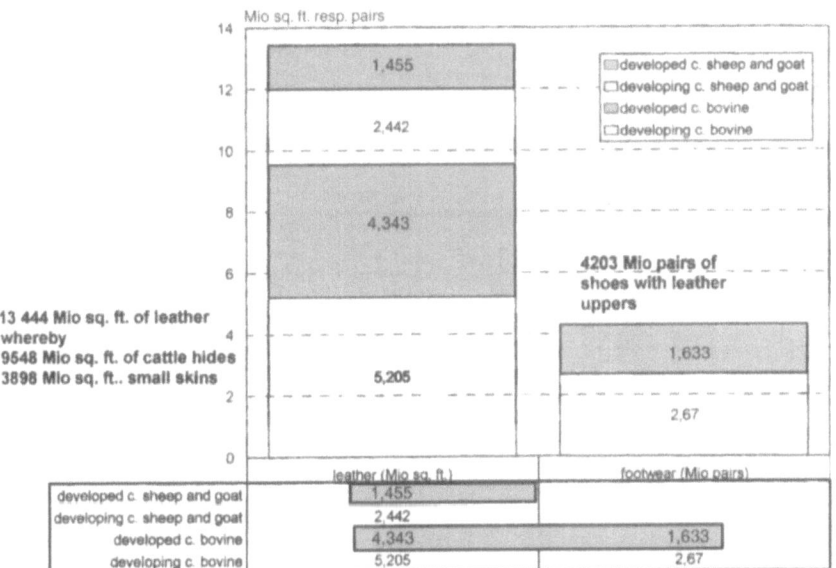

Fig. 1-4. Production of leather (cattle hides and skins of goat's and sheep) and shoes (figures for vegetable tanned leather not included – due to their minor importance) Developing countries in the nineties produced roughly half of world's production of leather and nearly two thirds of the world's shoe production

	1964	1974	1989/90	1994
India				1330
P R China	305	143	613	1423
TK	80	132	386	
PAK	120	174	333	
ROK	11	76	830	1449
JA	269	351	460	
E	209	456	586	
I	392	584	1723	1908
D	540	321	204	

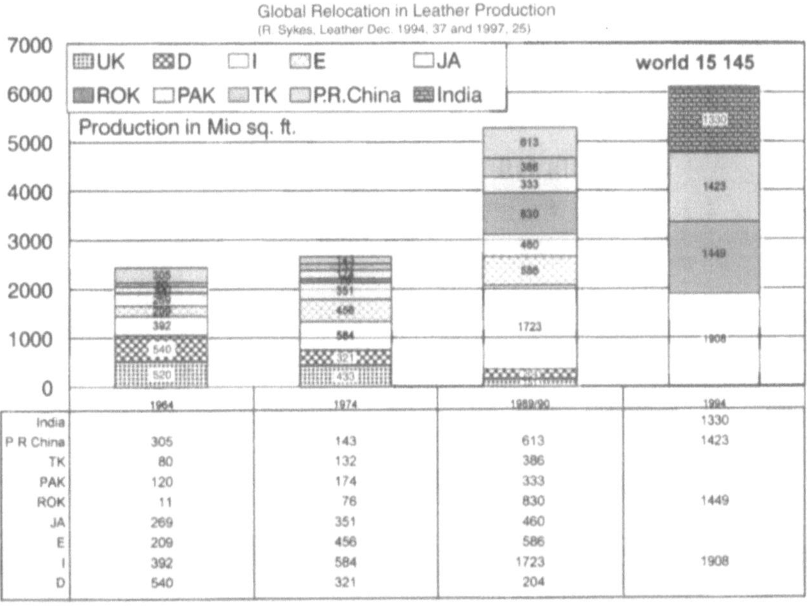

Fig. 1-5. Global relocation in leather production [1]. Apart from Italy, leather production in most developed countries over the last four decades has decreased. Korea (ROK), The People's Republic of China, and India e.g. were able to increase their production of leather remarkably

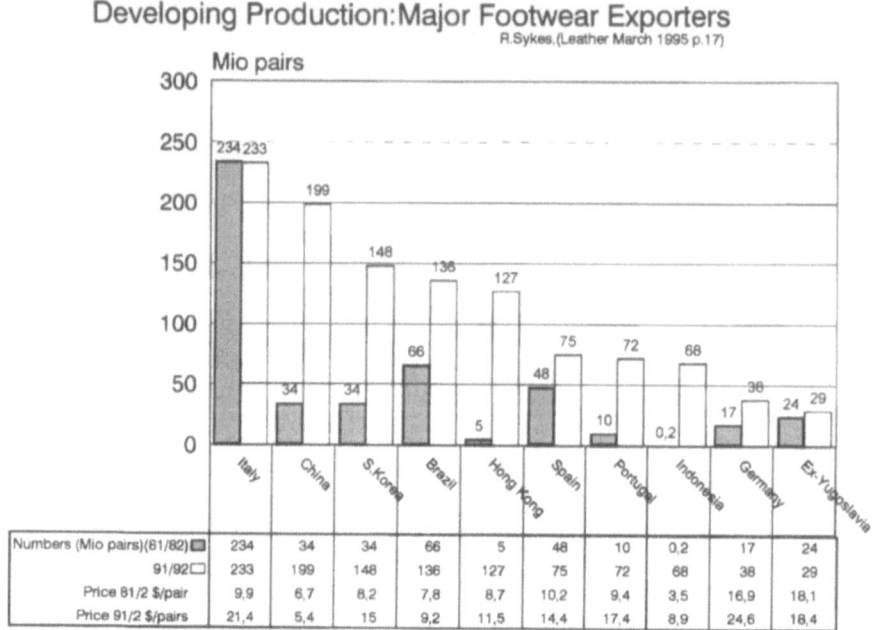

Fig. 1-6. Shoe production and price level of exported shoes of different countries – a comparison. The prices of shoes exported from Italy in the last 10 years have doubled. Italy was able to dominate the world's shoe fashion and maintain its high export rates. Apart from South Korea with special brands (children's and sport shoes) many other countries were only able to increase their exports by maintaining or lowering their price level

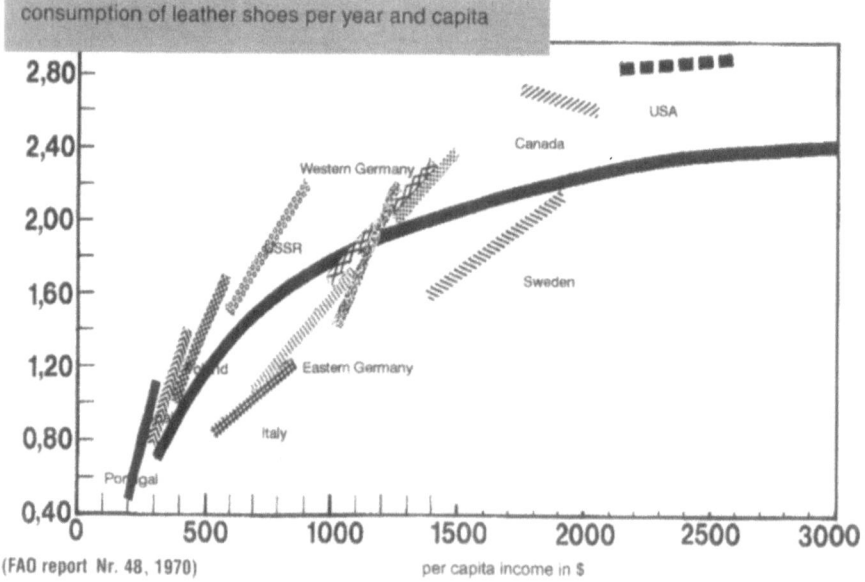

(FAO report Nr. 48, 1970)   per capita income in $

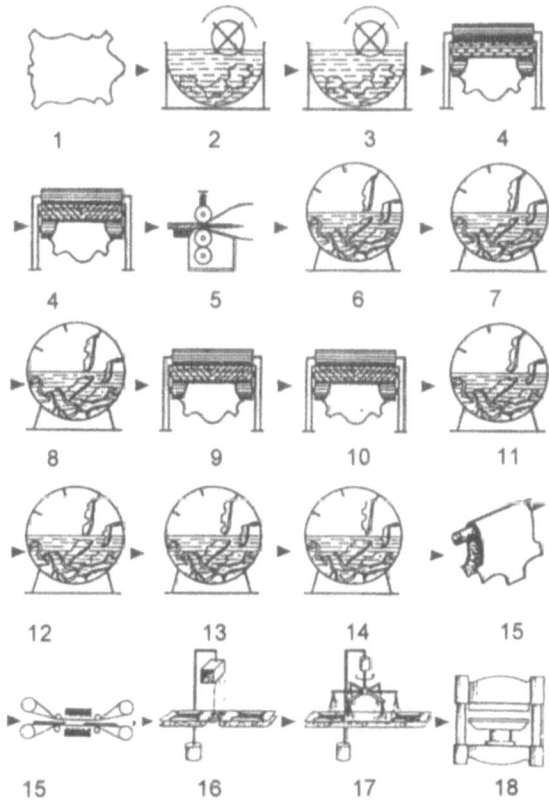

Fig. 1-8. The production of leather: Raw hides and skins are preserved in the slaughterhouse or by a hide broker by the addition of salt. By this process the hide (*1*) is to some extent dehydrated. The first step in the tannery is to soak (*2*) and wash the hide in the beamhouse. Then the material is treated with lime at pH 13 (*3*). By the addition of sulfide ions and by mechanical means (*4*) the hair is removed. The remaining flesh is removed (*5*). Then the hide is split horizontally to reduce its thickness (*6*). In the following deliming the pH is decreased to 5 (*7*). In the pickle (*8*) in an acid-salt mixture the pH reaches ca. 3. The material is then tanned with trivalent chromium salts (*9*), sammed (*10*), shaved (*11*) and neutralized to pH 4–5 (*12*), retanned (*13*), dyed (*14*), treated with fat liquors (*15*) and dried (*16*). This is the *wet end* of the leather production. In the *dry end* the leather is staked (*17*), finished in a variety of steps (*18,19*), ironed, embossed and/or plated (*20*)

Fig. 1-7. Shoe consumption and average income. An older investigation by UNIDO shows that high per capita consumption of shoes and leather goods depends on a high standard of living. This demonstrates that increased production of leather and shoes in developing countries is a contribution of the industrial development and export situation of developing countries. The main customers of the developing countries are the industrialized ones with their high standard of living. This development demonstrates also the intensive work by organizations like ILO (International Labor Organization) and UNIDO (United Nations Industrial Development Organization) to create jobs in the developing countries preferably by using natural sources (the local hides and skins). The labor intensive leather and shoe industry is suited very well for starting industrial development in a country

Finally, after all these economic statements, a short description of the process of tanning and finishing will be given [2–5]. Tanning and finishing of leather is a complex action whereby single pieces of skins or hides which differ according to race, sex, breeding method, seasonal killing time and treatment during and after the slaughtering of the animals are tanned, dyed and finished piece by piece. The transformation of animal hide or skin to leather is a mechanical and a chemical process [11]. Because the aim of this work is to describe the production of synthetic materials, the tanning, dyeing and finishing of leather is only shown in an extremely abridged schematic way (Fig. 1-8).

CHAPTER 2

Coating of Textiles

Since the earliest times, man has used leather and textiles to protect himself against low and high temperature, rain, sun, wind etc. Due to their permeability to water vapor both materials stand for a high wearing comfort. Leather protects extremely well against wind, high and low temperature but is not easy to clean. When wet, leather is heavy and very uncomfortable.

Textiles isolate against high and low temperature but rain is able to penetrate them easily and the protection against wind often is poor. Textiles can be cleaned easily.

For all of these reasons many trials have been done to obtain articles with a compromise of the good properties of both of them. By the impregnation of a textile with a resin and/or a hard wax, chintz and similar articles were created. Chintz has a special look, is less permeable to rainwater and protects against the action of wind. Articles made out of chintz, cretonne and similar products still have a textile feel and character. During the use of these materials the impregnation agent or the wax comes off and textiles of this type of minor use.

There are several ways to cover a textile with a polymer: By coating a textile and covering its surface totally by a polymer layer usually colored by pigments. The coating layer usually has a thickness of > 0.1 mm. In textile printing the surface may only be covered partially, normally by a pattern. This pattern may consist of dyestuffs or of pigments applied together with a binding agent. Usually, acrylate polymers are used as the binding agent. Textile printing is not considered in this book; only some applications using a printing technique (see Chap. 14).

One of the first methods used to improve textile substrates was the coating of the textile with a nitrocellulose product. This coating was a first transformation of a textile to a leather-like product. In 1864 an article looking like leather, an artificial leather, was created [1]. In the following years articles behaving more and more like leather were created; and the most essential point was that all of them were less expensive than the original model.

The most important polymers useful in the coating of textiles are polyvinyl chloride (PVC) (see Chap. 2.1) and polyurethane (see Chap. 2.2). For special articles, polyacrylates [3], ethylene vinyl acetate or butadiene copolymers [4, 5] are also used. Thinly coated raincoats are mostly produced by using silicones [2]. All these polymers have their own particular properties: PVC, most broadly used, is a cheap polymer, easy to process, needs a thicker coating and stiffens at low temperatures. Polyurethane (PUR) is easy to modify, more expensive, and also allows thin coatings with a high abrasion resistance. The price of poly-acrylate is between that of PVC and PUR and thin, less flexible coatings can also be produced with this polymer. A disadvantage may be the rather low cracking resistance at low temperatures. Silicones are expensive but they are suited for impregnations and thin coatings and they keep their elasticity even at low temperatures. For technical textiles, such as air bags, conveyor belts, etc. silicones have been frequently utilized as they are extremely well suited [6].

Due to the fact that in the main the application method depends on the polymer used for coating, the different application devices are described in the Sects. 2.1 and 2.2.

2.1
Coating of Textiles with PVC Polymers

Leather substitutes had a broad use in the 1940's. In central Europe during the war leather was not available in sufficient amounts. A normally dark gray colored homogeneous soft PVC foil (IGELITH®;· trademark of former IG Farben) was used in production as a leather substitute for raincoats.

On the other hand, the state-of-the-art in polymerization and application of the polymers, allowed the production of articles [1–3] which, also due to their low costs and good performance, were able to remain established in the market even when leather was again available in sufficient amounts [4]. Additionally, the synthetic substitutes were applied in fields such as for roofs for planes and automobiles, functional clothing and automotive interiors where, due to a lack of performance, leather or textiles were not or only partially suitable.

The most important polymer class for substitutes of leather are vinyl chloride homo- or copolymerisates [5, 6, 10] applied by direct or indirect coating (see Fig. 2-1).

Between 1950 and 1952 a substitute with a foamed, polymeric inner layer was developed [7]. The construction of this new article, a great step forward in the art of leather substitutes, consisted of a cheap knitted textile, a foamed layer of soft PVC and an abrasion-resistant top layer of homogeneous soft PVC [8, 9].

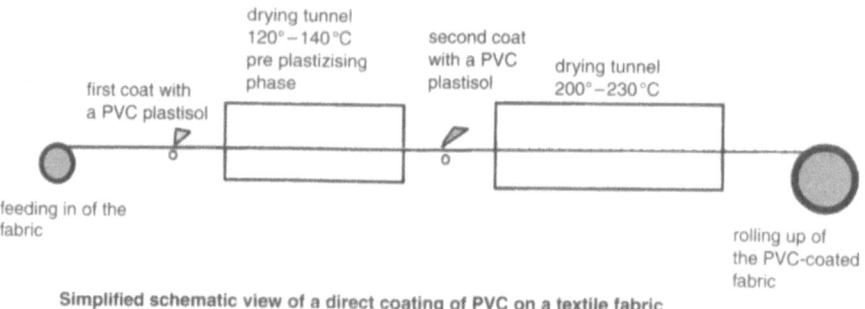

Simplified schematic view of a direct coating of PVC on a textile fabric
(Rollers for cooling – feeding in devices etc. are not demonstrated)

Fig. 2-1. Direct coating process of a PVC plastisol onto a textile substrate. A schematic view of a direct coating of a fabric with 2 layers of a PVC plastisol

The foamed layer is produced by blowing agents, i.e. chemicals decomposing at application temperature.

$NaHCO_3$
Sodium-bi-carbonate

$H_2N-CO-N=N-CO-NH_2$
Azodicarbonamide

Two blowing agents: Azodicarbonamide and Sodium bicarbonate.

This article could be produced at a very low price and had the look and behavior of leather. The only point missing was the lack of a certain water vapor permeability.

Due to the thermoplastic properties of PVC the coating of the textiles could be done in a machine where the polymers – together with all the other necessary ingredients – were melted and then transferred in one or more layers to a release paper, the textile is laminated in the melted mass and after cooling stripped off from the paper. This production process is not labor intensive and can be done at low cost.

To vary the color, the touch and the production procedure the coating needs additional products: pigments or dyestuffs [11], stabilizers [12], plasticizers [13] and foaming agents (see above) [14] as the most essential ingredients.

Additional products that may be added to the coating mix are (1) flame retardants (see Sect. 23.2) or (2) UV stabilizers.

The coating with PVC does not need any additional application medium such as water or a solvent. The coating is done by means of calenders and seems to be the most environmentally correct coating application procedure if the evaporation of the plasticizers is controlled. Unfortunately PVC belongs to the chlorine-containing polymers and, therefore, the polymer class itself is not regarded as a product which is compatible with the environment.

2.2
Coating of Textiles with Polyurethanes

Textiles, especially raised fabrics [12], can be coated by polyurethanes in a direct (Fig. 2-2) or an indirect process.(Fig. 2-4) [1, 9, 11, 14]. The polyurethanes may be applied in solution [6, 7], as high solids, by dispersion or by calandering.

Using the direct coating method, the PUR solution is normally directly coated onto the substrate by using a doctor blade. On the opposite side to the blade – and under the substrate – there is either nothing (knife on air), a roller (knife over roll) or a conveyer belt (knife over rubber blanket) (see Fig. 2-3). By the nature of the device used under the substrate the penetration of the polymer into the substrate is controlled. The deeper the polymer penetrates the substrate, the harder the resulting article will be, and the better the adhesion of the polymer to the substrate.

In the indirect or transfer process the coating is built up in an opposite way. At first the top layer is applied, then the intermediate, then the adhesive coat and

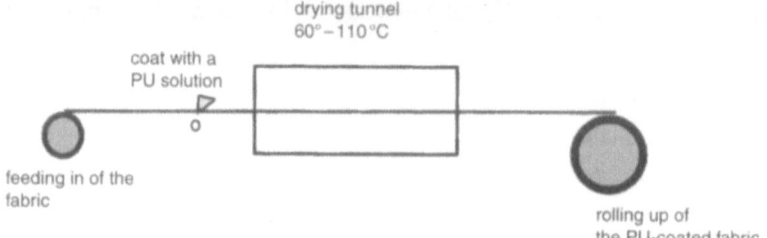

Simplified schematic view of a direct coating by a dissolved polyurethane on a textile fabric
(Rollers for cooling – feeding in devices etc. are not shown)

Fig. 2-2. Direct coating process [17]

finally the substrate. The coating material is applied onto a paper, itself coated with a silicone or polypropylene able to release the appropriate coating after drying and normally having an embossed structure. The first layer is usually, made from a hard and abrasion resistant polymer. The intermediate coat, if applied, is made out of a softer polymer. The softest polymer is the adhesive coat. If lower grades of textiles are used foamed intermediate coats help to equalize the surface of the coating [24] (see Chap. 11).

The different coats need to be dried prior to the application of the next one. If the coats are applied wet on wet, i.e. to apply a solution onto a layer still containing a lot of solvent, the polymer of the layer will swell. If a coating is applied onto a swollen polymer layer the final coated textile will not lay straight. Possible ways to avoid this are to dry the solvent out totally and/or to crosslink the polymer or to use chelating agents or metallic alcohols [9].

The engravings in the release paper, the kind of gloss or matting degree and the negative structure of it (see Sect. 3.2) is transferred to the coating. This kind

Fig. 2-3. Possible devices for direct coating

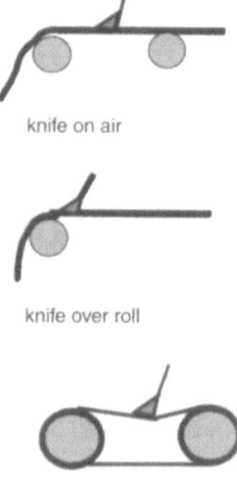

knife on air

knife over roll

knife over rubber belt

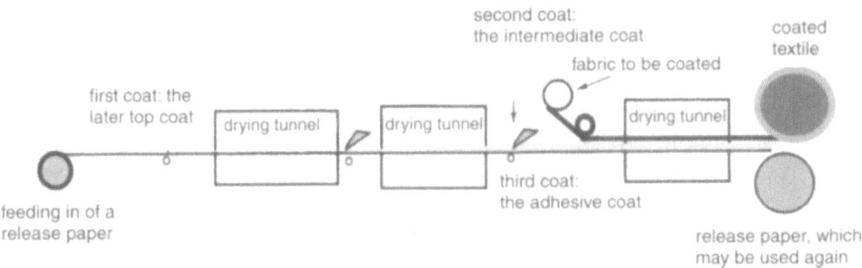

Simplified schematic view of an indirect coating of a textile fabric with a polyurethane solution or a dispersion; (Rollers for cooling – feeding in devices etc. are not shown)

Fig. 2-4. Indirect coating

of textile coating is used with polyurethanes, polyacrylates or similar products and mixtures thereof.

Very often the machinery used is able to allow an application by the direct and indirect coating alternately. Multipurpose equipment is also available [23].

When the process of PUR coating was first applied, the resulting material was only used for articles like shoes or bags – articles with a comparatively short time of use. At the end of the 1960s more and more uses, e.g. upholstery, were found. After a relatively short time the industry was faced with claims due to a lack in resistance to hydrolysis of the polyurethanes used [22]. Due to a decreasing resistance to hydrolysis, the tear strength and the abrasion resistance of the coatings are weakened and the coating no longer performs well. Today, we know that polyurethanes are also suitable for long lasting articles, so long as polyesters, like hexandiol-polyadipate, polycaprolactone, polycarbonate, etc. are used in the soft segments of the polymers (see Chap. 25). It is also possible to use polyethers in the soft segments. Polyurethanes containing polyethers in the soft segments are soft and resistant to hydrolysis. The tear strength of polyether-polyurethanes is less and they are more sensitive to oxidation than polyester-polyurethanes.

Nowadays the application of water-based systems is increasing slowly. However, most of the polyurethanes still used are applied as solutions in organic solvents, because the application of solvent-based products is easier and less energy consuming then having to evaporate water.

Also, for a long time, release papers, able to resist the attack of the water of a polyurethane dispersion, were not available. These papers could not be used as often as papers for solvent systems. During drying operations they lost their planarity and partially lost their releasing property. Today many producers of release papers also offer water-resistant release papers and, therefore, the coating with polyurethane dispersions also becomes of more and more importance.

To disperse a polymer in water, internal (see Chap. 25.5) or external emulsifying agents are necessary. If such a dispersion is used to get a film of a polymer, we need to realize that the water resistance of the polymer, due to the emulsifying groups still being present in the film, is weak. To improve the water resistance of the finished product it is necessary to crosslink the polymer [18]. There are several crosslinkers used in this field: isocyanates [26], epoxides, aziridine [25], N-methylol derivatives and carbodiimide group containing crosslinkers are used in industrial scale.

Usually crosslinking agents with three active groups are used. Tris-isocyanates, for example, crosslink polyurethanes via allophanate groups. The crosslinked polyurethanes do not swell as much in water or solvents as the ones which are not crosslinked.

Isocyanate crosslinking is not ideal for polyurethanes having carboxylic groups as dispersing medium. Aziridines, a product group with a certain toxicological potential, are better suited. For polyurethanes with carboxylic groups, carbodiimide crosslinking can also be used (see Chap. 25.5).

To avoid volatile organic liquids during coating operations, high solid systems, i. e. systems with low or no solvent content, may also be used [18]. High solid polyurethane systems normally contain a NCO prepolymer with blocked isocyanate groups which need a crosslinker based on a cycloaliphatic diamine (Eq. 2.2). The principle of high solid systems is to use oligomeric blocked compounds with a viscosity which can be managed at the temperature of application. After the addition of a chain-lengthening agent – also of low molecular weight and of low viscosity – the temperature is increased, the blocking agent is split off and the mixture reacts to the high molecular polyurethane – polyurea.

The mixture of the blocked prepolymer and the diamine has a pot life of more than 24 h at room temperature. After coating of the highly viscous reaction mixture, heating at more than 140 °C is needed. Then the blocking agent is split off and the prepolymer reacts spontaneously with the diamine. A polyurethane-polyurea is built up.

$$
\begin{array}{cc}
& \text{Me} \\
\text{\textasciitilde NCO} + \text{HO-N=C} & \longrightarrow \\
& \text{Et}
\end{array}
\qquad
\begin{array}{c}
\text{Me} \\
\text{\textasciitilde N-C-O-N=C} \\
\text{H O} \qquad \text{Et}
\end{array}
$$

| isocyanate prepolymer | and a blocking agent (in this example an oxime) | blocked isocyanate prepolymer |

$$
\begin{array}{c}
\text{Me} \\
\text{\textasciitilde N-C-O-N=C} \\
\text{H O} \qquad \text{Et}
\end{array}
+ \ \text{H}_2\text{N-X-NH}_2
\tag{2.1}
$$

temperature (ca. 150° C)   – $\begin{array}{c}\text{Me}\\ \text{HO-N=C}\\ \text{Et}\end{array}$ the blocking agent is split off and evaporates

$$\text{\textasciitilde NH-CO-NH-X-NH-CO-NH\textasciitilde}$$

polyurethane-polyurea

High solid system. The principle of high solid systems is to use oligomeric blocked compounds with a viscosity which can be managed at the temperature of application. After the addition of a chain lengthening agent – also of low molecular weight and of low viscosity – the temperature of the mixture with a rather long pot life is increased to split off the blocking agent. The blocking agent evaporates and the mixture reacts to the high molecular polyurethane-polyurea.

Solvent systems lose solvents during heating and their viscosity rises. High solid systems in contrast have a special property: At room temperature the viscous reaction mix becomes less and less viscous by heating it up The effect of lowering the viscosity by heating allows the reproduction of the finest details and is, for example, used in the manufacture of synthetic suede [19] where the reaction mix is applied on silicon rubber molds with a negative suede-like surface.

Besides a technical use to obtain an excellent abrasion resistance or cold and hot flexibility, reactive systems of polyurethane components are seldom used to coat textiles [29]. A reactive system of a a hydroxyl group containing, liquid polybutadiene, p-phenylene diisocyanate and expandable microcapsules can be used to coat woven, knitted or nonwoven base fabrics to obtain a good laundering and alkaline-resistant product [30].

Principally, a thermoplastic coating of textiles with polyurethanes is, in analogy to soft PVC, possible. The thermoplastic coating of textiles with polyurethanes is today only used for technical articles. Suitable films for this kind of coating are produced by calendering, extruding or using a melting process [16]. Films of polyurethanes produced by a thermoplastic process may be adhered to textiles via high temperature or by the use of an adhesive [15]. Thermoplastic polyurethane films containing a wax can be laminated to nonwovens via a microporous layer to avoid a non-textile break of the man-made leather [27].

Due to the high tensile strength of a polyurethane, thin films with sufficient properties can be applied. PVC must normally be applied as a thicker film to get the same level of physical properties [20, 21].

The definitions we have used up to now for the different coating systems are based on their technological items:

- *direct, indirect* process, *solvent, dispersion* and *thermoplastic* coatings are determined by the application method. The *high solid* process also is more determined by the application rather than by the chemical speciality.
  The following terms are more defined by their chemical properties then of the systems applied:
- The application of a solution of an oligourethane together with an isocyanate crosslinker is called a *two-component*-process [2, 5].
- Aromatic, high molecular weight polyurethanes [3, 8] being soluble in dimethylformamide (DMF) and/or methyl ethyl ketone may be applied alone, without additional reactive compounds. These systems are called *one-component* products.

The toxicology of DMF may cause problems (see Sect. 7.1 [3]). Solutions based on DMF are called solutions with a *hard solvent*. Since isophoron diisocyanate was introduced into the manufacture of polyurethanes in the 1970s people have learned to obtain solutions of these polyurethanes in mixtures of toluene and isopropanol [4]. These solutions are said to be "dissolved in *soft solvent*".

All of these different coating systems are not applied solely – mostly a combination of different systems is used:

- A top coat of a one-component-polyurethane may be applied to a textile substrate via a two-component adhesive coat. High solid top coats are always transferred to textiles by two-component adhesive coats.

Fig. 2-5. A fabric coated by
a polyurethane. The coated
surface (upper smooth part)
can be distinguished easily
from the textile substrate
(fibrous part)

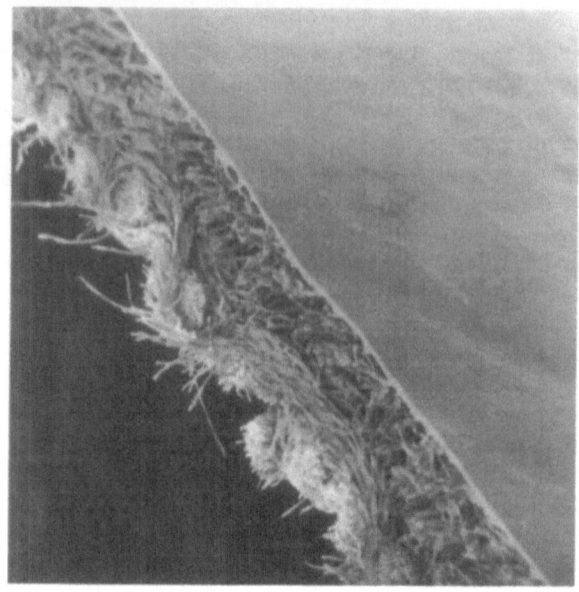

100 µm

- Intermediate coats with a foam structure [13] or backside coating of the
  textile as well [10] have also been reported.
- Polyurethanes have a certain water vapor permeability. The hard segments of
  the polyurethanes with their crystalline urethane or urea groups are responsible for the permeability which is higher than that of a PVC coating. Due
  to the fact that polyurethane films may be less thick than PVC films and
  that the water vapor permeability of thin films is better than of thick films (see
  Fig. 18-2) an additional contribution to the water vapor permeability is obtained. Nevertheless, the water vapor permeability of a fabric coated with a polyurethane (see Fig. 2-5) is not sufficient to guarantee a good wearing property.

Hybrid polymers, i.e. combined products consisting of elements of the polyurethane and the acrylate chemistry, play an increasing role in textile coating.
For instance, a polyol is reacted with an isocyanate and then with a (meth)-
acrylate also having at least two hydroxyl groups. If this is mixed with a reactive
diluent, glass powder containing silver ions as antibacterial agent, a coating
mixture is obtained with antifungal and antibacterial properties with excellent
adhesion to many substrates and high resistance to heat and light [28].

2.3
Economic Figures

Many coated textile substrates, membranes or nonwovens fall in the category
"technical textiles". Technical textiles are used in many fields such as agriculture,
building, clothing, ecology, geo, housing, industry, medicine, mobile industry,

packaging and sports. Technical textiles have grown much faster than the conventional ones: From 6 Mio tons in 1985, they grew to 9.3 Mio tons in 1995 and it is estimated that they will be at 13.7 Mio tons in the year 2005. The agricultural market including the fishing industry mostly needs ropes, twines, spunbonded or capillary nonwovens etc., i.e. products which are cheap to produce. In the building industry, membranes and other breathable products have a certain importance. Besides products consisting of glass fibers, PVC-coated polyester fabrics, sometimes polyethylene-coated products, dominate here. Breathable products are used as moisture barriers. In the clothing industry, threads and interlining materials are regarded as being technical textiles. In geotextiles products for separation, filtration and drainage cover the scope of this book. The home sector is dominated by floor coverings. Besides the use of conveyor belts in the industrial area, filtration is important in an industrial use. Medicine perhaps has the broadest use of water vapor permeable products, e.g. wound covers, protective textiles, products for hygiene purposes, dialysis, implants or artificial skin. Medical products must be biocompatible and non-toxic, etc., and easy to sterilize. It is estimated that this market will become more than 10% of all technical textiles used. The mobility market covers automotive, aviation and naval transport. Besides filters, carpets and seat belts mostly coated textiles are used in this area. The packaging area is dominated by cheap products which in the main are not permeable. For clothing to protect workers, and for sports and leisure, water vapor permeability is a must. Sports and leisure have an actual consumption of 240,000 tons of textiles with a value of 1.6 billion $ in 1995 and an estimated consumption in the year 2000 of 310,000 tons which will correspond to a value of 2 billion $. The market of technical textiles had a demand for nonwovens of 2.5 Mio tons in 1995 with a value of nearly 10 billion $ [2].

Up to the end of the 1980s the quantity of PVC-coated textiles overran the quantity of the leather being produced with 1.6 billion/m² by more than 10 times. Ca. 1 billion/m² of polyurethane-coated textiles were produced in addition. With a calculated average amount of 60 g solid polyurethane per m² more than 60,000 tons of polyurethanes were needed in the coating of textiles, 48,000 tons thereof for clothing and 80,000 tons for shoes. 200–300 Mio/m² of poromeric and hydrophilic materials were produced annually.

PVC-coated textiles, including planes and other technical textiles, cost ca. 1.60 DM/m². The quantity of ca. 16 billion/m² corresponds to a value of ca. 25 billion DM. Leather costs ca. 20 DM/m²; the value of leather being produced worldwide represents a value of roughly 32 billion DM. Polyurethane (PUR)-coated textiles cost ca. 5 DM/m² which adds to an annual production volume of ca. 5 billion DM and poromerics at a price of ca. 8 DM/m² to 2 billion DM.

Some of these PVC- and polyurethane-coated textiles are part of the so-called technical textiles. The market size of technical textiles is estimated to be 42 billion $ [1].

According to another publication (3 [4]) in 1983 only 2.4 billion/m² of coated textiles were produced. In 1992 355 Mio/m² of PUR-coated textiles were produced and 55 Mio/m² poromerics.

Kuraray actually currently produces 12.5 Mio/m² of Clarino® and it is indicated that this will increase to 17.5 Mio/m² by the year 2000 [3]. Teijin also reports an increase in its production of man-made leather from actually 12 Mio/m² by 13 %. The 12 Mio/m² consist of 4.2 Mio/m² of dry processed, i. e. conventional textile coating onto a polyester-substrate, and 6.4 Mio/m² of wet processed man-made leather, i. e. coagulated polyester and polyamide substrates [4].

According to a report in Japan, 41.6 Mio/m² of man-made leather were produced in 1996. In the same report it was stated that Kuraray estimates the worldwide demand for leather, i. e. natural, synthetic and man-made altogether, is approximately 1500 Mio/m². Natural leather accounts for 1200 Mio/m², synthetic leather is at 245 Mio/m² and man-made leather (leather substitutes which are poromeric) is only 55 Mio/m². Kuraray estimates the demand for man-made leather to increase to 90 Mio/m² in the year 2000. Besides Japan with 38, Korea will produce 14, Taiwan 12, Italy 8, China 11 and others 3 Mio/m² in 2000. Toray is the only producer with production in Japan (Ecsaine®) and outside, in Italy (Alcantara®). Alcantara is applied in Europe 60 % in upholstery and 20 % each in automotive and apparel. Mitsubishi produced 1.1 Mio/m² of Glore® in 1996, the only man-made suede consisting of acrylic ultrafine microfibers. Asahi's Lamous®, a man-made suede, amounted to 2.2 Mio/m² in 1997; the product is mainly used in the production of upholstery and automobile seats. Teijin's Airy® and Cordley® were sold in 1996 in a quantity of 9.2 Mio/m²; the main application was in the shoe sector. Bellace®, a product of Kanebo, is sold to 15 % in high tech fields as a polishing material for silicone, glass and metals. Belleseime®, another product of Kanebo, was sold in a quantity of 2.4 Mio/m² in 1996 and a 10 % increase in sales was estimated in 1997 [5].

There is a discrepancy between the different market figures. This discrepancy stems from the fact that Japanese sources regard as man-made leather products with textile substrates containing microfibers and poromeric polyurethanes whereby the polyurethanes are coagulated.

CHAPTER 3

Definitions

DIN 16 922 gives a definition of artificial leather as "textile and other substrates with or without a coating whose properties and/or surface appearance correspond to their use". This – technically orientated – definition covers all man-made, leather-like materials like homogeneous or foamed PVC coatings, polyurethane-coated textiles and poromerics [4]. In the main, man-made leathers are regarded as more sophisticated leather substitutes like coated nonwovens produced by a coagulation process.

Japanese authors describe poromerics, especially as long as they contain microfibers, as man-made leather and all others as artificial leathers. In Germany the definition synthetic leather is banned [3]: poromerics are not regarded as being leather being produced by a chemical synthesis.

Japanese authors very often use, in addition to the words artificial or synthetic leather, the production method as "produced by a wet" or "dry" process. *Wet process* means coagulation (see Chaps. 7 and 8), *dry* stands for normal textile coating (see Chap. 2).

Artificial leathers – according to the definition used in this work – stands for articles imitating the look and not the properties of leather. Leather imitations or substitutes imitate the original model in touch, feeling, and stress-strain properties.

Synthetic products being permeable to water vapor are either hydrophilic or microporous. They are named hydrophilic or poromeric (2 [7]). A microporous material is often called a *poromer,* a name coming from *por*ous and elast*omer*. In German, water vapor permeable products are called "*w*asser*d*ampf*d*urchlässig" or abbreviated WDD-articles. Consequently these products could be named in English as wvp-products, *w*ater *v*apor *p*ermeable products.

Nearly all synthetic products only partially offer the properties of leather. In Fig. 3-1 these properties are correlated with the technical efforts needed to achieve them. According to Fig. 3-2 the simplest property to imitate is the touch and the grain structure of leather: examples are textiles coated with foamed PVC or PUR. The handle or touch depends on the structure, hardness, matting or gloss effect of a surface. Therefore artificial products also need a "finish", a very thin top coat, of a dry polymer like cellulose-acetobutyrate, nitrocellulose or a polyamide.

A further improvement in touch or texture is achieved by using a raised fabric or a foamed intermediate coat. These achievements offer the touch, softness and a "full" feeling.

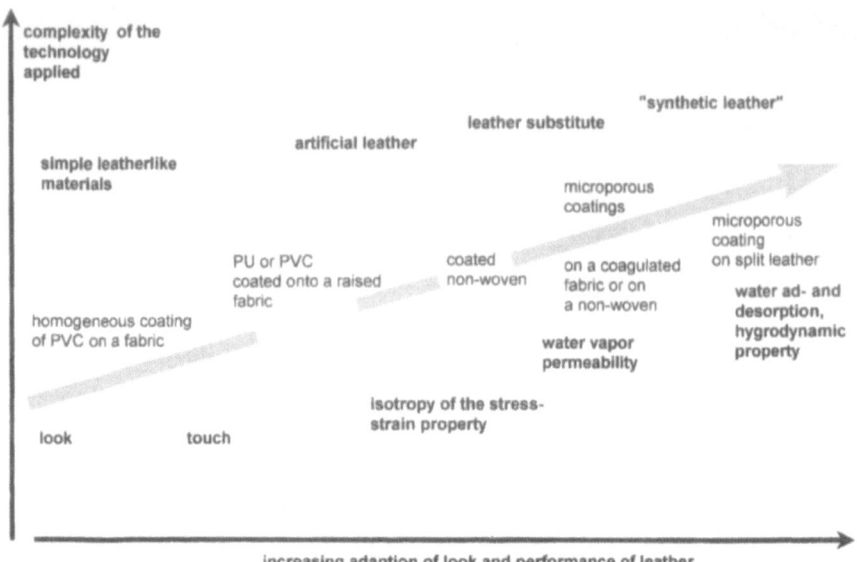

Fig. 3-1. Synthetic substitutes for leather: The necessary technology used to get the properties needed. The more the resulting product needs to be similar to leather, the more complex the technology needed

The stress-strain behavior of leather fits well with the properties needed for a shoe (Fig. 3-2). In contrast knitted fabrics are not suitable as the raw material for the production of shoes. Leather is elastic; the more the material is extended, the more power is needed to get the extension. These properties are needed to keep the shape of an article. Beside this the "set" is an additional factor for the wear comfort of the shoes. Knitted fabrics extend with almost no force at all. They are unable to keep the necessary form. Additionally fabrics and knitted textiles have an anisotropic stress-strain behavior: Normally fabrics extend differently in the weft or warp direction and knitted articles differ in the knitting direction or adverse to it. A decisive improvement in isotropy is offered by using a nonwoven as a substrate (see Chap. 21) [1, 2].

The products in Fig. 3-1, up to now, have not named the term water vapor permeability. If this property is desired, the required technology is more complex: instead of a simple coating machine a coagulation unit is needed. Nowadays inverse emulsions (see Chap. 9) or hydrophilic products (see Chap. 18) are also available, which can be applied on normal coating units to become a water vapor permeability.

New investigations into the influence of humidity and temperature of chrome tanned leather show a different behavior at low and high temperatures: Increasing humidity at low temperatures increases the thickness and the area of the leathers exposed – at higher temperatures the thickness also increases with growing humidity but at a loss in area. At temperatures up to 60 °C leathers become softer with increasing humidity – above this level of tempera-

Fig. 3-2. Stress-strain properties of different materials. In contrast to shoe upper leather, a PVC-coated knitted fabric does not resist any deformation. When the PVC-coated material is deformed the material almost does not resist the energy of deformation. When the stress is gone the material goes back to its original form; both lines are a hysteresis where both lines are more or less identical. To deform leather more energy is needed and the material does not go back to the original form. A rest of deformation is kept. The result is that a product manufactured by such a coated knitted fabric would be unable to keep its form

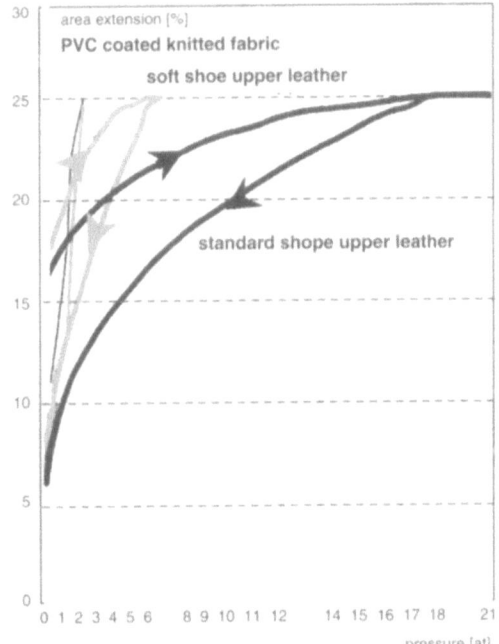

ture they become harder [5]. This effect is useful to produce a good wearing property.

Microporous products have pores connected to each other; the size of the pores is ca. 5–50 μm [4,5]. This guarantees a good water vapor penetration and no entry of liquid water, for example, in the form of rain. A microporous surface looks homogeneous to the eye without using a magnifying glass.

The next two properties of leather are (1) the ability to easily ad- and desorb moisture (see Chap. 5), and (2) the so called hydrodynamic property. Today these properties, together with the above cited ones, are obtained only by a mixed article with split leather as a substrate and a microporous or an appropriate homogeneous, hydrophilic coating (see Sect. 20.2).

CHAPTER 4

Water Vapor Permeability

Materials which are able to take in or let out moisture either need to be porous or selectively permeable. Membranes are products which are permeable to low molecular weight products or ions. In this work we will discuss mainly microporous or hydrophilic films, coatings and impregnations, i.e. materials permeable to water vapor. As will be seen later (Chap. 27) spin off materials, which were most probably originally designed as substitutes for leather, can be used for other purposes like filters, wound covers, slow release materials, etc. In general, all these kind of material are membranes – allowing defined products to permeate a barrier between an inside and an outside part of a technical product that they cover, include or contain.

Membranes, in nature also, are the most important part of a cell because they define the cell, its inside and its outside against its surroundings. "Membranes play a central role in both the structure and function of all cells ... they define compartments, each membrane associated with an inside and outside ... they also define the nature of all communication between inside and outside" [10]. Man-made membranes more or less do the same.

It is not only microporous products that have an ability for water vapor permeability. Water vapor transmission is regulated with microporous products by the partial vapor pressure difference between both sides. Water vapor passes through the microporous product by diffusion. Hydrophilic products are also able to transmit water vapor. The transmission of these products occurs by absorption and desorption of the water according to the difference in concentration of water vapor on both sides of the membrane. Besides water vapor other products are also able to penetrate microporous or hydrophilic products. Microporous products, for instance, are often used to filter particles according to the size of their pores. The transmission of ions or biomaterials may be improved by voltage, difference in pressure, etc. [9].

Microporosity is used in many articles (see Chap. 27). The best known article for a non-expert are filters. Microporous products that have been used for a long time by mankind are leather, earthenware and bricks.

An American publication [2] defines microporosity as "microporous films and coatings which have excellent tactile and tensile properties and have no visible pores".

The size of the pores of microporous products is, as mentioned, 5–50 µm. In comparison to that ultrafiltration needs 10^{-2} µm [3], microporous have $10^{-3} - 10^{-4}$ µm and macroporous ion-exchange resins have $10^{-2} - <1$ µm [3].

Microporosity is well known in rubber films [5] and polyurethane layers [4]. Microporosity is regarded as necessary for water vapor permeability [6, 7] and it also reduces the density of the polymer (see Sect. 10.1), increases its opacity and reduces tear strength and tensile strength of the polymer (Fig. 23-1).

Microporous Products: Comparison of the Structure and Properties

Leather is structured as follows: Fibers of a coarser structure on the under layer, the reticular layer, become continuously finer and finer extending to the surface, the papillary layer, of leather [1] (Fig. 5-1). Extremely fine fibers build up the surface, i.e. the grain of the leather [3].

Leather substitutes principally imitate the structure of leather [2, 6, 7] (see Fig. 5-2). A substitute differs in two points: the substrate, often a nonwoven, differs in structure and chemical composition from the top coat [9] and on leather the whole material has a homogeneous structure made from – tanned – collagen.

The stress-strain property, the water vapor permeability and the ad- and desorption of water of leather is connected with the structure of the collagen-fibers. Leather substitutes normally have a textile substrate and a polymeric coating.

Fig. 5-1. Leather from cattle – cross section (magnified 20 times) Leather is built of fibers consisting of tanned collagen, getting coarser from the surface (grain side) to the inside (fleshside)

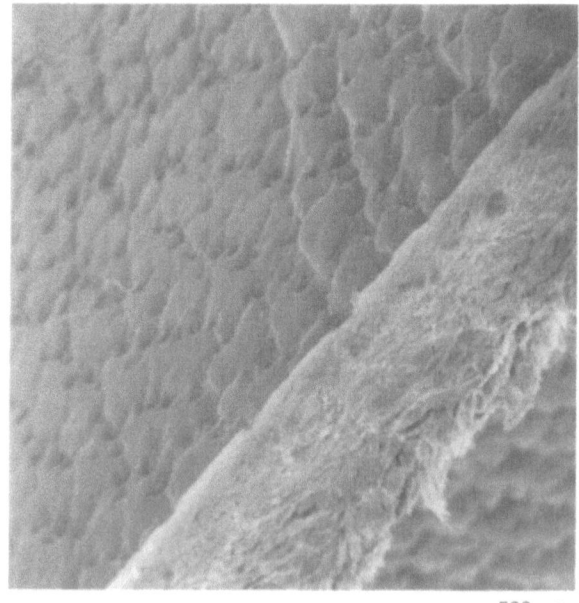

500 µm

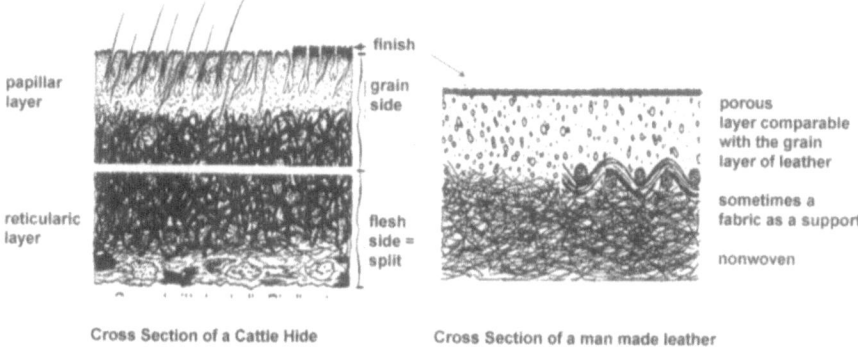

Cross Section of a Cattle Hide   Cross Section of a man made leather

Fig. 5-2. Structure of leather and a substitute – principle in composition

For shoe uppers nonwovens are used as a substrate. Only nonwovens have a kind of an anisotropic behavior. Woven and knitted substrates are suited for synthetic garments. For this purpose an anisotropic behavior is not so important.

Leather substitutes with a nonwoven as a substrate and a hydrophilic or microporous top layer are inhomogeneous in their composition. The nonwoven consists of synthetic fibers, like polyester or polyamide, and a polymeric binding agent like synthetic or natural rubber, polyurethane, polyacrylate etc. Leather and substitutes also contain sizing and coloring agents. In the case of a synthetic uppermaterial the nonwoven is covered by a polyurethane top layer. A typical microporous substitute is shown in Fig. 5-3.

Fig. 5-3. A substitute for leather with a microporous polyurethane topcoat (the upper layer with the fine pores) and a nonwoven as a substrate (under layer with fibrous structures)

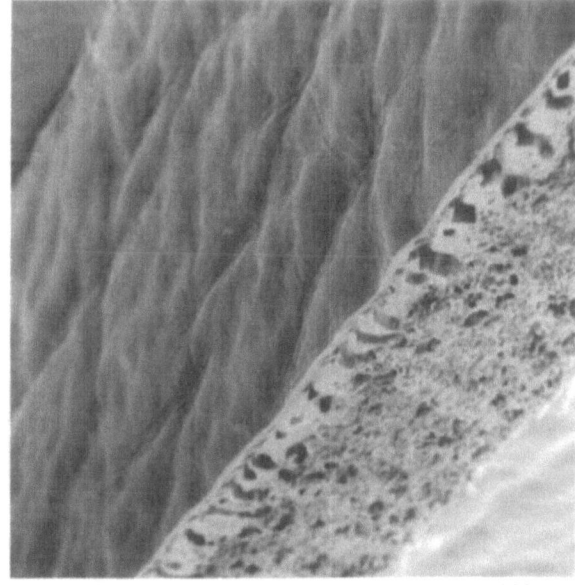

500 µm

Substitutes with a hydrophilic topcoat may have a construction like a normal polyurethane-coated textile (see Fig. 2-4). Hydrophilic products transport water vapor by ad- and desorption; they do not need a microporous structure.

In the last few years new articles have been developed; their construction differ from the above described ones: GORE-TEX® (trademark of Gore Ind.) contains a microporous layer and SYMPATEX® (trademark of Akzo) a hydrophilic one: The microporous or hydrophilic layer is situated between two textile layers. This construction is named "liner". Liners are used in sportswear and in leisure articles; they combine good water vapor transmission with a good protection against wind and cold or warm air and they are normally unpermeable against rain.

A comparison of different materials, now (1997) available on the market, has been published [15]. To date, no material has been marketed which does not need further improvement.

Microporous PTFE or hydrophilic polyester materials need a lamination to a fabric by an adhesive coat. The resulting bicomponent material loses smoothness and water vapor permeability. It seems to be difficult to get the desired size and quantity of pores in microporous films; this problem may be responsible for a possible lack in technical performance. Hydrophilic films may swell in the wet stage and, being wet, have an unpleasant touch.

Ordinary artificial leathers differ from the products mentioned above. Artificial leathers have a homogeneous or a foamed top layer on a textile substrate. The size of the pores of such a foamed top layer is much larger than those of a microporous film and the pores are closed. Therefore these foamed articles have no water vapor permeability at all or a very low one. Generally a PVC leather has a homogeneous top layer, which limits the water vapor transmission further (Fig. 5-4).

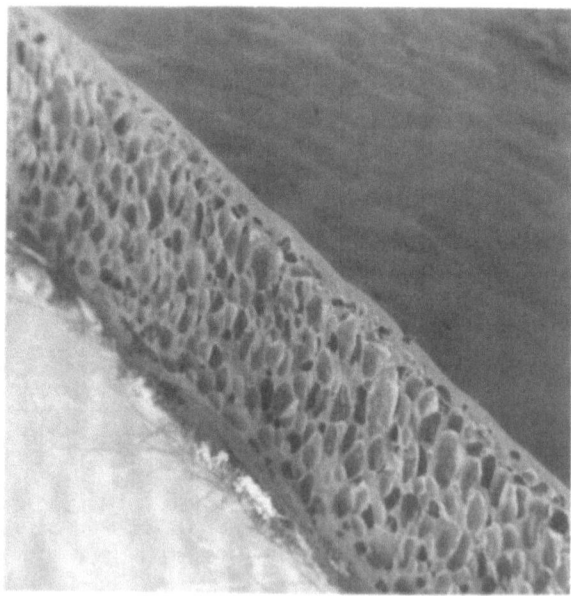

Fig. 5-4. An artificial leather with a foamed PVC layer. Three different layers can be distinguished: The textile substrate, the porous intermediate coat and the nonporous topcoat

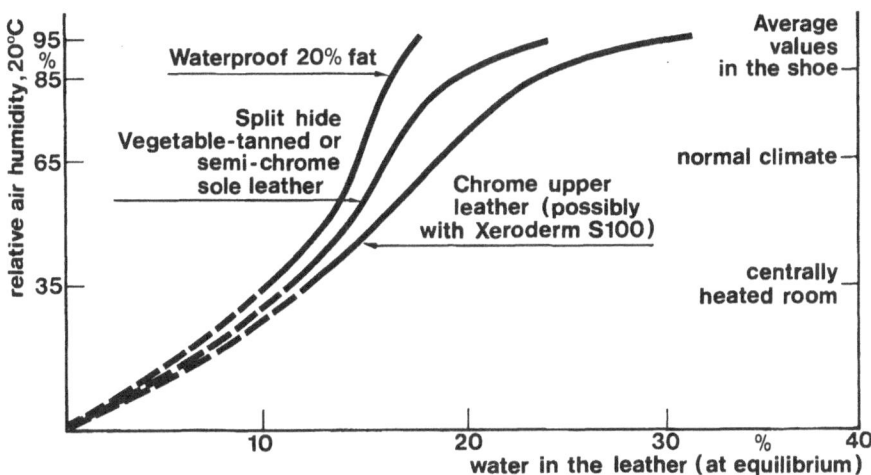

Fig. 5-5. Water vapor adsorption of different kinds of leather

The wear comfort of a shoe further depends on water vapor ad- and desorption [4] (Fig. 5-5). Hydrophilic materials offer excellent water vapor adsorption. In most cases this high water adsorption is connected with a low and slow, i.e. inferior, water desorption. Articles with a slow water desorption rate store the adsorbed water – wet shoes dry extremely slowly. Wet shoes also have a low insulation capacity against the cold: In addition to the wet feeling, feet feel cold in winter time and freezing may result. Besides the water vapor permeability the water ad- and desorption of the substrate is therefore also important [10].

Starch containing a PUR-dispersion may be transferred to a film. After the extraction of the starch a macroporous film is formed. It can be shown that the size of the pores determines the extent of water ad- and desorption. If the pores or the channels leading in them are under a particular diameter the water uptake increases due to capillary activity of the low sized pores. It is possible to describe this phenomenon by a mathematical formula [11].

The water vapor permeability of microporous films depends on the quantity of micropores. As will be shown later (see Sect. 10.1; Fig. 10-2), the amount of the added nonsolvent is responsible for the quantity and size of the pores: with increasing numbers and/or size of the pores the adhesion of a film with the pores, to a substrate, decreases. Microporous films, formed by a process of selective evaporation (see Sect. 9.2), are, by increasing the thickness, less permeable to water vapor [12].

Wearing tests with water vapor permeable materials in comparison to water adsorbing articles have been made: a non-permeable PVC coating on a water adsorbing cotton fabric was compared with the water vapor permeable GORE-TEX® membrane. The comparison was carried out at a low temperature (4°C) and a high temperature (60°C) as well as in wet (90% rel. humidity) and dry (10% rel. humidity) surroundings. The temperatures of the bodies of the subjects carrying out the tests were measured to study stress effects during the

Fig. 5-6. Absorbing water – change of the surface of leather. Leather increases its area by increasing the amount of absorbed water. This fits well with the wearing behavior of a shoe: During the daytime the foot increases its volume as does the shoe. After storage overnight the shoe loses absorbed water and decreases its volume again to the original shape. The feet during the sleeping period also decrease in volume. If a material has properties which differ, shoes made out of it transmit an impression as being "new" – a feeling people do not like

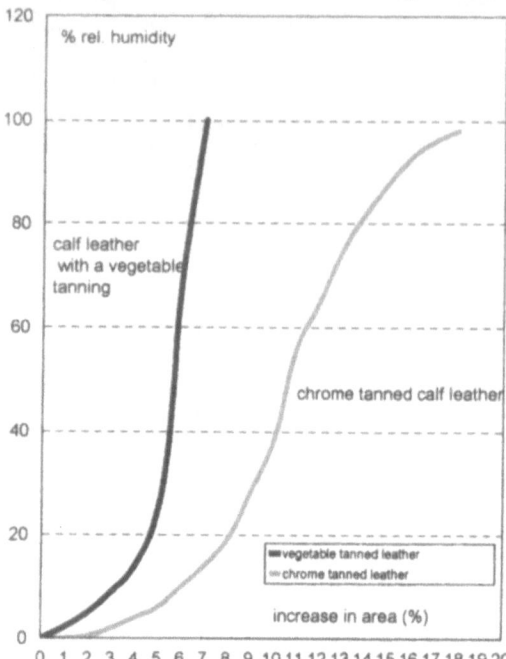

wearing trials. At the beginning of the wearing test no differences in the temperatures of the people were shown. After a longer wearing time of the products, the water vapor permeable material caused a lower increase in temperature and showed less stress to the person wearing those articles [13].

Water ad- and desorption should not result in a remarkable increase in thickness or area of the material exposed. A major problem of an average hydrophilic polymer is swelling upon exposure to water. Rain drops for instance may cause pustules if they come in contact with the surface of a coating with a hydrophilic polymer, which a customer would never accept. In addition these pustules have a low abrasion resistance – so the coating may be damaged easily.

Upon absorption of water, leather also increases its area slightly (Fig. 5-6). During the desorption of the water the area decreases to its original size. This change in the area correlates well with the person's wearing habits. During the daytime everybody's feet swell slightly. The change in the size of the shoes correlates with the increase in the volume of the feet [5]. During the night the volume of the foot decreases as well as that of the shoe, the shoe by desorbing the water. The next day the shoe fits the foot again.

Synthetic materials normally remain in the same size – every time they feel as if they were newly bought.

Synthetic materials do not all have the same deformatory behavior. Polyurethanes – due to their construction in segments – seem to be better suited to behave like leather than other polymers [8].

References to Part 1

1
Leather

1. FAO-Bulletin No. 48 "The World, Hides, Skins, Leather and Footwear Economy", Rome 1970, 1971 and "Pricing Policy for Hides and Skins", UNCTAD; CCP: HS/CONS 74/2 (Feb 1974)
2. H. Herfeld, „Die Bibliothek des Leders" Frankfurt; Vol. 1: H. Herfeld, B. Schubert, „Die tierische Haut"; Vol. 2: A. Zissel, „Arbeiten der Wasserwerkstatt bei der Lederherstellung"; Vol. 3: K. Faber, „Gerbmittel, Gerbung, Nachgerbung"; Vol. 4: M. Hollstein, „Entfetten, Fetten und Hydrophobieren bei der Lederherstellung"; Vol. 5: K. Eitel, „Das Färben von Leder"; Vol. 6: R. Schubert, „Lederzurichtung – Oberflächenbehandlung des Leders"; Vol. 7: H. Herfeld, „Rationalisierung der Lederherstellung durch Mechanisierung und Automatisierung – Gerbereimaschinen"; Vol. 8: L. Feikes, „Ökologische Probleme der Lederindustrie"; Vol. 9: H. Pfisterer, „Energieeinsatz in der Lederindustrie"; Vol. 10: J. Lange, „Qualitätsbeurteilung von Leder, Leder-fehler, -lagerung und -pflege"; Vol. 11: K. Mattil, W. Fischer, „Industrielle Fertigung von Schuhen"; E. Heidemann, "Fundamentals of Leather Manufacture", Darmstadt 1993
3. Stather F „Gerbereichemie und Gerbereitechnologie" Berlin 1967
4. O'Flaherty F Roddy WT Lollar RM "The Chemistry and Technology of Leather". New York 1956, 1958, 1964
5. Ullmanns Enzyklopädie der technischen Chemie, Vol. 16. Stichwort Leder, Weinheim 1978
6. Buljan J "Raw hides, preservation and trade", XXIII IULTCS Congress Friedrichshafen May 15–20, 1995
7. Sykes R (1995) Developing production. Leather, March, pp 17–22
8. Träubel H (1978) Die Entwicklung der Lederindustrie in aller Welt. Leder und Häutemarkt 37; (1979) Leather, July, pp 19–22; (1979) Das Leder 30:17; (1984) Tanning in industrialized countries. The Leather Manufacturer, May, pp 19–22
9. Träubel H Umwelt- und gesundheitliche Aspekte bei der Lederherstellung. Leder und Häutemarkt No. 26; 10.9.1993
10. Schweizer F:…"it sounds schizophrenic if tanneries situated in Germany working with a high ecological level are forced to finish their activities due to trickery of the public authorities", Leder und Häutemarkt Nr. 10 (19.5.95) p 5
11. Träubel H (1992) Leather. Chemtech, June, p 340

2
Coating of Textiles

1. Dingler's Polytechnisches Journal, vol. 171, p 312 citation according to Münzinger WH Kunstlederhandbuch, Berlin, 1950, p 3
2. Noll W (1968) Chemie und Technologie der Silicone. Weinheim, p 515
3. Hardtke G, Braeter K (1996) Beschichtung textiler Flächen mit wäßrigen Polyacrylat-Dispersionen nach dem Streichverfahren. Melliand, April, p 250
4. Rouette H-K (1995) Lexikon für die Textilveredlung, vols 1–3, Dülmen

5. Rouette H-K (1997) Trends in coating and laminating J Coated Fabr 26:241
6. Kubin I, Schreiber L, Djeha M, Dirschl F (1994) Silikonhaltige Spezialbeschichtungen auf wäßriger Basis für technische Textilien. Techn Textilien 37: Oct, p 149

2.1
Coating of Textiles by PVC Polymers

1. DRP 679 179 (Deutsche Celluloid Fabrik; 4.6.37/9.3.39)
2. DRP 685 839 (Kötitzer Ledertuch- und Wachstuchwerke; 8.7.37)
3. DRP 725 677 (I.G. Farben; 19.1.39/13.8.42)
4. E.g. for leather soles, see: Münzinger WH (1950) Kunstlederhandbuch. Berlin, p 24
5. Weiß F (1949) Die Verwendung der Kunststoffe in der Textilveredelung. Berlin, p 109ff
6. Spiess CH Leather manufacturing and related industries in Germany during 1939–1945. B. I. O. S. Report No. 27, pp 60, 78
7. Becker/Braun, Kunststoffhandbuch, vol. II-2. Polyvinylchlorid, München 1986, p 1100ff
8. Schmidt P (1963) Melliand 76, 186, 391, 1251, 1373
9. Schmidt P (1967) Beschichten mit Kunststoffen, München
10. Nass LI (1977) Encyclopedia of PVC, vols 1 and 2. New York
11. Leister K, Rösch G (1996) Farbmittel. Kunststoffe 86:965
12. Klaxmann J-D (1996) PVC-Stabilisatoren. Kunststoffe 86:987
13. Menzel B (1996) Weichmacher. Kunststoffe 86:992
14. H. Hurnik H, Facklam T (1996) Chemische Treibmittel. Kunststoffe 86:997

2.2
Coating of Textiles with Polyurethanes

1. Surveys: Becker-Braun Kunststoff-Handbuch, vol. 7 Polyurethane München-Wien 1993 chap 10.2, pp 621–630; Davies WD (1975) Urethane Coated Textiles. J Coated Fabr 4:205; Farkas F (1985) Plaste Kaut. 32:220; Farkas F, Stelczer T Kriterien für die Auswahl der Filmbildner zur Herstellung von Kunstleder nach dem Streich-Beschichtungsverfahren, Coating 4/1989, p 108–111; Gillibrand J (1974) Polyurethane coated fabrics. J Coated Fabr 3:55; Gasparrini FJ (1979) J Coated Fabr 3 and (1980) 105; Glenz O, Kassack F (1962) Über Textilbeschichtungen mit Polyurethanen. Melliand 42:323; Grant RR (1983) Beschichten und Kaschieren von Textilien. J Coated Fabr 12 No. 4; 196; Koch HJ (1970) Konstitution und Eigenschaften von Polyurethanen für die Textilbeschichtung. Melliand 1313; Koch HJ (1977) Polyurethan Textilbeschichtung, gestern, heute, morgen. Textil Prax Int 32 No. 3; 311, 323; Koch HJ (1971) Chemie und Anwendung von Polyurethanen in der Textilbeschichtung und Kaschierung. Melliand 1094; Koch HJ (1972) Neuere Aspekte der Textilbeschichtung mit Polyurethanen. Melliand 1272; Knafo G (1975) Structure-related properties of urethanes for coated fabrics. J Coated Fabr 2:142; Mann A (1995) Direct coating of urethanes on fabrics. J Coated Fabr October, 133; Pitzler G (1993) Beschichtung mit Polymerdispersionen und -lösungen. Coating 11:374; Popplewell D, Hole LG (1973) Urethane coated fabrics. J Coated Fabr 3:55; Santaniello A (1980) Chemie der lösemittelfreien Polyurethan Beschichtungen. J Coated Fabr 269; Weber KA (1968) Verwendung von Polyurethanen in der Textilindustrie. Textil-Praxis 525; Wittke W Beschichten textiler Flächengebilde, Coating 9/1988, p 340–345; Zinaman HJ (1979) J Coated Fabr 280; van Parys M Coating. Guimares, 1994; Wilkinson M (1997) A review of industrial coated fabric substrates. J Coated Fabr 26:87; a survey of machines and application devices is e.g. given in: (1963) Textil-Industrie 65:45; Zimmer JPM, Mayer K (1997) Versatile machine developments in coating technology. J Coated Fabr 26:188; Brocks J (1988) Beschichtungstechnologie – unterwegs zu neuen Ufern? Melliand p 66–70
2. JA 73 05 002 (Sun Star Chem. Ind. Co. Ltd.; 22.7.68/13.2.73)
3. DAS 1 112 041 (Bayer; W Thoma, O Bayer, H Rinke; 25.4.59/3.8.61)=FR 1 257 301=BE 590 114
4. US 3 678 011 (Allied Chemicals; JB Hino, B Taub, 13.1.70/18.7.72)

5. JA 73 39 829 (Kuraray Co. Ltd.; 23.6.70/27.11.73)
6. BE 594 629 (Chem.Werke Worms-Weinheim and P Spindler Werke KG; D Prior. 3.9.59; 1.9.60/30.9.60)
7. GB 1 098 505 (M. Noberasco; 12.3.65/10.1.68)
8. OS 2 116 162 (Hooker Chem.; Ch Wirth, G Albi; US Prior. 2.4.70; 2.4.71/14.10.71)
9. OS 2 442 686 (Dainippon Ink.; H Iwata et al.; JA Prior. 10.9.73 (=JA 101 234/73) 6.9.74/ 27.3.75)
10. GB 1 270 848 (Nairu-Williamson Ltd.; K Norcross, H Gofts; 5.8.69/19.4.72)
11. GB 1 050 459 (Kohkoku Kagaku Kogyo KK; K Takase; 11.12.63/7.12.63)=JA 48 857/63
12. US 3 574 106 (Plymouth Rubber Co. Inc.; LD Bragg; 2.10.68/6.4.71)
13. BE 736 798 (Pirelli; A Angioletti, G Ferrante; IT Prior. 3.8.68; 30.7.69)=IT 19 797/68
14. JA 73–18 803 (Toray Ind. Inc.; 4.11.68/8.6.73)
15. OS 2 145 510 (H. Pannenbecker; 11.9.71/22.3.73)
16. OS 1 469 575 (Toyo Rubber Ind. Co. Ltd.; H Matsushita et al.; JA Prior. 27.12.63 (=JA 70 432/63; 29.9.64/2.1.69)=FR 1 414 241)
17. Schröer W (1987) Die Beschichtung von Textilien mit Polyurethanen, Conference 11.12.1986, Zürich; Textilveredlung 22:459
18. DE 3 313 237 (Bayer, W Thoma et al.; 13.4.83/18.10.84); DOS 2 814 173 (Bayer; W Thoma, G Berndt, J Pedain, W Schröer, W Kling; 1.4.78/11.10.79)
19. DOS 3 004 327 (Bayer, F Komarek; 6.2.80/13.8.81)
20. Lasman HR (1975) Vinyl and urethane coated fabrics for shoe uppers. J Coated Fabr April, p 256
21. Brandt TJ (1974) Urethane coated fabrics for shoe uppers. J Coated Fabr July, p 3
22. Morley DJ, Symonds WE (1977) Polyurethane coated upholstery fabrics. Textile Institute and Industry April, May and June
23. Offermann P, Janssen B, Giessmann A (1996) Universal Beschichtungsanlage für die Textilindustrie. Melliand 794
24. Conway R (1995) How to achieve added value by coating. J Coated Fabr 25:69
25. Pollano G (1997) Crosslinking with aziridines. Polym Metr Sci Eng 77:383
26. Shaffer M, Wicks D (1997) Two-component waterborne polyurethanes. Polym Mater Sci Eng 77:377
27. JA 92 48 891 (Kuraray Co. Ltd.; S Kaneda, H Shunji, K Hirai; 14.3.96/22.9.97)=JA 96–57 438
28. PCT WO 97 46 627 (DSM NV; Japan Synthetic Rubber Co. Ltd.; Japan Fine Coating Co. Ltd.; T Takahashi et al. 5.6.96/11.12.97)=JA 96–143053
29. (1993) Polyurethane Handbook. Munich Vienna New York, p 456
30. JA 10 58 576 (Asahi Kako KK; H Matsunami et al.; 23.8.96/3.3.98)=JA 96–241 186

2.3
Economic Figures

1. Anon. (1997) Weltmarkt für technische Textilien 42 Mrd. $. Techn Text April, 112
2. D. Ridge et al. "The world technical textile industry and its markets: prospects to 2005"; a report prepared for the Techtextil Messe Frankfurt GmbH; April 1997
3. Anon (1997) Kuraray plans man-made leather sales expansion by 40%. JTN March, p 32
4. Anon (1997) Teijin increasing man-made leather output by 13%. JTN March, p 32
5. Anon (1997) Man-made leather grows in production and demand. JTN June, 66

3
Definitions

1. Herfeld H, Königsfeld G (1965) Das Leder. 16:229 and (1968) Leder- und Häutemarkt 20:154 (techn. Beilage)
2. van Vlimmeren PJ (1974) Leder- und Häutemarkt (Gerbereiwissenschaft und Praxis) Mai 20:86
3. (1977) Das Leder 150; Aktenzeichen (BGH I ZR 152/75)

4. Katsumi Hioki Leather-like materials. In: Kirk Othmer encyclopedia of chemical technology, 4th ed, vol. 15, New York; Kruse H-H (1990) Leather imitates. In: "Ullmann's encyclopedia of industrial chemistry" 5th ed, vol A15, Weinheim
5. Scheibe R, Wolff H (1997) Eigenschaftsänderungen von Chromleder bei Klimawechsel. Das Leder 6/7, 134

4
Water Vapor Permeability

1. Everett, Stone (eds) (1958) The structures and properties of porous materials. London
2. US Patent 3 000 757 (DuPont; RA Johnston et al.; 28.1.57/19.9.61)=GB 849 876; DAS 1 419 147 (DuPont; EK Holden; 17.3.59/6.3.69; US Prior. 25.3.58); DAS 1 419 148 and DAS 1 419 149 (DuPont; 17.11.59/13.12.68)
3. Saier HD, Strathmann H (1974) Untersuchung des Membranbildungsmechanismus asymmetrischer Filtrationsmembranen. Chemie-Ing-Technik 46:109; Scheuble P (1970) Zur Bildung von asymmetrischen Zelluloseacetatmembranen zur Meerwasserentsalzung. Dissertation, Aachen; Strathmann H, Saier HD (1973) The formation mechanism of asymmetric reserve osmosis membranes, 4th Int. Symp. on Fresh Water from the Sea, vol. 4, pp 381–394; Saier HD, Strathmann H (1975) Asymmetrisch strukturierte Membranen-Herstellung und Bedeutung. Angew 87:475; Oehme Ch, Martinola F (1973) Removal of organic matter from water by resinous adsorbents. Chem and Ind 823; Martinola F, Richter A (1970) Makroporöse Ionenaustauscher und Adsorbentien zur Aufbereitung organisch belastete Wasser. Jahrbuch vom Wasser 37:1
4. DAS 1 110 607 (DuPont, RA Johnston et al. 28.1.58/13.7.61, US Prior. 28.1.57)
5. DRP 275 697 (M Wünschmann et al.; 30.10.13/23.6.14)
6. DIN 53 333 and IUP 15; see (1961) Das Leder 12:86
7. Spiers CH (1967) Leather 165:389
8. Kesting RE (1971) Synthetic polymeric membranes, New York
9. GB 1 493 822 (Inmont Corp.; 19.9.73/30.11.77)=US 73 398 696 and US 74 47 4406
10. Gennis RB (1989) Biomembranes, New York

5
Microporous Products: Comparison of the Structure and Properties

1. Stather F (1967) Gerbereichemie und Gerbereitechnologie, Berlin
2. Träubel H (1975) Leder und seine Substitute. Leder 1
3. Graßmann W (1961) Kollagenforschung unter dem Gesichtswinkel der Praxis. Leder 165; Stirtz M (1975) Beitrag zur elektronenmikroskopischen Struktur der Rindhautepidermis. Leder 155; Zahn H, Wortmann F-J (1997) Kollagen als hydroplastisches Material Das Leder May, pp 110–116
4. Diebschlag W (1972) Klimatechnische Untersuchung der physiologischen Zusammenhänge im Funktionssystem Fuß-Schuh unter Verwendung verschiedener Schuhschaftmaterialien, Dissertation (Marburg/Lahn); Diebschlag W, Müller-Limmroth W (1971) Leder- und Häutemarkt 23:212, 260; Müller-Limmroth W, Diebschlag W (1971) Das Leder 22:1; idem (1974) Schuh-Technik 315; Müller-Limmroth W et al. (1975) Behaglichkeit in der Fußbekleidung. Leder- und Häutemarkt 330; Seligsberger L "Comfort factors in leather footwear", US Army Natick Laboratories, Report No. 17 August 1963, Natick/Massachusetts, USA; Diebschlag W (1975) Das Dehnungsverhalten verschiedener Schuhschaftmaterialien sowie deren maximale Anpreßdrucke auf den Fuß beim Gehen. Das Leder 26:7; Diebschlag W (1973) Vorzüge und Nachteile von Leder und Synthetics bei Verwendung als Schaftmaterial. Leder- und Häutemarkt 25:682; Herfeld H, Königfeld G (1965) Über einige grundsätzliche Vorteile von Leder gegenüber Austauschstoffen bei Verwendung für Schuhe und Bekleidung. Leder und Austauschstoffe III, Das Leder 16:229; van Vlimmeren PJ (1974) Kraft-Dehnungs-Eigenschaften und Formstabilität von Leder und synthetischen Materialien. Leder- und Häute-

markt May, 86; Brooks FW, Mitton RG (1968) Wear trials for the comparison of leather and synthetic shoe upper materials. J.S.L.T.C. 52:42; Pepper KW (1966) The challenge of synthetics to leather. Chem and Ind 2079; Schubert R (1975) Untersuchungen über das Wasserdampfspeichervermögen von Leder und anderen Schuhmaterialien. Das Leder 26 (1975) p 1–6; W. B. Beeinflußt der Schuh die Gesundheit. Schuhtechnik 66 (1972) p 472–473

5. Martinelli B (1957) Chem Rundschau 466

6. Saitoh Y, Takayama G (1982) Artificial Leather. Nippon Gomu Kyokaishi 55(3) (1982) p 47–52

7. Karasek O, Hadobas F (1982) Die Entwicklung von poromerischen Kunstledern mit hohem Tragekomfort. Schuh-Technik December, 1071

8. Kellert H-J, Reich G (1978) Aufbau von Poromeriks als Schichtenverbund und deformationsmechanisches Verhalten der Aufbaukomponenten. Leder Schuhe Lederwaren June, 270; Reich G (1991) Leder und synthetische Austauschmaterialien heute – eine vergleichende Betrachtung. Leder- und Häutemarkt 43(2):1 and 43(5):6

9. Whittaker RE (1972) Structure and viscoelastic properties of poromerics. J Coated Fibrous Mat July, p 3

10. Jansen Y, Rouette HK (1992) Vorbehandlungseinflüsse von Polyestersubstraten auf den Feuchtetransport von Zeltgeweben mit kompakter oder mikroporöser Polyurethanbeschichtung. Melliand, September, 748; Jansen Y (1990/1991) Feuchtetransport durch kompakte oder mikroporöse Polyurethanbeschichtung auf Polyesterzeltgewebe in Abhängigkeit von der Vorbehandlung. Diplomarbeit, Krefeld

11. Buschmann HJ, Schollmeyer E (1992) Kinetische Untersuchungen zur Aufnahme und Abgabe von Wasser durch makroporöse PUR-Folien. Melliand, September, p 745

12. Becher D, Barthau R Herlinger H (1995) Polymer-Faser-Haftung bei Beschichtungen. Melliand, April, 263-

13. Kenney WL, Hyde DE, Bernard TE (1993) Physiological evaluation of liquid-barrier, vapor-permeable clothing ensembles for work in hot environments. Am Ind Hyg Assoc J 54:397

14. Zorn B (1982) Einige grundsätzliche Bemerkungen zur Hydrophobierung von Leder. Das Leder 33:79

15. Painter CJ (1997) Waterproof, breathable fabric laminates: a perspective from film to market place. J Coated Fabr 26:107

Part 2

Ways To Create Water Vapor Permeability by Elimination of Solid or Liquid Products

Pores can be obtained after the removal of soluble or combustible substances present in a material. The substances should not be miscible with the material. After elimination of these substances more or less large pores remain in the material.

CHAPTER 6

Ways To Create Microporosity

Leaching of salts is the best known process for creating microporosity. This process to remove included products has been used for a long time: straw or salt is incorporated into a raw mass like clay. Then the clay is formed into bricks or plates and baked. Straw is eliminated by burning, salt must be leached. Another example is the production of diaphragms: Salt is incorporated into cement, then a plate is formed, hardened and afterwards the salt is leached. Diaphragms suitable for the electrolysis of sodium chloride to produce chlorine and sodium hydroxide are produced by this method [1]. In both cases articles with pores are created.

Today leaching of salts or other materials is broadly used to create microporosity. The only major differences are in the art and manner of the leached material.

Examples to create microporosity are:

- leaching of salts (Chap. 7),
- leaching of organic materials, like solvents (Chap. 8),
- or leaching of oligomers and polymers (Chap. 8.6)

The following methods differ:
- evaporation (Chap. 9),
- blowing (Chap. 11),
- controlled melting processes, sintering (Chap. 12),
- perforation (Chap. 7), and
- stretching of crystalline polymers (Chap. 16).

Many of these processes are not carried out singly – often a combination of the methods is used. There are known ways of leaching a solvent together with a salt, or an evaporation process which is followed by leaching, etc.

CHAPTER 7

Elimination of Solid Particles – Especially Leaching of Salts

A solid product is incorporated into a solution of a polymer. The incorporated solid substance must not be dissolved by the solvent used for the polymer. After impregnation, coating or film forming and drying the solid substance is extracted with a non solvent for the polymer [1–5, 13, 14]. At every place that the incorporated product has been, pores are created.

Polymers [7, 57] which are suited are: polyurethanes [40], polyamide [5, 8, 17, 19], soft PVC [10, 12], rubber [9, 51, 54], copolymeres of olefinic monomers or their derivatives [6, 16, 30].

This method of leaching salts is mainly used for solutions of polymers. Plastisols [17, 26], which may contain additional solvent [11, 23, 33, 44], can be used in the leaching process too. A plastisol is a suspension of PVC in powdered form in a plasticizer or a mixture of plasticizers.

A special method of leaching is the incorporation of a solid substance in a thermoplastic material on a calender [22–25, 47, 49]. After film forming the substance may be leached out especially under tension. Polyurethanes are especially mentioned [50] for this method.

Thermoplastic polymers, such as PVC plasticizer-plastisols, are treated with a water-soluble substance like sodium chloride, sucrose, PVA, starch, etc. and transformed into a film. Then the material is treated with a swelling solvent, the water-soluble substance is leached, then the foil is heated to eliminate the swelling agent [62].

Polyolefins may be treated with a plasticizer like dioctyl phthalate, transformed to a foil, then treated with water to eliminate the plasticizer at least partially. The microporous foils are suited to the filtering or packaging sector [63].

Leaching may also be used for porous adhesive coats [15, 34, 41]. An impregnation of nonwovens [42, 45, 46] or fabrics [56] will become porous after leaching out incorporated substances.

Besides salts like ammonium chloride [22], sodium chloride [36, 42, 43, 48], sodium hydrogen carbonate other electrolytes have been mentioned. Oxides [7] are also suited. Organic substances like caprolactam [59], sugar [16, 42] and urea [42] are suited for a leaching process. Silica particles in PVA films can be leached by lithium, sodium or potassium hydroxide [64].

High molecular products [28, 31, 34, 38] can also be leached according to this method: polyvinyl alcohol [13, 20, 27, 29, 37, 52, 55, 61], polyvinylpyrrolidone (5 [63]), carboximethyl cellulose [34, 42], powdered cellulose [56] polyacrylate [42, 45] and starch [9, 33, 51, 53]. Sometimes these polymers also assist in the application of polymers by modifying the viscosity, etc. [60]. The leaching of starch may be assisted by the addition of an enzyme in the leaching bath [51, 53, 70].

PVC-plastisols are mixed with a water-based solution of polyvinyl alcohol to get a water-in-oil emulsion. The emulsion is transformed into a film, heated to gel the PVC and polyvinyl acohol is leached out afterwards [72].

Natural polymers [59] like starch, casein, or gelatin may be leached in textile jet-dyeing machines by the assistance of sodium sulfosuccinic ester of 2-ethylhexylalcohol as an emulsifier [58]. Clay [24] can also be leached.

Not only powders but also specially formed components like fibers [21, 35, 44, 55] are suited for the leaching process. Water is the mostly used leaching agent. Acids to assist the leaching of carbonates of earth-alkaline metals [56] or emulsifying agents [41] may be added to water. Organic substances like wax may be leached with solvents [18].

To assist the formation of pores, blowing agents may also be added [9,45]. Thermoplastic compounds like PVC-plastisols are treated with salts, emulsifiers and a polymer, able to build a "channel" in the film. After leaching of the salt a microporous sheet is formed [65].

Borax, which is able to split off water when it is warmed up [32], is also named as an additive.

Leaching may also be used in an indirect process [67]: microporous films can be produced by coating a releasing surface with a salt-containing polyurethane solution in DMF. After evaporating the solvent, transferring the film onto a textile substrate and stripping the releasing surface, salt is washed out. A microporous coating is then produced [69].

A two-component polyurethane mixture with an excess of isocyanate is coated on a releasing surface, then a polyurethane solution in DMF containing a leachable substance is applied. The polyurethane is coagulated and the substance leached. The isocyanate excess in the first coat improves the adhesion of the first to the second coat [68].

Nonporous films of a polyurethane containing keratin or leather powder are treated after application with an enzyme, a protease, to create microporosity [73–75]. After film forming, the incorporated substances are surrounded by the polymer. It is difficult to get the leaching agent into contact with the incorporated substances. Extending and stretching of the films during the process improves the leachability of the incorporated substances – even an electric field may help [5,9].

Water-based mixtures may also be applied: a polyurethane dispersion and a PVA solution in water are mixed and coated onto a nonwoven textile. After evaporation of the water, polyvinyl alcohol is extracted from the coating which becomes microporous [71].

The leaching technique is also suited for impregnations: Textile substrates may be impregnated with a polyurethane solution in an organic solvent containing 1–20-µ particles of sodium or potassium chloride, sulfate or carbonate. Preferably the solution should contain an alkylene oxide adduct of a phosphorus ester. After evaporating the solvent, the salt is leached [66].

Advantages of the leaching method are that leaching can be broadly used – almost no special equipment is needed, and it is easy to incorporate salts or other substances. Disadvantages are: the leaching liquid contains the leached products; only manufacturing plants situated on the coast are able to elute salt and to get rid of the spoiled solutions. Special permission from the local authorities is required which today is no longer available. Besides the evaporation of water there exists almost no other treatment of the solutions to eliminate salt and this method cannot be run economically.

The size of the particles incorporated determine the size of the pores. An intensive grinding and bolting of the salts is necessary. The biggest disadvantage is to dissolve the substances incorporated after the films or coatings are formed. Even stretching operations during the leaching process are not able to help to leach out the incorporated substances completely. Microphotos of articles being produced by this method, many times, show traces of incorporated salts.

Products which have been produced by this method are HYDROLETTE® and LUDOLETTE® (both trademarks of Göppinger Kaliko).

7.1
The Coagulation Process

Coagulation is the method mostly used today to produce microporous sheets or impregnations [1]. Coagulation means an internal precipitation of a dissolved polymer by a nonsolvent [8]. The coagulation process is often applied with polyurethanes dissolved in dimethylformamide (DMF). DMF is hygroscopic so the precipitation occurs gradually by absorption of water by the DMF.

A DMF solution of a polyurethane is transferred to a film which is exposed to a humid atmosphere. The solvent DMF, by gradually absorbing water, becomes less and less soluble for the polymer. The polymer dissolved in DMF remains in the continuous phase, the nonsolvent penetrates this phase and separates in it in the form of discontinuous droplets. After washing and/or squeezing out the solvent, the polymer remains in the form of a film with fine interconnected pores.

A similar result is obtained if a freshly prepared film, impregnation or coating is dipped into a bath consisting of a DMF/water (80:20) mixture. The material then passes batches of DMF/water mixtures with increasing water and decreasing DMF content until it is finally washed by water alone.

If a polymer/solvent mix is treated with a nonsolvent in a way that it precipitates the polymer as a powder then the discrete particles are unable to form a film [14]. These powders are – besides controlled melting (see Chap. 12) – useless for water vapor permeable materials and they will not be mentioned again.

To accelerate the time of coagulation a limited amount of nonsolvent can be added to a polymer solution. Exceeding this amount the polymer precipitates; a film cannot be made anymore. It is possible to determine the ideal amount of nonsolvent which may be added by a titration process [19]: 50 g of a 5 % elastomer solution in DMF should be titrated by stirring with a water/DMF mixture (1:1) until turbidity occurs. The amount of water/DMF mixture in ml needed to get this turbidity is regarded as the point of turbidity. This point gives a hint as to how much water could be added to a polymer solution to accelerate the coagulation. Coagulation under the influence of added nonsolvent is described in detail in Chap. 8.4.

The principles of the coagulation process in the manufacture of an artificial glazed kid leather were described as early as 1913 [4].

The coagulation process may be applied in the production of films, coatings and impregnations. The polymer is dissolved [3] or even produced [17, 18] in dimethylformamide (DMF), dimethylacetamide (DMA) or tetrahydrofurane. Substrates may be coated or impregnated [5] with these solutions. Then the substrate is treated – possibly in several steps – with a nonsolvent [6]. The nonsolvents should not dissolve or even swell the polymer and should mix well with the solvents used. The nonsolvent [16] most often used is water. The treatment of the film etc. with a nonsolvent may be stopped when ca. 98 % of the solvent is washed out. Polymers [2] suited are polyethylene, polypropylene, polyamide [13], polystyrene, vinyl (co)polymers [9, 15, 20], polyurethanes [8, 11, 12] (see Chap. 8) etc.

It is also possible to get hydrophilic fibers by a coagulation process. Fibers can be produced by the addition of a major part of tertaethylene glycol to a solution of polyacrylonitrile in DMF and spinning the polymer at a temperature just below the boiling point of the solvent. These fibers have an increased water retention and a core-shell structure [26].

Mechanical devices against spoiling of the surface of the coagulate by dust [21], by treating the wet coating with a knife to get a smoother surface [22] or to coat the backside of the coated substrate as well [23] have been published. To accelerate the coagulation speed the possibly warmed nonsolvent water is directed towards the coated surface at high speed [24]. Equipment for the coagulation process of polyurethanes is shown in Fig. 7-1.

With an increasing temperature a polymer becomes more and more soluble in solvents. This effect is described by the Gibbs–Helmholtz-equation. At a higher temperature it is sometimes impossible to get micropores with the coagulation process. This point also has to be taken into consideration in the drying

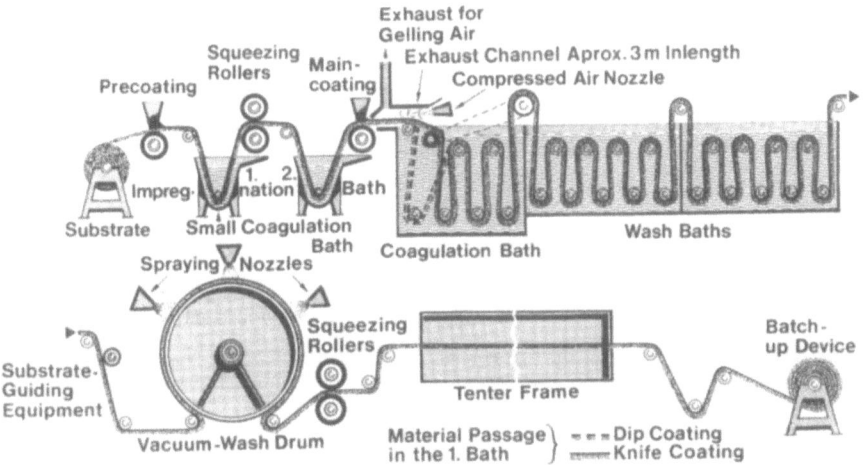

Fig. 7-1. Schematic representation of a coagulation equipment (8 [13])

Fig. 7-2. Microphoto of a polymer film with micropores produced by a coagulation process

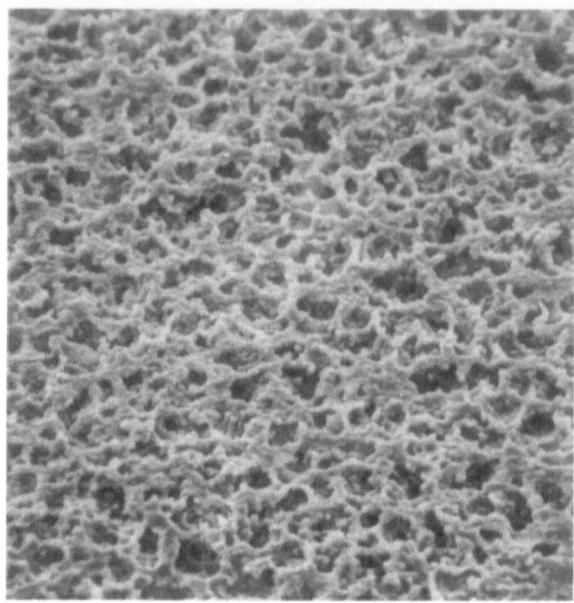

1 μm

process. As long as there is too much solvent present in a film or coating, a microporous structure may collapse during the drying operation. This may occur by the increased solubilizing power of the solvent at a higher temperature and due to the softening effect of the solvent still present in the polymer. The production parameters, like temperature, humidity of the nonwoven bonded by a coagulated polyurethane and the thickness of the coating, can be optimized by the help of a mathematical equation [25].

The adverse effect to heating may be used by the following method. A solution of a polymer in a weak solvent is cooled down. The weak solvent will then become a nonsolvent. The polymer coagulates. The solvent which now is a nonsolvent [7, 10] is removed at this low temperature. Improvements, e.g. to get a dust-free immersion of the coated textile into the coagulation bath [21], or using a special knife to get a smoother surface [22], have also been published. Figure 7-2 shows the typical structure of micropores produced by a coagulation process.

Advantages of the coagulation process are:

– A great variety of polymers can be used. A lot of articles can be produced, and
– The necessary solvents and nonsolvents can be recovered easily and reused.

Disadvantages of the coagulation process are:

– The mostly used DMF is expensive and abortive, and
– The necessary investment for the equipment to recover the solvent is high due to the fact that DMF, due to bacterial action, develops corrosive formic acid. Therefore stainless steel is necessary.

7.2
Coagulation of Polyamide and Other Polymers

The coagulation of polyamide to coat or impregnate [4, 16] substrates is done by the following processing steps: polyamides belong to a group of polymers which are not easy to dissolve. Besides formic acid, methanol together with a metal chloride of an alkaline group element or a mixture of different solvents is able to dissolve polyamides. A solution of a suited polyamide [26, 31, 37] is, for example, prepared in lithium chloride/methanol [8, 10, 17, 22, 36, 38, 40], lithium chloride/cyclohexanol [46], methanol/methylene chloride/water or formic acid [3, 28]. This solution is used to impregnate [30, 32, 41, 42] or coat a substrate. Then the substrate is dipped in water to coagulate the polyamide. As previously described, water penetrates the solution of the polymer – slowly doing so – the polymer precipitates with fine water droplets included which results after drying in a microporous structure. Solutions of polyamides can be used for coatings and impregnations of textile substrates [4, 10].

N-Hydroxymethyl derivatives or the N-methoxymethyl ethers [9, 18, 23, 39] of polyamides are more soluble than the polyamides themselves. These derivatives of polyamides may also be used in the coagulation process.

To increase the speed of coagulation it is possible to evaporate a part of the solvent [27] or to add minor amounts of a nonsolvent before coating or impregnating [24, 33]. If the coagulation mixture is cooled prior to coagulating (7.1 [7]), the polymers coagulate easier [10, 35].

Mixtures of polyamides with other polymers [15] may also be used in the coagulation process. Polyvinyl alcohol [1], polyvinyl butyrate [25], polyurethane [5, 7, 14, 20], cellulose xanthogenate [11] or butadiene–styrene copolymers [21] may be added.

A polyamide dissolved in a water-based solution of hydrochloric acid may be coated on a textile substrate. The coating is coagulated in a sodium hydroxide solution [48].

Leachable salts (7 [6]), crosslinking agents like maleic or citric acid [12], or isocyanate prepolymers [19] can also be used as additives in the coagulation of polyamides. To get a smooth surface volatile solvents may be added [13].

Nonwovens containing fibers of polyamide may be treated with a solvent for the polyamide; after the coagulation the nonwoven is bonded by the coagulated polyamide [29, 43–45]. In this special process polyurethanes [34] may also be present. Polyvinylidene fluoride is dissolved in acetone. The solution is coated on a releasing conveyor belt, immersed in an acetone/water bath separated from the conveyor belt and then washed with water [47] to get a microporous structure.

CHAPTER 8

Coagulation of Polyurethanes

The polymer most often used for the coagulation process is polyurethane. Nearly all leather substitutes sold in the world consist partially or totally of polyurethanes. Therefore a whole chapter is dedicated to the coagulation of polyurethanes.

In 1951 a coagulation process for a polyurethane was published for the first time [1]. Since 1954 [36] additional process steps have been developed to produce microporous polyurethanes [32].

The structure of the soft segments of the polyurethanes influences the coagulation of the polyurethanes remarkably. Soft segments with a higher hydrophilic property result in polyurethanes with a better coagulation behavior. The structure of the hard segments is of minor importance [33].

Only linear polymers have good solubility in organic solvents. Crosslinked polymers are normally insoluble in solvents; due to a certain decomposition crosslinked polyurethanes can be dissolved in heated DMF because hot DMF is able to split branched polymer chains which results in a linearity of the polymer-chains. The NCO/OH ratio determines the linearity of the polyurethane chain. A ratio NCO/OH <1 results in easily soluble OH-terminated polyurethanes [35], NCO/OH >1 results in crosslinked, poorly soluble ones (details see Chap. 25.1).

The following principally different process types are known:

- *Microporous films* may be produced by applying a polymer solution onto a stainless steel band or a glass fiber fabric coated with a fluorine polymer (TEFLON® by DuPont) dipped in water, stripped off the band or fabric, washed and dried.
- The polymer solution may be applied in a *direct coating* process onto a fabric or a nonwoven and coagulating in a bath of water as a nonsolvent [20, 23, 24, 30]. After washing and drying the coated substrate is ready to sell or only needs an additional finishing step (see Chap. 22). This method is especially suited if the substrate is treated with the coagulation solution too. Sometimes it is difficult to get a smooth surface. Air bubbles may show as open pores on the surface. This may be a disadvantage of the direct coating process.
- The *indirect process* involves a releasing fabric which is, usually, coated with a thin polyurethane layer, then with a second top layer or finish. After a short drying of the first layer, the polyurethane solution in DMF is applied and a textile laminated into the wet solution layer. After coagulation, washing

and drying the coated finished article is stripped off the releasing substrate. Materials manufactured by this process have an extremely smooth surface. It is possible to use a silicon rubber mold as the releasing substrate (8.1 [26]). It is also possible to wet the releasing substrate with water, before coating with the solution and coagulating at higher temperature [27]. This process is much more complicated than the others mentioned; therefore, it is not normally used.

- *Impregnating* textile substrates is used technically. Fabrics [29, 31, 34], knitted textiles or nonwovens [25] may be impregnated with solutions of polyurethanes, able to be coagulated, dipped in water, washed, dried and buffed. This impregnation process of nonwovens will be discussed in detail in Chaps. 17.1 and 19.

- To simplify the process modification of the polyurethanes during the coating and coagulating process has also been reported: Dissolved polyurethanes may be treated with an isocyanate. A part of the polyurethane then is no longer soluble in the solvent and precipitates. This step [18] facilitates the coagulation process. Coagulating in sodium sulfate solution [26] has also been reported. Additives help to speed up the coagulation [13, 14].

Finished articles, such as shoe uppers etc., may also be produced in a coagulation process [19]. Besides one article, PORVAIR®, foils need to be laminated onto a textile substrate.

- Foils are usually laminated by a spraying, printing or coating process onto fabrics, split leather or nonwovens. Laminating does not result in extremely soft articles. Soft articles will be achieved more easily by impregnating of nonwovens.

The first synthetic leather substitute produced in a coagulation process was CORFAM® (DuPont Fig. 8-1 [2]). Corfam consists of a nonwoven impregnated by a coagulated polyurethane, a woven textile interlayer and a coagulated polyurethane top coat. Other materials have a similar composition. Only the textile interlayer is a special feature of Corfam. The production of Corfam was discontinued in the early 1970s, as well as the production of XYLEE® (Glanzstoff [3]), DESMODERM® foil (Bayer, Fig. 8-2 [4, 15, 16]) and other materials [10]. CLARINO® (Kuraray, Fig. 8-3 [5, 10]), PORVAIR® (Porvair) [17], and the artificial suede articles ECSAINE® (Toray) [6], ALCANTARA® (Iganto, Fig. 8-4 [5, 7, 8]), ASTRINO® (Kuraray, Fig. 8-5 [9, 11]) and JANECK® (Dai Ichi Kasei, Fig. 8-6 [12]), as well as polyurethane products for the coagulation process like DESMODERM (Bayer) and ESTANE® (Goodrich), are still on the market. Besides these articles there are also others that have been around for a longer time now, produced by chemical or fiber producing companies, as well as special brand names of coating companies (see Chap. 29.1).

A patent application was published with the aforementioned composition, i. e. a combination of a nonwoven containing microfibers with a microporous [21] or a homogeneous [22] top layer was claimed recently.

Fig. 8-1. CORFAM®. Three distinct layers can be seen: The base structure is an impregnated nonwoven, then a fabric can be distinguished (the 4 regular round spots corresponding a thread in warp direction) and the microporous top layer

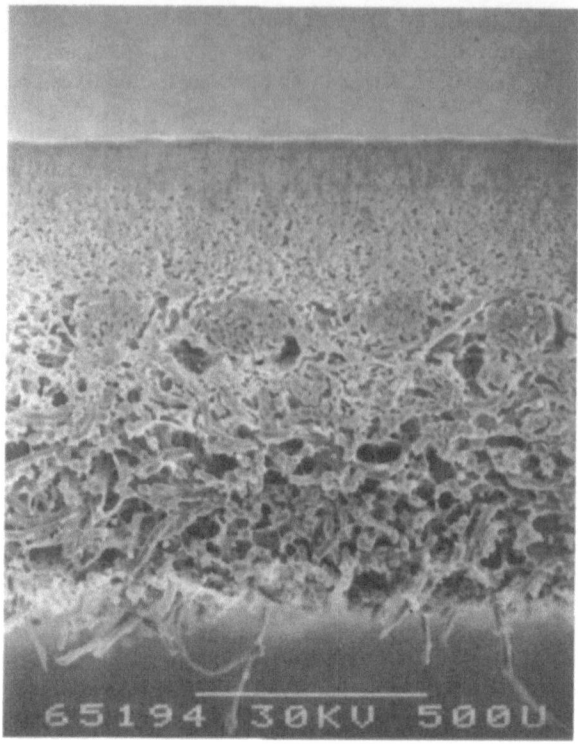

Fig. 8-2. DESMODERM® foil. The microporous polyurethane foil has no textile substrate because the foil should be used to laminate fabrics or split leather

Fig. 8-3. CLARINO®. Two types of Clarino, one with a coarser pore structure in the top layer, the other with a rather regular pore structure. The materials resemble Corfam – only the fabric intermediate layer is missing

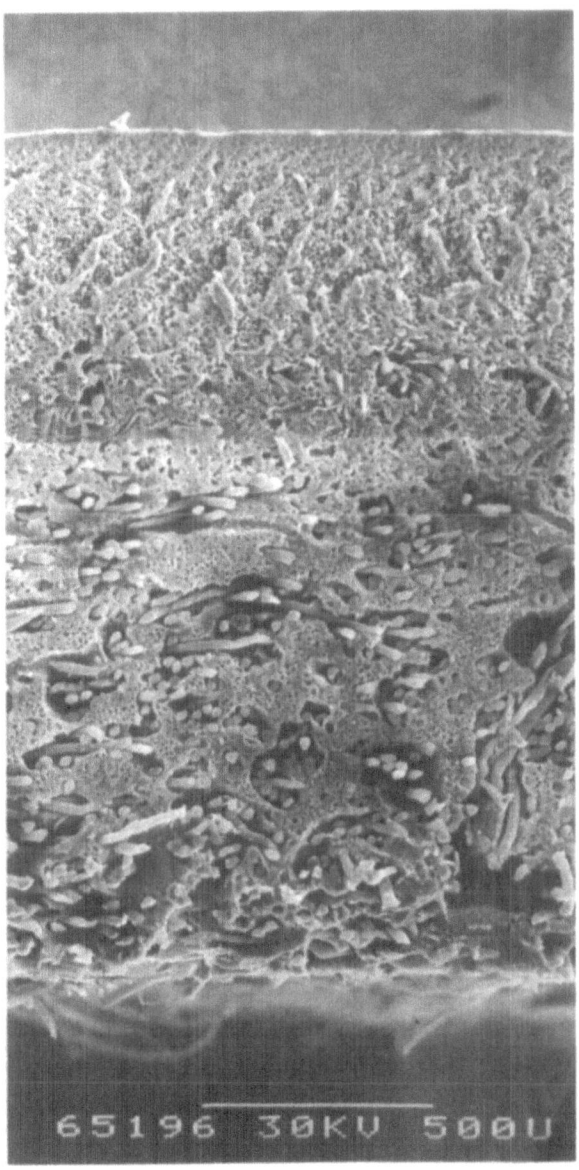

Fig. 8-4. ALCANTARA®.
Alcantara as a synthetic
suede-like material which has
no coated, closed surface.
Its surface consists of microfibers. The whole material
consists of a nonwoven bonded by a coagulated poly-
urethane

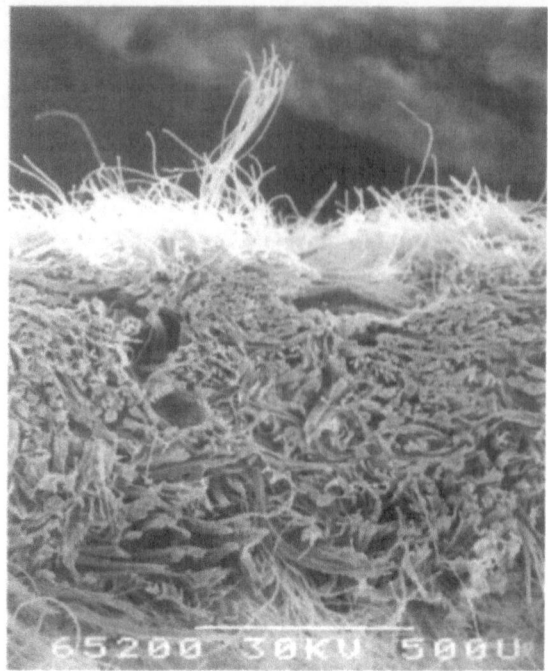

Fig. 8-5. ASTRINO®

Fig. 8-6. JANECK®

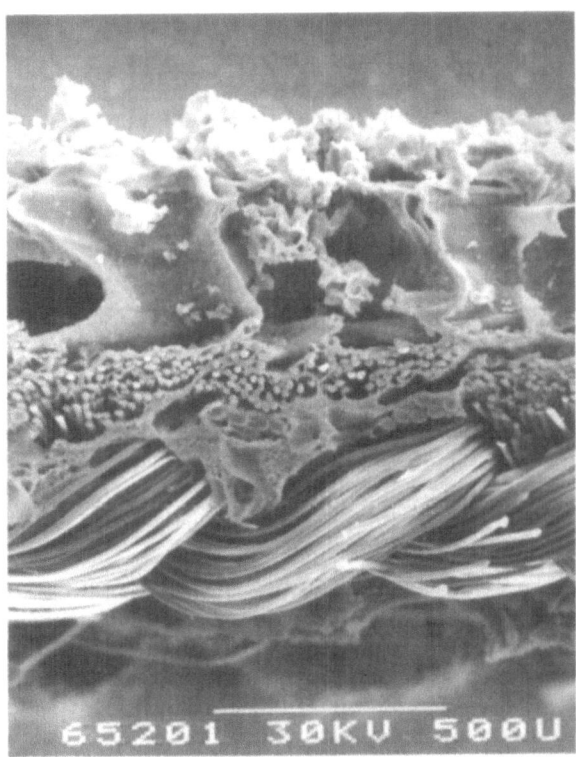

8.1
Coagulation of Polyurethanes Without Additives

As shown in the previous chapter, basically no additives are needed to get poly-urethanes coagulated. Nevertheless, in some cases, it is impossible to get poly-urethanes coagulated at all or it is difficult or not fast enough. In many cases the leaching of the solvent needs too much time. If films are dried while still contain-ing some solvent, then the micropores will collapse or be damaged by this operation.

The easiest way to improve the speed of coagulation is to change the tempe-ratures of the system. Increasing the temperature between 80 °C and the boiling point of DMF [1], water vapor [26], hot [2] or a heated (up to 70 °C [3]) non-solvent bath may be applied. Heating the solution, adding nonsolvent and cooling down [25] is also a method for improved coagulation.

Cooling down the coagulation solution to 5 °C [4] or below 10 °C and adding calcium chloride or sodium sulfate [5] also improves the coagulation. Even the cooling of a coagulation solution after being coated onto a substrate, by means of solid carbon dioxide or liquid nitrogen [6] has been reported.

Usually it is easier to get a polyurethane coagulated in a substrate [29] rather than as a film or coating. Impregnation of a nonwoven followed by a direct coating – both by means of a polyether – polyurethane (but of different polyethers) – and mutual coagulation results in a leather substitute with an extreme softness [30].

Besides dimethylformamide and dimethylacetamide, polyurethanes are soluble in dioxane, pyridine and tetrahydrofuran. Nonsolvents are water, methanol and ethanol – even acetone has been mentioned as a nonsolvent.

Special methods:

- Mixtures of polyester– and polyether–polyurethane [7].
- Evaporation of the solvent after coating and before coagulating by heating with a lamp in a water/methanol medium [8].
- Using a fluorinated polyurethane [27] or other special polyurethanes such as aromatic ones [15, 16].
- Stepwise coagulation of thin layers of polyurethanes [9], or of an aromatic polyurethane in the base and an aliphatic one in the top coat [17, 18]
- Using solutions of different viscosity [10].
- Applying polyurethanes of different hardness and mutual coagulation [11].
- Coating one side of a porous substrate and treating the opposite side with nonsolvent [12].
- Wetting the substrate before coating with a nonsolvent [13, 14] (see also Chap. 8.3).
- Mixing with a water-soluble polymer, coagulating, washing out the water-soluble polymer and crosslinking with toluylenediisocyanate [24].
  Special methods of impregnations:
- An impregnated cotton fabric can also be coated with a polyurethane solution then impregnated and coated, coagulated, crosslinked with formaldehyde and finished with a wax [20].
- Impregnating a NYLON® fabric coagulating and then buffing [22, 32, 33].
- Pretreating textile substrates with paraffin wax or an emulsion before impregnating it with a DMF solution of a polyurethane and coagulating [31].
- Treating a NYLON® fabric with a fluorocarbon (ASAHIGARD®) then with a polyurethane for coagulation [21].
- Impregnating a polyolefin or polyester nonwoven [27].
  Surface treatments during coagulation:
- With embossed or glossy rolls to obtain a special surface [19], or with a negatively structured releasing layer [23].

8.2
Coagulation of Polyurethanes and Additional Small Sized Solids

Small sized solids or small sized products with a high surface area added to a polyurethane solution prior to coagulation help to speed up the coagulation and washing process. Powdered, crystalline cellulose [29 – 31, 45], other cellulose types [32, 45, 64, 68], cellulose in combination with a polyoxyethylene–dimethyl siloxane [75], hydroxyalkylene cellulose, ethers of cellulose [33], ground paper fibers [48], insoluble starch [32], fibers of collagen [34, 63], chrome tanned collagen particles [88] and leather [35, 36] may be added as coagulation accelerators.

To adjust the viscosity of the impregnation or coating solutions, and to assist the coagulation, bentonites etc. are added [85]. The salts added should have a particle size of 30 – 95 µm [90] or 6 – 30 µ [91].

Inorganic compounds such as small sized silicium dioxide [37 – 41, 65], silicates [39, 43, 44, 46], silicic acid [71, 76, 94], calcium carbonate [42 – 45, 60], magnesium oxide [43, 45, 46], zinc oxide [43, 44], zinc carbonate, calcium and magnesium chloride [83], titanium dioxide [44], cerium oxide [57], zirconium carbide [73] and mixtures of carbon dust with a cellulosic slurry [47] are suitable additives. Inorganic salts may be added in the form of a slurry [82]. Special chlorides of metals [iron(III) chloride, chromium(III) sulfate or zinc chloride] added to a dissolved PUR in DMF after coagulation and exposure to heat and pressure develop special colored effects [89].

Tensioactive substances which are able to crystallize under the conditions of a coagulation have been reported as additives [49]. A cloth is impregnated with a water-repellent agent then with a polyurethane dissolved in dimethylacetamide (DMA) and then coagulated. The polyurethane solution contains 1 – 400 % of an inorganic salt to obtain a soft water vapor permeable material [74].

So called metal soaps like aluminum, calcium or zinc stearates are mentioned to improve the touch and the water impermeability of the finished article [50].

Very effective additives are polyurethane dispersions [58] with a cationic character [51] or dispersions of poyurethane – polyureas [52]. These dispersions – after being mixed into a DMF solution of a polyurethane – polyurea – build up fine microstructures [53 – 55] preforming the desired micropores.

If additionally, anionic, synthetic tanning agents are added [51,52], then these synthetic polycondensates being polyanions precipitate the cationic polyurethanes to a microstructured matrix.

methylene-bis-naphthalene-sulfonic acid
dispersing agent

tanning agent consisting of 4.4'-di-oxy-di-phenyl-sulfone condensated with formaldehyde and naphtalene-sulfonic acid

Different kinds of polycondensation products

Under the influence of this matrix the polyurethane coagulates easier. Also polyurethanes containing built in silicon groups dispersed in water may be disposed in a polyurethane solution in DMF to improve the coagulation [56].

$$
\left[\;
\begin{array}{ccc}
\text{alkyl} & \text{alkyl} & \\
| & | & \\
\text{---Si-O-}\!-\text{alkylen-Si-O} & \text{---C-N-polyurethane-chain} \\
| & | & \;\;|\!|\;\;| \\
\text{alkyl} & \text{alkyl} & \;\;\text{O H}
\end{array}
\; \right]_{\geq 2}
$$

A polyurethane with internal silicone groups

The structure of small sized products also plays an important role, i.e. natural or synthetic fibers, like pulp for the production of paper [70, 92], are also suited to help coagulation. Fibers or fibroids (0.5–1.5 mm) of PVC [27], polyamide, cellulose [79], polyester, polyurethane etc. [28] may be added. The impregnation of textile substrates may, therefore, also be regarded as a method to improve the coagulation conditions.

Specially suited *small sized products* are:

- aliphatic polyurethanes [81], polyureas produced by the polyaddition reaction of H12MDI with H12MDA in the form of finely dispersed polyurea particles in a polyurethane solution [69],
- a mixture of polyurethanes with different viscosity [19] using an indirect process [20],
- mixtures of water-soluble cellulose derivatives together with metal salts of sulfuric acid [93],
- a polyamide-6 fabric pretreated with fluorocarbon and with a solution of a polyurethane in DMF together with silk [61] or keratin [62].
- *impregnations* of porous substrates [3–5] possibly containing shrinkable fibers [6–8, 12],
- hydrolyzing a polyester substrate by an alkaline treatment, impregnating with a polyurethane solution in DMF, coagulating, drying and buffing [87],
- assisted possibly by vibrations [9] or light vacuum [5],
- of collagen-containing nonwovens [10, 11],
- substrates with cellulose-fibers [14–18, 92],
- nonwovens [59, 77] with microfibers [see Chap. 12.1] [66, 67, 72, 80, 84, 86]
- coagulation in a bath of pH 13 [13],
- impregnation of a polyester fiber nonwoven with a polyether-polyurethane [1, 2] or using a mixture of a polyether with a polyester-polyurethane [21], polyureas [22], addition of polyglutamic acid [23], or
- of fabrics together with a nonwoven [24–26] or adhering two nonwovens together by coagulation [78].

8.3
Coagulation of Polyurethanes – Addition of Other Polymers

The addition of other polymers to a solution of a polyurethane(polyurea) in DMF in most cases also helps to increase the speed of coagulation. This method may be combined with the addition of salts or other coagulation additives. Therefore there is often no strict distinction between all the methods described in Chap. 8.

The polymer most often added to polyurethanes is polyvinyl chloride [1, 3, 18, 19, 29]. PVC is especially suited in the production of synthetic suedes [2, 7]. It may also be added in the impregnation of nonwovens or to bond fabrics [5] to a non-woven in the form of a microporous adhesive.

At most 50 % of PVC (based on the weight of the polyurethane) is added. Mixtures of PUR and PVC in polyamide-6 fabrics coated by coagulation were investigated. There seems to be an optimum with PUR/PVC mixtures of 80:20 for abrasion resistance of the coated fabric; crumple resistance and peeling strength were better with lower PVC contents [38]. Besides PVC, polyvinyl alcohol [10, 15], possibly hydrolyzed [12] vinyl chloride/vinyl acetate or vinyl alcohol copolymers – sometimes as a mixture with silica [6, 8] – also help to coagulate a polyurethane. Even mixtures of polyurethanes with polyurethane-polyureas coagulate quicker than the polyurethane or the polyurea alone [9]. Thiolignin [13], esters of polyglutamic acid, methyl cellulose, hydroxyethyl cellulose [16], polyethylene glycol [15], polysiloxane [27] possibly bearing reactive groups [30], solutions of polyamides in lithium chloride/DMF [28] are suited additives too. Polyurethanes mixed with a copolymer of acrylonitrile and maleic anhydride and with polyisobutylene are suited for the impregnation of non-wovens consisting of collagen and polypropylene fibers [17]. Acrylonitrile [22], acrylonitrile/acrylic ester copolymers containing functional amino, alkoxyamino or carboxylic groups [31], poly(meth)acrylates [3, 26], poly(methyl)styrene [4], acrylonitrile/maleic anhydride, fibroids of cellulosic compounds or of butadiene/acrylonitrile, PVC, polyvinylidene chloride [20], vinyl copolymers containing carboxylic groups [32] or polyamides [2, 21–25], silicones [33] and dimethylpolysiloxane [27] are suited polymeric additives for polyurethanes. Maleic acid–vinyl ether copolymers [36] also are suited additives; after coagulation they are leached partially or totally.

Polyacrylic acid partially or totally neutralized as additive to a polyurethane dissolved in DMF assist in the rheological behavior of the coating solution [34]. A textile substrate is printed with a solution of a polymer immiscible with polyurethanes. After drying of the printed product a PUR solution is used to impregnate the substrate and coagulated. The microporous polyurethane only fills spaces uncovered by the printed polymer [35].

Polyurethane solutions, possibly containing another polymer like a vinyl chloride copolymer or an acrylate are coated in different layers on a substrate and then coagulated. The polymer may be leached out afterwards [37].

Solvents which can be used in the coagulation process are – besides dimethylformamide (DMF) or dimethylacetamide (DMAc) – acetone, tetrahydrofuran, acetonitrile or dimethyl sulfoxide. Nonsolvents besides water are methanol, ethanol, ethylene glycol or glycerol [11].

8.4
Coagulation of Polyurethanes – Addition of Nonsolvents

The addition of a nonsolvent to a polyurethane solution in DMF may be applied together with the previously mentioned additives or methods.

There are three principle differences in how the nonsolvents may be applied:

Method 1   The nonsolvent is applied as a vapor or as an aerosol to the substrate already coated with the polyurethane solution in DMF.

Method 2   The nonsolvent is added to the PUR solution prior to coating or impregnation of the substrate.

Method 3   The nonsolvent is applied to the substrate prior to coating or impregnation.

Method 1. This method is generally used in such a manner that the coated substrate passes a channel containing an atmosphere of saturated water vapor. Sometimes water is sprayed onto the coated substrate [8]. By the slow action of a nonsolvent, the solvent absorbs the nonsolvent and gradually changes its solubility parameter. The solution of the polyurethane becomes viscous and less and less stable, the polyurethane is then in a gel-status. If this gel is immersed in water, the polyurethane coagulates and is washed out as previously mentioned.

If the polyurethane solution is applied onto a textile substrate, the nonsolvent may also be applied on the reverse [1, 2]. If the polyurethane solution also contains PVC [3], the treatment with water vapor should be ended when the content of water in the solution has reached 3.7 – 4.5 % [3]. 1 – 50 m/s is the optimum speed of the water containing air moving over the polyurethane coating [4] or over a film on a releasing surface, like release paper [5].

The treatment with (water vapor containing) air at 80 – 100 % relative humidity may also be applied at a temperature of 75 – 100 °C. This treatment should last 0.1 – 30 min [6]. Advantageously the polyurethanes have a higher temperature than the humid air [7]. Nonsolvents named are – besides water – usually ethylene glycol and glycerol.

One method consists of the application of a polyurethane solution together with a nonsolvent by mutual spraying. By this method the polyurethane coagulates in situ – there is only an additional drying step needed [10, 11].

Method 2. As mentioned in Chap. 7.1, the ideal quantity of nonsolvent to be added to the polyurethane solution may be determined by a titration. It is possible to add such a quantity of water to a PVC/polyurethane solution that the solution starts to become opaque or to develop micelles. Thereby the coagulation speed is accelerated [12, 17]. A possible way to find out the ideal quantity of nonsolvent [13] to be added was previously indicated (see Chap. 7.1).

Volatile nonsolvents which evaporate easier than the solvent DMF may also be used [26, 40]. Examples for this method are ethyl acetate in combination with sodium bicarbonate for the suede production [27] and tetrachlorodifluoromethane or trichlorotrifluoroethane [28] and white spirit [29]. Using volatile solvents the coagulation may be carried out in boiling water.

Sometimes a quantity of a nonsolvent, such as water, ethylene glycol, *tert*-butanol, acetone, hexane or toluene [16], is added so that a serum and a polyurethane gel separate. The polyurethane gel is applied to a substrate and coagulated [14]. The coagulation of the gel fraction may also be achieved in water containing more than 5% of lithium chloride or bromide, sodium bromide, or magnesium or calcium chloride [18].

The combination of adding a nonsolvent with a leachable solid is shown in a process wherein a solution of a polyurethane or a polyurea is mixed with magnesium or calcium carbonate, zinc oxide or clay to get a microporous coating [9].

The addition of 0.1–1-mm fibers together with water, methanol or ethanol to a polyurethane solution to improve the coagulation has also been reported [30]. In addition it is possible to coagulate a polyurethane solution containing fibers in nonsolvents [20].

If the addition of a nonsolvent is further assisted by a filling agent and a vinyl copolymer the coagulation is further improved [15]. A typical coating composition consists of 15–35% polyurethane, 20–80% solvent, like DMF, 5–30% of a nonsolvent and 0–15% of an additional polymer like PVC, ABS, SBR or ABR polymerisate [21, 22]. Similar to this formulation are additions of a thickening agent like carboxymethyl cellulose, polyacrylic acid in the form of a sodium salt or polyvinyl alcohol [23]. In a special method cyclohexanone is added. Cyclohexanone is a solvent for PVC and claimed to be a nonsolvent for polyurethane [24]. In another case 1–10% of the nonsolvent silicone oil is added; silicone oil is a nonsolvent for polyurethane as well.

Another method is the precipitation of the polyurethane by a nonsolvent and the use of the polyurethane powder for coating [31, 36].

Method 3. Textile substrates like fabrics or nonwovens may be treated on the surface with the nonsolvent water then coated or impregnated with a polyurethane solution and coagulated [32]. The indirect process can also be used: a polyurethane dissolved in DMF is applied on a silicone rubber belt. Then a fabric, humidified with 30–150% of water, is laminated into the thin PUR film prior to coagulation [39]. The coagulating liquid, water, may contain a tensioactive product. Such a method is suited to the coagulation of a knitted textile consisting of 70% polyamide and 30% PVC fibers [33]. Cotton fabrics may also be pretreated in an impregnating device with a DMF/water mixture [34] then with a polyurethane solution which is coagulated later. Ethanol, polyethylene glycol [35], methanol, propylene glycol, and toluene solutions of a paraffin are suited as added nonsolvents [37].

8.5
Coagulation of Polyurethanes – Addition of Emulsifying or Polar Auxiliaries

The size of pores is of particular interest in the performance of synthetic suedes and at the grain break of artificial leather. Tensioactive materials and/or alcohols help to control the size and shape of pores. $C_8 - C_{39}$ alcohols like stearic [46], cetyl alcohol [1, 2, 5] or esters of carbonic acids of such alcohols [13, 36], laurylbetain [42], castor oil [25, 34], mono-, di- or triesters of glycerol or sorbitol [7, 10, 39, 40], esters of α-ω-glycols with 6–22 methylene groups, ethers of C_1–C_{28} alkyl groups, $C_8 - C_{28}$ ketones and aldehydes [2–4], long-chain aliphatic carbonic acids [5, 11] are suited and influence as an additive to a polyurethane solution the improvement of the pore structure of the coagulated product. A combination of sorbitol monostearate with the oleic ether of polyethlene glycol or 2-(2-hydroxy-5-butylphenol)benzotriazole is an additive for the solution to be coagulated [9].

The pore size of the microporous products can be regulated by type, the manner and the quantity of the additives and by the temperature of the coagulation bath [48].

A treatment of nonwovens with fatty acid amides and water-soluble polymers prior to impregnation and coagulation of the polyurethane results in very soft man-made leathers [41].

Metal salts, like Mg, Ca, Fe, Al, Ba, Zn salts, of long-chain carbonic acids like lauric, stearic or naphthenoic acid [6] are suited additives.

In addition to the pore-regulating agents, other polymers like polyacrylonitrile and nonsolvents may also be added [12].

The organic or tensioactive substances are added in an amount of 0.1–30 % in reference to the quantity of the polyurethane used. Polar organic and emulsifying agents may be used together [14, 26] or an anionic and a nonionic emulsifier can be used together [29] to obtain a polyamide-6 or a polyester fabric coated by the coagulation process.

Anionic tensioactive substances may have sulfonate or sulfate groups: a typical example is sodium bis(2-ethylsulfosuccinic acid) [22]. They may be used as additives to polyurethane solutions which also contain protein or a powder of cellulose [27] or a silicone [31]. Another example of an anionic tensioactive substance is an alkylfluorinated product like C_8F_{17}-SO_3M (M = monovalent ion) [23]. These fluorine-based emulsifying agents may be added to a polyurethane together with diethylsulfosuccinic acid [24]. Anionic tensioactive substances may have a high molecular weight. Anionic tensioactive substances in polyurethane solution can be leached out during the coagulation process by the help of an electric field [32].

Nonionic tensioactive substances are polyethylene or polypropylene glycol [15], polysilicones [15], block copolymers of 80 % polyethylene oxide and 20 % ethylene oxide [16]. Ether, ester, amines, urethanes and amides with C_4–C_{30} alkyl or alkylene groups [17] or phosphoric acid esters of C_2–C_{30} alkyl or alkylene groups [18, 19] or trinonylphenyl phosphite [30] are polar compounds.

Not only anionic or nonionic [43] but also cationic tensioactive substances like octadecyltrimethylammonium chloride [20, 21] are suited additives.

Nonionic tensioactive substances, dyestuffs and/or waterproofing agents can be applied together [28]. A water-repellent man-made leather can be achieved by treating a fabric with a fluorocarbon and then a coating of a polyurethane solution in DMF together with a tensioactive compound and coagulating afterwards [37]. Pretreatments of textile substrates with emulsifying or tensioactive substances prior to coating and coagulating help in getting better porosity [38].

Hydrophilic substances like polyethylene glycol ether are suited additives in polyurethane solutions in DMF, resulting, after coagulation, in finer pores [47].

$$R\!\!-\!\!O\!\!-\!\!\left(CH_2\text{-}CH_2\text{-}O\right)_n H$$

$R = C_{4-18}$ Alkyl
$n = 2 - 20$

Polyethylene glycol ether

Tensioactive compounds based on polyaddition products of polyethylene-polybutylene adipate with MDI and polyethylene glycol–polypropylene glycol ether are also good coagulation additives [45].

Additionally the polyurethane solution, fatty acids, alcohols, amines, alkylphenols, etc., may be added to the coagulation liquid to get better coagulation conditions [44].

8.6
Coagulation of Polyurethanes – Addition of Leachable Solids

To improve or accelerate the coagulation, leachable substances may be included in the coating or impregnation solution. The removal of these soluble substances included in coatings or films can be carried out together with the coagulation process itself. Soluble substances in coagulated coatings result in big pores; this method is often used in the production of synthetic suedes. Leachable substances are:

– salts,
– organic substances of low molecular weight, or
– polymeric organic substances.

A disadvantage of this procedure is the increasing concentration of the leached substance in the coagulation bath. The recycling of the solvent in the coagulation bath may be difficult due to the substances being present in it. Normally the solvent is distilled. It is necessary not to use leachable substances which may hinder the distillation of the solvent. As shown previously (Chap. 7) it is difficult to leach out the substances enclosed in a film or coating. Leaching processes therefore often mean low production speed and large production vessels due to the high quantities needed of the washing liquids.

Leaching of Salts. Additives most often used are salts of strong inorganic acids [1]. Salts are ammonium, sodium or potassium chloride or sulfate [18, 46]. The particle size of the salt should be $\leq$ 0.105 mm – or better $\leq$ 0.075 mm [8, 10] or 20–25 μm [4–6]. Sodium bicarbonate or pyrophosphate [11] or magnesium or aluminum sulfate [3] may also be used.

Coatings of textile substrates or releasing paper may be done by a solution of a polyurethane containing 10–50 % of a mineral salt like sodium or potassium chloride or sulfate.

Salts like copper nitrate sprayed onto a coated textile followed by a coagulation at 80 °C result in a substrate for synthetic suede [19].

Bentonite has also been mentioned as a leachable compound [62]. A special effect – corresponding to the addition of a nonsolvent (see Chap. 8.4) – is the addition of salts containing crystalline water, like aluminum sulfate octadecahydrate [9].

A combination of sodium chloride with added water and coagulation at 50 °C [15–17] may be used to produce microporous films [17].

To achieve extreme softness in combination with large pores, 300–1500 % of sodium sulfate may be added to a DMF solution containing 10–50% polyurethane. This solution is coated and coagulated –then again a coating with a salt-free solution is carried out which is coagulated thereafter [12].

Besides salts the coagulation solution may also contain silica at a pH of 2–5, polycarbodiimide like polytriisopropyl benzocarbodiimide [13] to stabilize the coating to hydrolysis.

The coating of textile substrates may be done in two steps: The first step with a polyurethane solution containing a lower and the second one with a higher salt content. The first step is done with a polymer/salt ratio of 0.6:1 to 2.5:1 and the second one with a ratio of 3:1 [7]. Foils produced according to this method are useful in the production of shoes [61].

Polyurethane solutions in aliphatic ketones, esters or cyclic ethers containing ammonium or metal salts of an inorganic acid may be coated on textile substrates and coagulated [1].

A nonwoven consisting of polypropylene fibers is coated with a polyurethane dissolved in DMF then covered by a salt. After coagulation and elimination of the salt, big pores remain on the surface. This method is especially suited in the production of man-made suedes because big pores are created [51].

A mixture of a leachable salt like sodium or potassium chloride or sulfate together with stearic, palmitic or myristic acid achieves coagulated coatings which may be buffed afterwards to get a synthetic suede [14].

During coagulation and leaching of salts the substrates may be dyed in the same working step by the addition of a 1 % solution of an azo, triphenylmethane, anthraquinone or azine dyestuff [20].

A solution of a polyurethane containing an inorganic salt can also be sprayed onto a substrate in several layers prior to coagulation and leaching of the salt [2].

Leaching of Organic Compounds of Low Molecular Weight. Organic compounds of low molecular weight are powdered lactose [11], urea [21] or thiourea [22, 23].

Metal salts of carboxylic or sulfonic acids able to be leached during or after the coagulation [48], hydrophobing agents or nonionic tensioactive substances [49] which may be crosslinked afterwards by an organic polyisocyanate [50] stabilize and help to improve the pore structure.

Temporal solvents like caprolactam, 3,5-xylenol or dichlorobenzene dissolve a polyurethane at a higher temperature. By cooling after application the polyurethane becomes insoluble in the temporal solvent and by this phenomenon the polymer coagulates. The nonsolvent at ambient temperature must be leached finally by a real nonsolvent like water or alcohol [24].

Leaching of High Molecular Organic Compounds. Powdered gelatin applied on the surface [25], vinyl ether–maleic anhydride copolymerisates [26], partially saponified polyacrylonitrile or polyvinylacetate [27], hydroxyalkyl cellulose [28], carboxymethyl cellulose [27, 29, 54–56], starch, dextrin [29], polyvinyl alcohol, sodium acrylate or methoxymethyl cellulose [31] are suited products to be added to a dissolved polyurethane which may be leached after or during coagulation.

A textile substrate is washed with a nonsolvent or a tensioactive substance, then treated with a water-soluble polymer like PVA, sodium polyacrylate or carboxymethyl cellulose (CMC), impregnated with a polyurethane solution in DMF, coagulated and washed with water to leach the water-soluble polymer [58].

To achieve reserved spaces on a surface, products like polyvinyl alcohol [30–32], methyl cellulose [33], natural rubber, polyacrylate, starch, casein or carboxymethyl cellulose may be applied on locally limited areas. The polymers may be mixed with the polyurethanes in the form of fine fibers [57]. During coagulation the applied polymers are leached [34].

The impregnation of textile substrates like nonwovens is also improved if the polyurethanes applied contain leachable polymers. A nonwoven is impregnated with a polyurethane dissolved in DMF also containing polyvinyl alcohol, polyvinylmethyl ether, casein [35], carboxymethyl cellulose [36, 37], a sodium salt of polyacrylic acid [60] or dextrin [37]. After coagulation and leaching the fibers of the nonwoven do not stick together and the impregnated nonwoven acquires an improved softness.

It is also possible to add the second polymer (e.g. carboxymethyl cellulose) to the bath of the nonsolvent and to coagulate a textile substrate impregnated with a polyurethane in that bath, dry and leach the second polymer after these steps [53].

A modification of this process consists of the steps of producing thin coatings on polyamide or polyester fibrids with leachable polymers like polystyrene, producing a nonwoven out of these fibrids, treating the nonwoven with polyvinyl alcohol, dissolving the polystyrene in trichloroethylene and impregnating with a polyurethane solution. During coagulation the polyvinyl alcohol is also leached [38–45]. Nonwovens consisting of microfibers of polyethylene–polyoxyethylene modified polyamide-6 were impregnated with a polysiloxane. During this impregnation polyethylene or polyoxyethylene is leached, then a swelling agent for the rest of the fibers is applied and the nonwoven is pressed. Afterwards an impregnation with a polyurethane containing polyoxyalkylene groups dissolved in DMF, and coagulation is carried out [47].

A special case to eliminate an added compound consists of the following method: two polymers of different thermal stability like polyurethane and PVC are coagulated in glycerol at such a temperature that the less stabile polymer is decomposed [52].

8.7
Coagulation of Polyurethanes – Addition of Blowing Agents

To increase the diameter of the pores, blowing agents are added to a polyurethane solution to be coagulated later. As a blowing agent sodium bicarbonate is normally used. For example, a coated nonwoven is coagulated in a bath of hot water. In hot water sodium bicarbonate splits off carbon dioxide. The carbon dioxide increases the pore size of the coagulated polyurethane [1–3]. The effect is especially useful in the production of artificial suede leather.

8.8
Coagulation of Polyurethanes by Reacting Prepolymers of Polyurethanes

Polyurethanes still containing some free isocyanate groups are soluble in DMF. It is possible to use these solutions for the coating or impregnation of textile substrates. During coagulation the free isocyanate groups react with water developing carbon dioxide [1] to a polyurethane–polyurea [Eq. (8-1)].

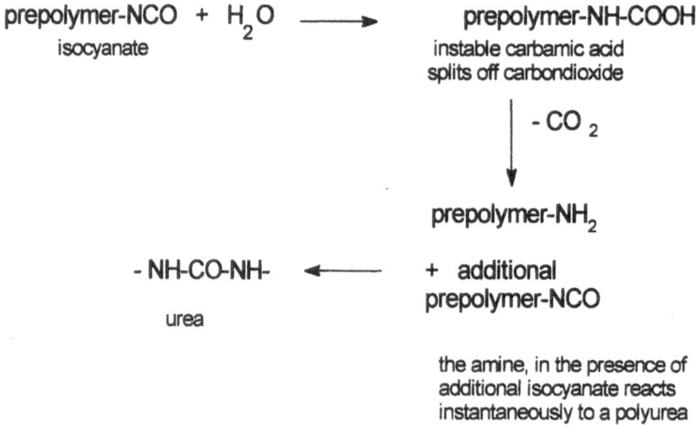

prepolymer-NCO + H_2O $\longrightarrow$ prepolymer-NH-COOH
    isocyanate   instable carbamic acid
                                                    splits off carbondioxide

$\downarrow$ - CO_2

(8.1)

prepolymer-NH_2

- NH-CO-NH- $\longleftarrow$ + additional
                                 prepolymer-NCO
    urea

the amine, in the presence of
additional isocyanate reacts
instantaneously to a polyurea

Reaction of isocyanate groups with water

It is possible to dissolve products with amino groups in a solution of a polyurethane. These added amino-containing products are able to react in the same manner as described in Eq. (8-1). The only difference is that no carbon dioxide develops [2 – 4, 8]. This kind of reaction is described in detail in Chap. 25.1.

As well as additives with amino groups, polyurethanes containing isocyanate groups can be used to modify the microporous product later. The addition of crosslinking agents or monomeric isocyanates like toluylenediisocyanate (TDI) is also possible [5, 7].

A crosslinking of a polyurethane dissolved in DMF may be carried out by UV rays as long as the polyurethane is in the gelating state [6].

CHAPTER 9

Microporosity by Evaporation of Volatile Products

If a polymer is dissolved in a volatile liquid and this liquid is removed by spraying, the polymer forms fine fibers which may be deposited on a releasing surface as a fine, microporous fiber fleece (see Sect. 9.1). In this case the molecular weight of the polymer should be rather low so that the polymer is able to dissolve in the volatile liquid. Another possibility of creating micropores is to mix a nonsolvent into the polymer solution in the volatile liquid. If this nonsolvent is also able to evaporate and has a lower volatility than the solvent, micropores are produced in a similar way to the coagulation process.

9.1
Spraying of Polymer Solutions

Slightly unstable solutions of polymers in volatile liquids can be transformed into microporous fiber fleeces by spraying with air gun which has fine nozzles. A presupposition for this is a solvent which evaporates under spraying conditions almost totally [1, 2, 11, 13, 15]. An example of such a solvent is tetrahydrofurane. A solution of a polymer in tetrahydrofurane is sprayed with an air gun in such a way that the solvent evaporates almost totally. Fibers are formed in the air stream. These fibers still containing solvent deposit on a release paper or a glass plate in the form of a fleece of fine fibers. The fibers stick together due to the remaining solvent still in them. By a plating operation the density of the fiber fleece can be increased; the water vapor permeability is decreased by such an action. With the naked eye the microporous structure cannot normally be seen, the surface of such a fleece is smooth. The fleece looks similar to products processed from fibrids [8] (see Sect. 21.1). A typical picture of a fiber fleece produced by a spraying method is shown in Fig. 9-1. If the spraying is carried out in an atmosphere of diminished pressure the evaporation is accelerated [17].

The solution may contain an additional crosslinker prior to spraying. The crosslinker builds up a polymer of a higher molecular weight during storage or drying [3, 5]. To reduce overspray the spraying may be carried out in an electrostatic field [19]. If a gas is dissolved in the solution the particles sprayed out build pores additionally [9, 16].

Not only solutions may be used, mixtures of a solution of a polyurethane with a dispersion of the same polymer may also be used [6, 7].

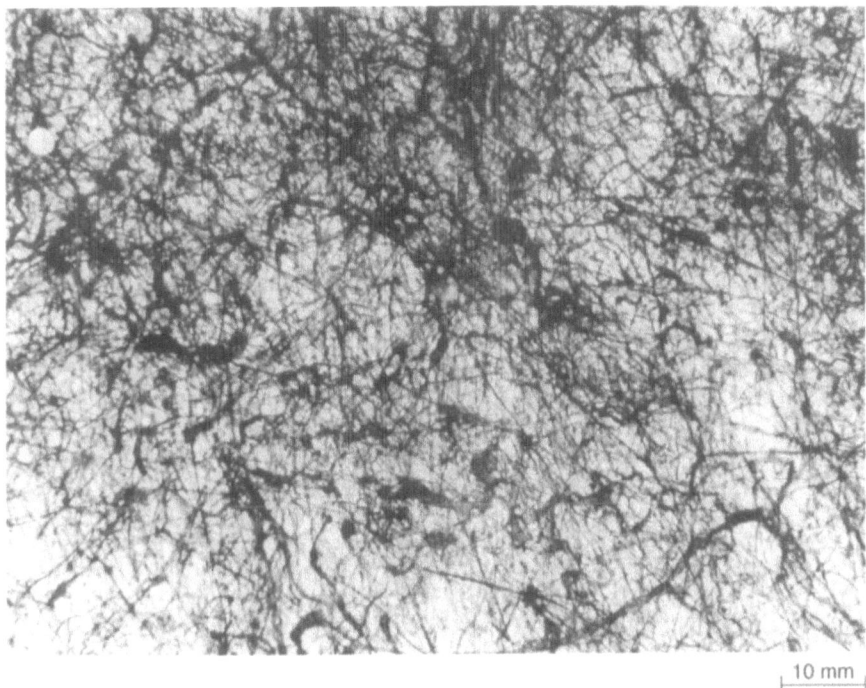

Fig. 9-1. Microporosity by spraying. By the spraying operation and evaporation of the solvent the dissolved polymer produces fibroids which are disposed as a fleece like structure

In addition to polymers, prepolymers like isocyanate prepolymers dissolved in methylene chloride may also be sprayed on microporous substrates in the presence of water. During and after the spraying the water reacts with the isocyanate prepolymer to form a polyurea [18]. Water may be present as a warm humid atmosphere in which the spraying is carried out [13]. Prepolymers containing isocyanate groups in a mixture with N-alkoxymethylated polyamides in a DMF solution may be used as a spray coat for nonwovens [21].

If the solution contains additional polymer fibers, extremely porous products result [12]. Ultrafine polycarbonate fibers, possibly containing a binding agent as well, are sprayed onto a nonwoven to become a microporous surface [20].

In a special case the spraying is carried out without solvents by using melts which are sprayed [4, 10, 14] (see Chap. 21; spun bonded nonwovens).

Spraying is not carried out in a technical way due to the fact that only solutions of a rather low concentration can be used. The concentration of a polyurethane in tetrahydrofurane normally does not exceed 15% solids, otherwise the viscosity would be too high for spraying.

The spraying is done with a high quantity of air. It is very difficult to recycle the solvent out of the spraying air. Polyurethanes are normally not soluble in volatile liquids. Tetrahydrofurane, and to a certain extent acetone, has the appropriate solubility. Both products build highly explosive mixtures with air –

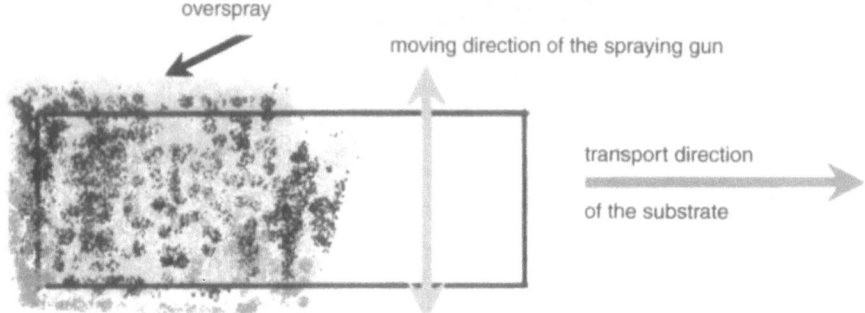

overspray

moving direction of the spraying gun

transport direction

of the substrate

The rectangular part symbolizes the substrat. To
obtain an even thickness of the coating the spraying
gun needs to pass over it. The overspray is a loss in
product.

Fig. 9-2. Overspray. The substrate is moved in transport direction, the gun moves in a perpendicular position to the transport direction. The spraying gun moves back and forth over the substrate to be coated. In the endpoints just before changing the direction again the velocity of the gun is reduced thereby increasing the amount of sprayed product because it is difficult to close and open the nozzles of the gun to avoid differences in thickness. Therefore the endpoint of spraying is outside the edges of the substrate to be coated. The overspray usually is a loss in product

so a low concentration of the solvent in the air is needed. Strict care must be paid to this low concentration to avoid explosions.

Another negative point is overspray. If a coating is carried out by spraying normally, more than 25% of the coating material is not deposited on the substrate to be coated. An overspray is necessary to get the same thickness on all parts of the coating. Therefore the spraying gun not only moves above the substrate but also over it. The point of return is very often at the edges of the coatings or outside of it (Fig. 9-2). The only broader technical uses of the process in an industrial way was the spraying of adhesives on microporous films to laminate them to a substrate by punctual adhesive points.

9.2
Selective Evaporation Process

In solutions of polymers in a volatile solvent a nonsolvent is mixed by thorough stirring. The nonsolvent is less volatile than the solvent. Each solution of a polymer can be mixed with a maximum quantity of a nonsolvent – specific for the polymer and the solvent used – just enough to prevent the precipitation of the polymer. The amount of nonsolvent added should be as high as possible to get the maximum in microporosity. This amount corresponds to a quantity just high enough so as the remaining polymer still stays in solution without precipitating. The solution becomes opaque. The polymer/solvent/nonsolvent mixture is then used to coat a textile substrate, transferred to a film or to impregnate nonwovens etc. Then the organic solvents are allowed to evaporate. The evaporation number of the solvent used should be lower than the evaporation number of the non-solvent [2].

With a solvent of a lower evaporation number than that of the nonsolvent, the concentration of the solvent in the polymer/solvent/nonsolvent mixture decreases. The more the solvent evaporates, the less soluble the solvent/nonsolvent mixture becomes for the polymer and a polymer matrix is formed finally with the nonsolvent in an inner phase. As in the coagulation process, finally the polymer includes almost only nonsolvent which – also after evaporation – leaves micropores [9]. The first time the selective evaporation process was ever described was with vinyl polymers dissolved in acetone and ethanol as the nonsolvent [1, 3, 10, 16].

Good evaporating solvents are tetrahydrofurane, butan-2-one, acetone etc. – slow evaporating nonsolvents are C_1–C_{10} alcanols, paraffins and water. The concentration of the solvent mixture contains normally 3–25 % polymer and 5–70 % nonsolvent. Besides PVC, polystyrene are also polymers well suited for this process.

Polyurethanes can also be used as a polymer [17]. If crosslinking products, like trifunctional isocyanates, are added, the polyurethane during evaporation becomes additionally insoluble in the solvent mixture [5] and the formation of micropores is improved further. If the polyurethane contains perfluoroalkyl groups, water may be used as a nonsolvent [23].

Polyurethanes may be synthesized in a solvent/nonsolvent mix like water/methyl isobutyl ketone. The resulting dispersion may be applied as a coating or a film; after selective evaporation micropores are created [24]. Microporous adhesive coats may also be obtained by such a dispersion [15, 19]. Evaporation may be achieved by microwaves [21]. Composed products containing a fiber substrate and a microporous impregnation as well as a coating produced by such a dispersion can be used as a leather substitute [26].

The spraying technique can be used for polyurethanes dissolved in a solvent/nonsolvent mixture as well. The solutions are sprayed onto a release paper, a substrate is adhered to the wet film and – after evaporation of the solvent and nonsolvent – the composed material is stripped from the release paper [27].

A variation of the selective evaporation process consists of a method of dissolving a polyurethane in dioxane or tetrahydrofuran and mixing the solution with a melted alkane carbohydrate melting in a range of 95 to 175 °C. The resulting mixture is cooled to 39 °C, coated on a cotton fabric and the solvents are evaporated at room temperature during a 16-h period [25].

A special application of the selective evaporation process consists of mixing polyvinyl fluoride with γ-butyrolactone or dimethylacetamide at a temperature where polymer and added swelling agents and/or solvents are able to mix homogeneously. Then this mixture is extruded and cooled down. The added solvent separates in the polymer matrix. Then the film is stretched and the solvent evaporated. The resulting films are microporous and have a smooth surface [11]. Besides polyvinyl fluoride, other polymers like the polyamide of hexamethylenediamine and adipic and sebacinic acid and the mixture of polyamide-6 or -6.6 with polycaprolactam have been mentioned. In these cases polypropylene glycol or glycerol as a latent solvent was used [12]. Cellulose acetate may be processed to a microporous film in an analogous process by dissolving it at 80 °C in diethylene glycol ether, and cooling down after film formation and evaporation [13].

Besides coatings and film formation, the selective evaporation process can be used for impregnation. In a patent application a nonwoven was impregnated and coated as well by a mixture of a polyurethane and polyethylene glycol ether in butan-2-one, emulsifying agent and water as a nonsolvent [28].

To get the selective evaporation done properly the following factors are important:

- The volatility of the solvent and the nonsolvent should be different. Solvents should evaporate faster than the nonsolvents [7,14].
- The nonsolvent is defined as being unable to be adsorbed easily by the polymer by swelling. This can be proved by putting a nonporous polymer film into the (non)solvent to be tested for 24 h at 20 °C. During this time the film should not absorb more than 50 % of its weight by swelling of the solvent to be proved. If the weight increase is less than 50 % the solvent tested may be used as a nonsolvent [7].
- Additionally, the solubility-parameter of a polymer and of a solvent may be advantageous. Similar solubility parameters of polymers and solvents are a sign that the solvent dissolves a polymer easily. Different parameters show nonsolvent effects [4].

Polymers, as other products, are more soluble at higher temperatures than at lower ones. This effect is described in the equation of Gibbs–Helmholtz [8]. Therefore the mixing of the solution with the nonsolvent should be done near the evaporation temperature [2, 7]. If the mixing temperature is much lower than the temperature of evaporation, the added nonsolvents are able to swell or even to dissolve the polymer and, therefore, the selective evaporation will not result in micropores. A variation of this effect may be used in the following procedure: a polymer is dissolved at a high temperature in a potential nonsolvent or a mixture of a solvent with a nonsolvent Then a coating at this temperature is carried out, the coating is cooled down and the solvent (mixture) evaporated which – at this temperature – does not dissolve the polymer any longer by sublimation [18, 20].

The amount of added nonsolvent defines the amount of micropores and the grade of water vapor permeability [6]. In Fig. 9-3 this effect is demonstrated. The density of the polymer film decreases with increasing amounts of added nonsolvent during manufacture (Fig. 9-4).

The films (see Fig. 9-5) and coatings produced by the selective evaporation process – as long as organic nonsolvents are used – look like films produced by the coagulation process (see Fig. 7-2).

The selective evaporation process is advantageous because:

- it may be applied on normal coating equipment used in textile coating, and
- the amount of the microporosity desired may be adjusted easily by the amount of nonsolvent added.

Fig. 9-3. The amount of added nonsolvent defines the microporosity. In this figure a low boiling petrol fraction (in ml) in increasing amounts was added to 25 g of a 10 % solution of a polyurethane in tetrahydrofurane. Then the solution was poured onto a glass plate to evaporate selectively tetrahydrofurane and a low boiling petrol fraction. With each amount of a low boiling petrol fraction a film was produced. Increasing amounts of a low boiling petrol fraction resulted in films with increasing water vapor permeability. The water vapor permeability of the resulting films was measured according to DIN 53333

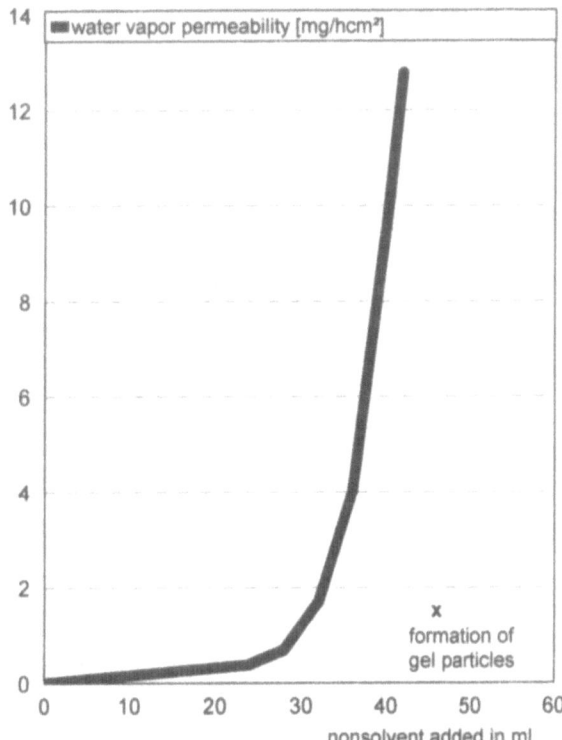

Fig. 9-4. The density of the resulting film also decreases with increasing water vapor permeability

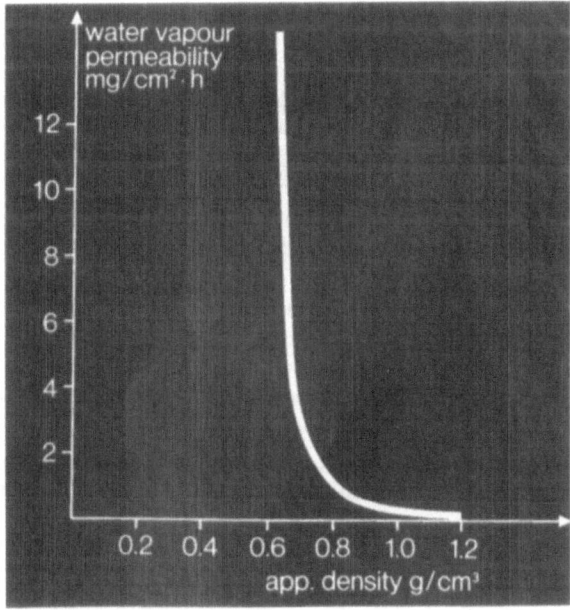

Fig. 9-5. Micropores of a film produced by the selective evaporation process. The pore structure of a film obtained by selective evaporation is very similar to the structure of a coagulated polymer

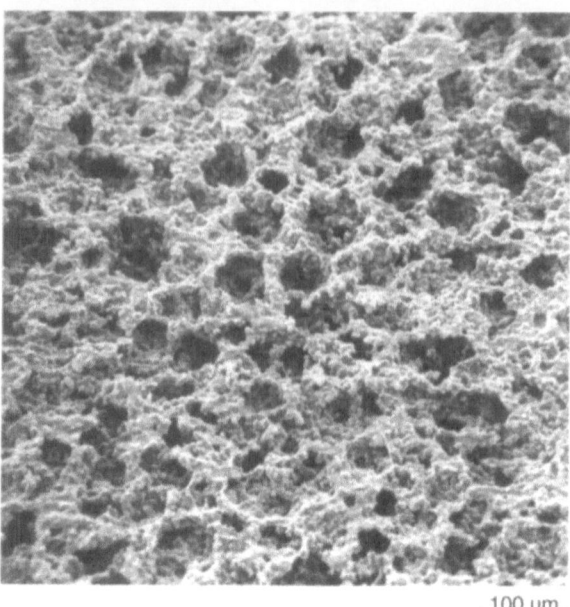

100 μm

The selective evaporation process is disadvantageous because:

– high amounts of solvents and nonsolvents are needed (with water as a nonsolvent higher drying energies are needed), and
– organic solvents and nonsolvents should be recovered from the drying air which is – sometimes – technically difficult and expensive.

The process is used on a technical scale.

9.3
Selective Evaporation Process of Dispersed Polymers

In the previous chapter *solutions* of polymers were treated with a nonsolvent. In most cases only a slight flocculation of the polymer or a turbidity of the solution occurred when the dissolved polymer was mixed with a nonsolvent. Mostly the polymers remained in a dissolved stage.

In this chapter dispersions of polymers are treated to obtain a microporous structure. The polymers are usually dissolved in a solvent. By the addition of a nonsolvent a dispersion is obtained. Solvent and nonsolvent are not or only partially miscible with each other. Similar to the selective evaporation process described in the previous chapter the dispersed polymer is able to form a continuous film or coating. If the solvent and the nonsolvent are immiscible, this is usually the case if water is used as a nonsolvent, two types of dispersions may be created:

– oil-in-water-type, and
– water-in-oil dispersion (Fig. 9-6).

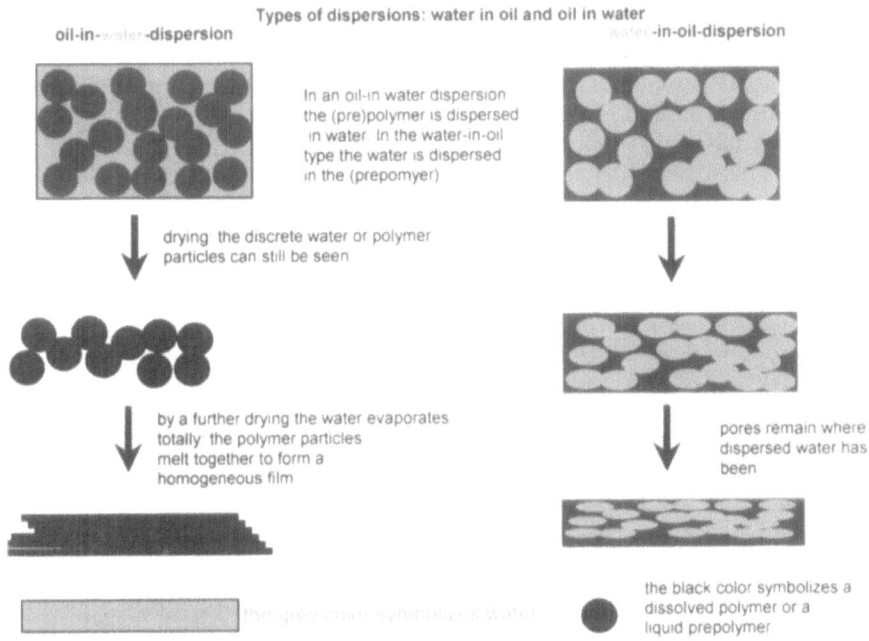

Fig. 9-6. Types of dispersions: oil in water and water in oil

The difference in film-forming properties of an oil-in-water to an inverse, water-in-oil dispersion is that in an oil-in-water dispersion the polymer is in the discontinuous phase and water in the continuous phase. During drying the polymer stays in separated droplets, which under drying conditions melt together to give a homogeneous film. In a water-in-oil dispersion the polymer is the continuous phase which includes finely dispersed, separated water droplets. As long as under drying conditions the forming polymer film is weak enough to enable water to evaporate, small channels are built and water may evaporate. The pores and the channels of evaporating water are the base of the microporous structure (Fig. 9-6).

Due to the fact that in most patent applications no clear distinction between homogeneous solvent/nonsolvent mixtures and solvent/nonsolvent dispersions was done, some publications are mentioned here which could also have been described in the previous chapter.

In a PVC-plastisol a polyvinyl alcohol solution in water is stirred in to get a water-in-oil dispersion. After evaporation of the water, a microporous coating or film is created [5]. Instead of a polyvinyl alcohol solution, a polyurethane dispersion in water may also be used to get an inverse dispersion. This dispersion can be applied onto silicone rubber molds, e.g. of a shoe upper, which under the influence of high frequency gelates. At the same time water evaporates to form a microporous shoe upper. Instead of the polyurethane dispersion, a water-based PVC dispersion may be used in a similar manner [7]. Other thermoplastic polymers, plasticizers and water can be used as a coating mixture to get microporous coatings [2].

A PVC plastisol containing water can be coated on a release paper and a fabric is then applied at 140 °C to get a microporous coating [9, 24].

A water-based dispersion containing 20–60 % PVC is treated with 50–100 parts of a plasticizer (per 100 parts of PVC). Then 10 parts of polyvinyl alcohol are dissolved in the mixture and some PVC powder is added. A textile is coated with the resulting mixture and at 150–170 °C the mixture is plasticized and the water is evaporated [20] to form a microporous PVC coating.

PVC powder may be dissolved in cyclohexanone. The solution is then treated with a poor solvent, like naphtha. For 100 parts of solution 300–400 parts of the poor solvent are used. Water, as well as a powder of diatomaceous earth, is further added. The mixture is extruded or calendered to microporous plates [19].

With the help of a water-based polyvinyl alcohol solution, water is emulsified into a PVC plastisol and dimethylol urea as a crosslinker for polyvinyl alcohol is added. The resulting emulsion is transferred to a film becoming microporous by selective evaporation and having included crosslinked polyvinyl alcohol [3].

In PVC or polyethylene dispersions a dyestuff dissolved in paraffin or a plasticizer is stirred in. After formation of a film the mixture is plasticized by heat. The dyestuff is incorporated in the film. Under pressure the dyestuff is liberated onto the film-surface [22]. Materials according to this principle are also suited as carbonless copying papers (see Chap. 22).

Plasticized PVC needs a gelating step at around 140–160 °C to get a good performance of the material. Water boils at 100 °C. Therefore it is difficult to create micropores by the evaporation of water and to also get good gelating conditions. It is therefore assumed that the processes described previously are not applied technically. Other polymers not needing special thermal steps are better suited for this kind of process.

Water-based dispersions of a polymer are treated with a solvent not miscible with water and which boils at a higher temperature than water. The resulting emulsion can be coated onto a nonwoven to get a microporous coating [23]. A dispersion of a polyurethane in a solvent which is miscible with water, like butan-2-one or methyl isobutyl ketone is treated with a water-based chitosan solution in the presence of an emulsifying agent. to become a water-in-oil dispersion. After application on a textile substrate a water vapor permeable material results, able to be used for apparel or for medical uses [6].

If a water-based dispersion of an acrylonitrile–butadiene rubber is stirred into a solution of a polymer like a dicyanodiamide–formaldehyde condensate, microporous coatings can be achieved [1],

A similar process is to use a dispersion of a polyurethane in a poor solvent, like butan-2-one, in which a polymer latex, like acrylic ester copolymerisate, is stirred. After application and selective evaporation of the solvent and water a microporous coating is achieved [8].

A latex of a polyurethane or an acrylonitrile–butadiene copolymer is mixed with a nonsolvent, like mineral oil, not miscible with water and boiling above

100 °C. A thickening agent, like methyl cellulose, is also added to the mixture. This mixture can be used for microporous coatings. The mineral oil should evaporate at a slower rate than water [14]. A latex is mixed with a hydrocarbon which boils above 204 °C as a nonsolvent; 0.3 – 3 parts of the nonsolvent are used for every part of dispersed polymer. This water-in-oil dispersion can be used to produce microporous films [13].

N-Alkoxy derivatives of a polyamide can be dissolved in an alcohol. A polyamide consisting of hexamethylenediamine and adipic acid is N-methoxymethylated, which is dissolved in ethanol, mixed with a nonionic emulsifying agent and then treated with a water/alcohol mixture to get an emulsion. This emulsion can be used to impregnate fabrics; the impregnation is able to adsorb water easily [10].

The polymer best suited for this kind of process is polyurethane [26].

A nonsolvent is dispersed in a solution of a polyurethane. The dispersion may be used to get microporous impregnations, films or coatings [9].

Polyurethanes containing polyethylene oxide groups are especially suited to obtain water-in-oil-emulsions because they may be easily mixed with water due to the high hydrophilicity of polyethylene oxide: 60 – 600 % water can be stirred into a solution or suspension of a polyethylene oxide containing polyurethane in THF or butan-2-one. The resulting dispersion is applied onto a nonwoven or a release paper to get a microporous coating or a film [11] after selective evaporation of the solvent and water. Polyethylene oxide groups containing polyurethanes may be applied together with polyurethanes not containing hydrophilic groups as water-in-oil-emulsions for microporous coatings [12].

A polyurethane consisting of up to 70 % of polyoxyethylene glycol is dissolved in DMF or dioxane. The solution is treated with water to produce a water-in-oil dispersion. Textile substrates are impregnated with such a dispersion [18]. Water-based dispersions of polyurethanes containing polyoxyethylene gylcol ether groups and/or carboxylate and/or sulfonate groups are treated with organic nonsolvents which boil above 220 °C. The resulting oil-in-water dispersion may be used for the production of microporous coatings [29].

Water-in-oil emulsions may be coated onto a microporous coating prior to being produced by wet coagulation to get softer and also better dyeable leather substitutes [28].

An insoluble, but swellable, polymer, dispersed as a latex, is stirred into a solution of a polyurethane in butan-2-one. The insoluble polymer is an acrylate–, vinylacetate–, butadiene–styrene- or an acrylonitrile copolymer. The resulting water-in-oil dispersion is coated onto textile substrates. After evaporation of butan-2-one and water a water vapor permeable microporous coating is achieved [16]. These emulsions may also be used to impregnate a nonwoven [17]. Dyed polyester fabrics are treated with a dispersion of a polyurethane in a toluene/butan-2-one/water mixture to get – after drying – a water vapor permeable material [27].

Two polyurethanes, dispersed in water, having opposite charges are separately transferred into a water-in-oil emulsion. A textile substrate is coated in two steps with the two latexes [25]. One polyurethane contains cationic groups, the anionic one carboxylate groups.

Nonwovens or other textile substrates are treated with a dissolved or dispersed polymer and cooled to a point where the solution solidifies. The polymer is crosslinked by high energy radiation. After warming up, the rest of the solvent or dispersing liquid is evaporated [4].

A composite material is claimed for the manufacture of sports shoes consisting of a nonwoven, containing polyester, rayon or viscose fibers, and a polyurethane dispersion, which becomes microporous, after a selective evaporation [30].

CHAPTER 10

Reactive Processes

All the processes described up to this point are based on high molecular polymers. In this chapter processes using oligomeric or low molecular "pre" -polymers are described. The polymers are polyadded or polymerized to a product of a high molecular weight during their application. A polycondensation which is less often published to produce microporous articles exists, however too [5].

In the previous chapters polymers were sometimes treated with crosslinking agents to increase their molecular weight or to improve their applicability behavior. In particular in the selective evaporation process crosslinking agents help to diminish the solubility of the polymers in the solvent mixture during the application. Therefore, the formation of micropores is assisted by crosslinking.

Examples of a polymerization during application are:

– Prefabricates of microporous ion-exchange resins are, for example, produced by polymerizing a mixture of styrene and divinyl benzene in the heterogeneous phase, like water [25].
– Panels, films or shaped articles may be produced by a water-in-oil emulsion [4] of a mixture of styrene and methyl methacrylate, forming it, polymerizing and evaporating the water formerly included [7, 24].

For the most part, polymerization in a heterogeneous phase results in articles with rather a high hardness; therefore, this kind of process is only described here in these few examples.

The process mostly used for man-made leathers is the polyaddition reaction. A polyaddition reaction in a homogeneous phase and then transforming the polyurethane into a dispersion is well known [3], as well as the polyaddition in a heterogeneous phase by using NCO prepolymers and amines for the same purpose [2]. Both processes yield an oil-in-water dispersion, i.e. a dispersion of the polymer in water. A polyaddition reaction in a nonsolvent like ligroin, a low boiling petrol fraction also yields an oil-in-water dispersion [26].

A cellular substrate, e.g. a polyurethane foam, or fibers of polyacrylonitrile impregnated with a polyacrylate latex may be coated with a reactive polyurethane mixture [15]. As a polyurethane mixture ADIPRENE® L 1671and MOCA® (3,3'-dichloromethylenebisaniline) (both trademarks of DuPont [1]) may be used. This kind of surface treatment probably results in a porous coating [12].

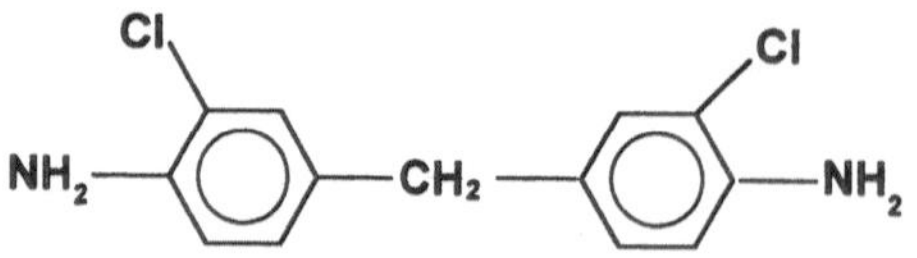

3,3'-di-chloro-di-4,4'-di-amino-diphenyl-methane

The chlorine atoms in the ortho position to the amine groups hinder the reactivity of the amines with the isocyanate groups. Therefore, these amines may be mixed with an isocyanate prepolymer without a spontaneous reaction occurring as, for instance, with 4,4'-diaminodiphenylmethane [Eq. (10-1)].

OCN ------------ NCO + H_2N-R-NH_2 →

compound with   diamine
isocyanate groups

------- NH-CO-NH-R-NH-CO-NH -------   (10-1)
       polyurea

Formation of a polyurea

The reaction of an isocyanate with an aliphatic amine is extremely fast, with an aromatic it is still very fast. To control the reaction of an isocyanate with an amine it is possible to use a derivative of it. Amines react with a ketone to form a ketimine [see Eq. (10-2)]. The rate-determining step in an amine-isocyanate reaction then is the hydrolysis of the ketimine back to the amine and ketone. By that, the extremely fast reaction of an aliphatic amine with an isocyanate may be slowed down and can be controlled [16]. This polyaddition principle is suited for the coating of fabrics [8, 16, 22].

Ketoximes are also suited to block an isocyanate group [see Eq. (2-1)]. In the presence of an amine and at a temperature of >140 °C, ketoxime is split off and

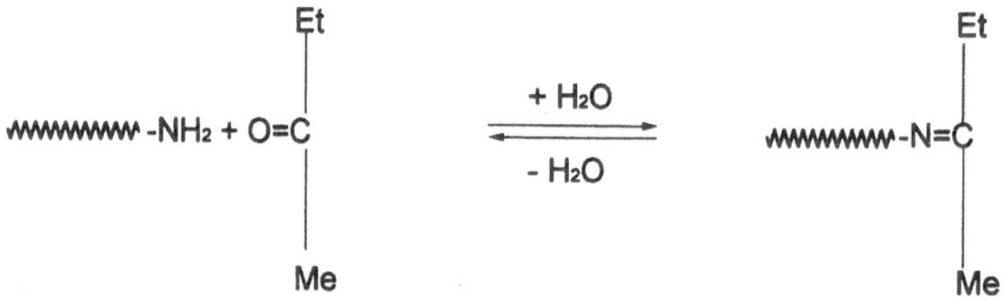

                                                                              (10-2)

Amines react with ketones (e.g. butanone-2) to oximes. This reaction is an equilibrium, which in absence of water, is on the side of the oxime.
In presence of water the equilibrium is partially on the side of the free amine – adding of water reforms the amine again. Even small quantities of free amines react spontaneously with an isocyanate added to the oxime/amine mixture. The reaction of amines with isocyanates reduces the quantities of amines in the equilibrium, amine is further reformed under liberation of water which catalyzes the reformation of amines further.
By this effect the extreme fast reaction of an amine with an isocyanate can be controlled because its speed is determined by the reformation of the amine in the equilibirium.

the isocyanate group is liberated, able to add to an amine to form a polyurea. So a blocked prepolymer reacts at a temperature of 80–200 °C with an amino alcohol [17] to form a coating mixture. A solution of a polyurethane or an isocyanate prepolymer, blocked by a ketoxime, is coated onto a release paper. Then more than 50 % of the solvent, e.g. toluene or butan-2-one is evaporated and the film can be used to coat a substrate [18].

Other coating compounds are mixtures of polyesters, toluylene diisocyanate and leather or cork dust [11]. Isocyanate prepolymers, together with a glycol and a polymer containing hydroxyl groups, like a partially hydrolyzed PVC, may be applied onto a release paper and then laminated to a textile substrate [13, 14].

Coating of a fabric with an isocyanate and a polyamide, modified by an isocyanate, dissolved in a solvent, only partially dissolving the mixture, result in coatings which also should have some porosity [6].

A mutual spray application of a prepolymer and an aromatic diamine results in porous coatings too ([20, 27], see also Chap. 11.2).

Another way to obtain a porous coating is the application of an isocyanate prepolymer onto a textile substrate followed by the reaction with the humidity of the air [21, 23].

10.1
Polyaddition of Emulsified Oligomers and Monomers with Nonsolvents in the Inner Phase

In this chapter only reaction processes using polyaddition to form polyurethanes or polyureas are discussed. To get a microporous structure, monomeric or oligomeric precursors of the polyaddition reaction are treated with a nonsolvent. After the polyaddition and evaporation of the nonsolvent, the resulting polymer may have the desired shape directly with a microporous structure. If the components have a high reaction rate, the precursors are mixed directly beforehand to give them the desired shape.

A typical reactive emulsion contains between 25 and 300 % of nonsolvent; suitable nonsolvents are, for instance, aliphatic hydrocarbons [2].

As nonsolvents, organic compounds may also be used which are able to dissolve the precursors of the polyaddition reaction. These organic compounds, during the polyaddition, should gradually become a nonsolvent for the final polyurethane(urea). A mixture of toluene and heptane, for example, is able to dissolve a polyhydroxylic compound together with an isocyanate or a prepolymer. After the addition of MOCA® (DuPont) the mixture is used for a coating. At the moment the polyaddition reaction is complete, the solvent mixture then has nonsolvent characteristics and micropores are formed [20].

A typical procedure is to dissolve a mixture of a polyester and a glycol in a rapidly evaporating solvent and to emulsify a dispersion of an isocyanate in that mixture. The reacting mixture is used to form a film or a coating and the solvents are evaporated [1]. It is also possible to eliminate part of the (non)solvents by pressing the resulting film or coating between rollers [7].

In a precursor solution of a polymer, 60–200 % a nonsolvent is stirred in. The precursor solution builds the continuous phase and the nonsolvent the discontinuous one. Then after formation of the coating or a film, the emulsion is hardened by polymerization and evaporation of the solvent and the nonsolvent [9]. It is possible to create thick layers of such mixtures. These thick layers may be cut into thin layers prior to evaporation of the solvents [11] or to form finished articles like shoes from them [12].

The reaction may be carried out as a two-step process. First by means of an emulsifier [15, 16] a nonsolvent is dispersed in a solution of a prepolymer [14, 17]. Then the chain extender is also dispersed in. This reaction mixture is used for coating or film forming in which during the polyaddition reaction solvents and nonsolvents are evaporated [3 - 5]. This process can be carried out with a low amount of solvent as long as isocyanate prepolymers are used which have a low viscosity. The low viscosity is achieved by mixing the prepolymers with a plasticizer as a thinner. As a nonsolvent water can be used. The chain extender is dissolved in water also. After formation and evaporation of water microporous films or coatings are produced [22].

If solvent/nonsolvent-mixtures are used which are immiscible with each other, an emulsion is created at the moment the nonsolvent is added to the solution. Therefore liquid emulsions of reactive components with a volatile solvent in the continuous phase and nonsolvent in the discontinuous phase are used for coatings. During the reaction time a release paper is placed on the surface of the coating to keep the solvent/nonsolvent mixture in the forming coat for a longer time. Afterwards the release paper is removed, and the solvents are allowed to evaporate. By this method pores on the surface of the coating are created [7, 8, 13].

Polyisocyanate, solvent, catalyst, water and a Zerewitinoff active compound are reacted under gel formation. The solvent of the resulting gel is removed and the product is used as an insulation material [21].

Oligomer products having amino groups, like aminopolyethers, can be reacted with low molecular diisocyanates to form microporous polyureas according to the method described in this chapter [6]. This method may be applied together with a polymer latex able to coagulate under the reaction conditions [10].

Figure 10-1 shows a film produced by a polyaddition reaction with a nonsolvent as the discontinuous, inner phase. The pore structure differs to a large extent in comparison with the micropores formed by a coagulation (see Fig. 7-2).

- Micropores created by a coagulation look like a microfoam.
- Micorpores created by a nonsolvent in the inner phase look different from those formed by coagulation. If water is used as a nonsolvent the micropores have a spherical structure.
- The water vapor permeability of microporous films created by nonsolvents in the inner sphere generally is lower than by coagulation. If water – instead of an organic nonsolvent – is used, the vapor permeability may be even lower. Due to the high surface tension of water, spherical droplets are formed in the inner phase and channels of water connect the droplets. Only the channels are responsible for the transmission of water vapor. By the use of emulsifying agents the size of the channels can be increased.

The advantages of a reactive emulsion process with nonsolvent in the inner phase are:

Fig. 10-1. Typical structure of a film produced by a poly-addition reaction with water in the inner phase. Water with its high surface tension forms spherical droplets in the emulsion which also keep their structure in the finished polymer

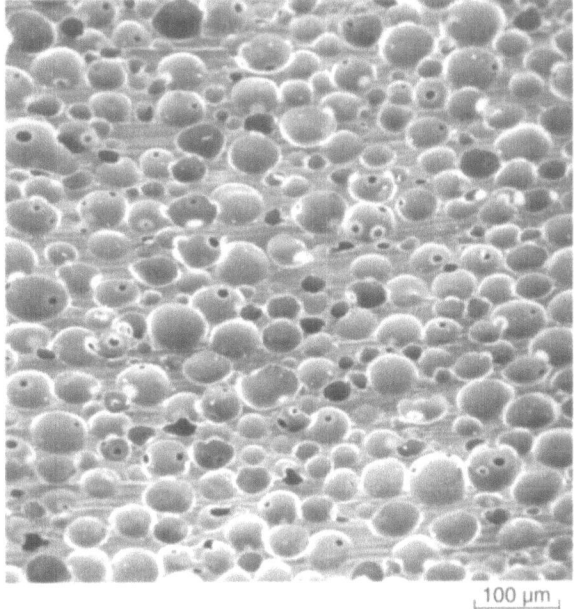

100 µm

- Few or no amounts of organic solvents are necessary.
- The physical properties of the products are high due to the fact that a high molecular weight or even a crosslinked polymer can be synthesized which can never be achieved in a process for a dissolved polymer. Even polyurethane-polyureas which may – as high molecular products – be dissolved only in hard solvents like dimethylformamide or -acetamide can be produced by this method.
- The process needs one working step less: the isocyanate prepolymer reacts directly with the amine, so there is no need for a further dissolving step.
- In correspondence to selective evaporation, the amount of microporosity can be controlled easily by the amount of the nonsolvent used; this is advantageous.
- Due to the fact that increasing porosity is combined with decreasing physical properties, the films and coatings should only have a range of water vapor permeability which fulfills the demands and not more.

The process of polyaddition with a nonsolvent in the inner phase has the following disadvantages:

- Contrary to reactions with dissolved polymers in this process the physical properties of the films and coatings can only be seen after the application is done. Additionally, a storage of 1 – 2 days is necessary to get the right level in physical properties. There is almost no way to correct the physical properties after the application.
- The process needs dosing pumps which must work in a precise manner.
- The chemistry of polyaddition is only partially done in a chemical plant: the main step is done by the applier.

Fig. 10-2. Microporosity of a polyurea as a function of the quantity of water used. The microporosity is measured as water vapor permeability; × ml nonsolvent are used for 50 ml (10 mmol NCO) of a prepolymer (NCO content 3.6 %) of hexamethylene-diisocyanate and polypropylene glycol ether (hydroxyl number 56) dissolved in benzene. Hydrazine was used as a 1 M solution in water. The dispersion was prepared by a high speed stirrer; it was poured onto a glass plate as a film with 0.3-mm thickness. The water vapor permeability was measured after the evaporation of benzene and water. For each different quantity of the nonsolvent water a film was produced and the water vapor permeability measured according to DIN 53333

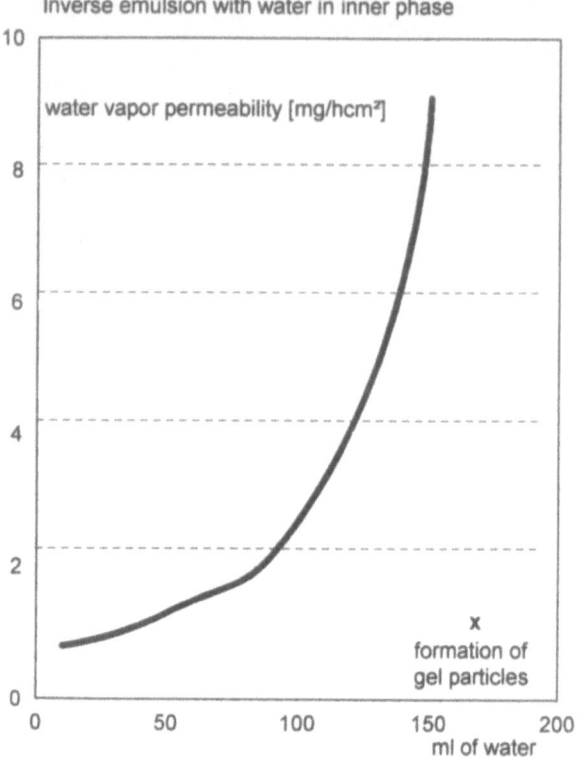

The water vapor permeability depends on the amount of nonsolvent used (Fig. 10-2). There are products on the market being produced according to the process described in this chapter [18, 19].

Normally a mixing device [3] is used which is well known in the production of polyurethane foams. Figure 10-1 shows the typical structure of a film produced by a polyaddition reaction with water in the inner phase.

10.2
Polyaddition of Dissolved Polymers

An alternative to the process described above, where all products are dispersed in an organic- or water-based medium, is to dissolve the starting materials in an appropriate solvent or a solvent mixture. The only stipulation for the solvent or solvent mixture must be that the polyurethane(urea) by the ongoing reaction should become increasingly insoluble in the solution. If the solvent or the solvent mixture is present in the right quantity, i.e. a quantity where the forming polyurethane(urea) is able to include the (non)solvent in a kind of a water-in-oil emulsion, then the same conditions exist as above. The solvent mixture of solvent and nonsolvent is in correlation to the mixtures used in the selective evaporation process. After final elimination of the solvents, micropores remain in the polymer matrix.

Usually, a mixture of the precursors of a polyurethane or a polyurea are dissolved in the solvent (mixture). Directly after the addition of a catalyst or a crosslinker the solution remains clear. The clear solution is used for forming a film or a coating. After a few moments, the clear solution shows a beginning turbidity. It must be applied in this stage [1, 9, 11]. Due to the fact that the viscosity then starts to rise and the product also starts to solidify, this process – technically – should be carried out continuously.

The behavior of the products synthesized by this process is similar to that of products produced by the selective evaporation process. The amount of water vapor permeability can be governed by the quantity of the nonsolvent added (see Fig. 9-3).

The process is suited for the glycol as well as for the amine polyaddition reaction. In the glycol reaction the polyol, i.e. the polyester or polyether, glycol and isocyanate can be mixed in one step. Usually, a two-step reaction is needed for the amine addition. The formation of an isocyanate prepolymer must take place prior to the amine polyaddition, because the amine reaction with an isocyanate reaction is so rapid that no controlled formation of a polymer can take place in the presence of hydroxyl groups (see Sects. 25.1 and 25.2).

To determine if a special solvent or solvent mixture is suited for the process, the solubility parameter of the polyurethane(urea) desired and the liquid medium at the temperature of application should be different as much as possible. The evaporation numbers of the solvents used must differ in the same way as in the selective evaporation process.

A comparison of the polyesters to be used show that, for example, a polyethylene glycol adipate needs a solvent less lipophilic than polybutylene adipate to dissolve the starting materials and become a nonsolvent for the resulting polyurethane. For the first polyurethane(urea), toluene is suited, in the second one, xylene may be used to achieve microporosity.

Similarly, the isocyanates should also be taken into consideration: the hard segments arising from 4,4′-diisocyantodiphenylmethane (MDI) are more rigid than those from toluylenediisocyanate (TDI). MDI-containing polyurethane(urea)s need less lipophilic solvents than TDI-containing ones.

Figure 10-3 is a photograph of a film produced by polyaddition in a solvent mixture. Apparently, there is no difference seen between this structure and the structure of a film produced by coagulation or selective evaporation (see Figs. 7-2 and 9-5).

It may be helpful to add a small quantity of a good solvent, like DMF, to the solvent mixture which boils at a higher temperature than the other solvents present in order to obtain a smoother surface and a better spreading of the products during reaction [10, 12].

Besides processes running at higher temperatures, the polyaddition can also be carried out at lower temperatures. In this case the solution of the precursors for a duroplastic polymer and their reaction partners is cooled to a temperature below the melting point of the solvent. The solvent should melt between 50 – 150 °C. The forming polymer includes the solvent in the crystallized state; the

Fig. 10-3. Film produced by
a polyaddition in a solvent
mixture

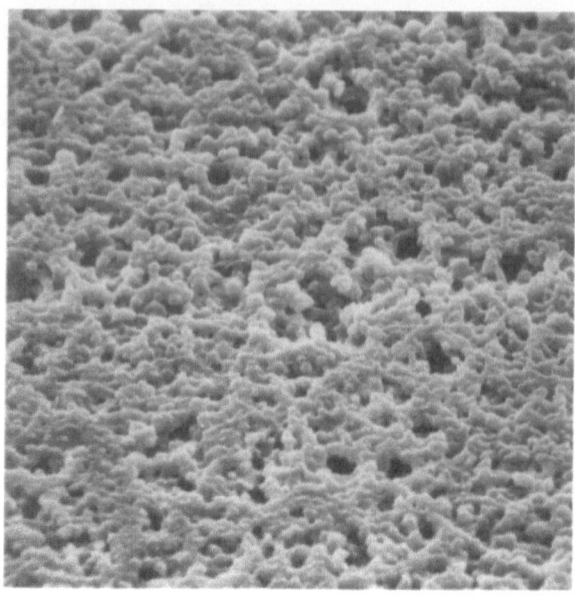

20 μm

crystals can be removed by sublimation. The remaining pores on average have a
diameter of 20 μm. As a polymer, isocyanate prepolymers together with a polyol
or a polyamine or compounds with olefinic unsaturated groups and an alkyl
halide together with a polyamine are suited [13].

An isocyanate prepolymer is produced by mixing polytetramethylene glycol
ether, a diol and an isocyanate in a solvent. After the addition of a diol, water or
an amine, coatings are prepared which are resistant to alcohols, hydrolysis or
oleic acid [15].

Polyethylene glycol, triol and MDI are transferred to a prepolymer. Naphtha
is added to the prepolymer solution. Drying in the air results in a porous product. The spinning of such a prepolymer into a nonsolvent containing a chain extending agent is also possible. The products, water swellable hydrogels, may be
used in the production of wound plasters [14].

Using silicone rubber molds of prefinished articles, like a shoe upper, products looking like a prefinished article can also be produced by this method [4].
Microporous top layers or topcoats can be achieved by applying acetone solutions of prepolymers. Acetone becomes a nonsolvent during hardening of the
prepolymer [16]. This process has the same advantages and disadvantages as the
similar process in emulsion (see Chap. 10.1).

Products which may be applied according to this process [2, 3] are LEVA-
CAST®-adduct and -hardener [6], NEPTAN® [8], MARSIPOL® [7], and ICAP-
PLAST® [5].

10.3
Bonding of Nonwovens by Reactive Processes

The chemistry discussed for films and coatings (see Chaps. 10, 10.1 and 10.2) can also be applied to bond nonwovens. Reactive solutions of isocyanate-, amino- and/or hydroxyl-group-containing compounds in solvent/nonsolvent mixtures can be used to bond nonwovens (see Chap. 21).

The aqueous reaction of isocyanates is used in the bonding of nonwovens, for example, by the addition of an isocyanate prepolymer based on a polyester to impregnate the textile substrate. By the aqueous reaction of the prepolymer polyureas are formed [1, 5]. A chain-lengthening by the amine is done in the following way: A nonwoven impregnated with an isocyanate prepolmer is dipped in water which may contain a diamine [2] or may be treated on one side with a water-based diamine solution [10]. Besides solutions, emulsions of these reaction partners may be also used [3]. This impregnation may be modified in such a way that prior to the production of the nonwoven, the fibers are treated with prepolymers, solvents and plasticizers, and then dried in humid air [9].

The impregnation may be carried out with a solution of a prepolymer and a chain-lengthening agent [8] whereby the polyaddition is done in the nonwoven [6]. Following this impregnation a coating is achieved on a release paper by a solvent/nonsolvent solution of an isocyanate prepolymer and a diamine. After evaporation of the solvent mixture the impregnated nonwoven obtains a microporous film [7] and the composite material is pressed and stripped off the release paper. For instance, a dispersion of a polyisocyanate, polyoxypropyleneglycol, butandiol, acetone, paraffin oil and starch may be used to bond nonwovens [12].

Ketoxime-blocked isocyanate prepolymers may be mixed in solvent/non-solvent mixtures with an aromatic diamine like MDA. A nonwoven is treated with this mixture and then exposed to higher temperatures to de-block the isocyanate groups and to bond the nonwoven [11]. These types of bonded nonwovens may be used as diaphragms [4].

Reactive solutions of components containing at least two isocyanate, OH and – possibly – NH groups are each coated and impregnated onto nonwovens to obtain a microporous leather substitute [14].

The following process is a combination of bonding and laminating: reactive mixtures of isocyanate prepolymers and diamines are sprayed onto layers of fibers. Prior to hardening the next layer is put on followed by a spraying and an additional fiber layer and so on [13].

10.4
Polymerization of Emulsified Polymers

Vinylic monomers as such or in solution may be transferred in an inverse emulsion by stirring water into them. So acrylates, such as butyl acrylate, acrylic acid and acrylonitrile, are transferred to water-in-oil emulsions, a polymerization is started and after elimination of the liquids – mostly by evaporation – microporous products are produced. This kind of reaction may be used for impregnations [2, 10] or films [14].

Water is stirred into olefinic unsaturated compounds, able to polymerize, a fabric is impregnated, the polymerization is started and water is evaporated [12]. The same can be done to produce a coating [13]. Instead of acrylates, mixtures of styrene with propylene glycol ether and ethylene glycol ether may also be used [4].

A solution or dispersion of monomers in water or a solvent is cooled down to solidify the liquid medium then the reaction is started, e.g. by radiation. After the polymerization is finished, the liquid medium is eliminated [6]. An improvement to the process is to add the (redox) catalysts needed, in two separate steps and add one component at the moment the crystallization starts. This kind of polymerization is carried out in a fibrous substrate [1].

Isocyanate prepolymers dissolved in dioxane can also be frozen and polymerized to microporous products [3]. A polyamide of 40 % hexamethylene adipamide and 60 % caprolactam is dissolved in water/methanol, ethyl acrylate, methyl acrylate, glycidyl acrylate, etc., together with sodium stearate. Azoisobutyric nitrile is added. After forming to a film, polymerization is carried out and methanol and water evaporated to form microporous products [5].

A method to further improve the mechanical properties is to use hydroxyethyl acrylate, hexylmethacrylate and 1,6-hexandiol dimethylacrylate together with $F(CF_2)_{12}(CH_2)_{12}H$ and to start the polymerization by means of UV rays in a matrix. The fluorine compound is extracted afterwards with hexane [8].

10.5
Polymerization Reaction of Dissolved Polymers

It is also possible to create a microporous structure by polymerization in solution. This process is similar to a polyaddtion, which is carried out in a solvent or a solvent mixture, whereby the precursors of the polymer are soluble (see Chap. 10.1). Vinylic compounds, like styrene, vinyl chloride, acrylonitrile, etc., preferably with at least 10 mol% of a crosslinking monomer, like divinylbenzene, are polymerized in 50–2000 % of a solvent or a solvent mixture. The solubility parameter of the solvent or the solvent mixture should be + 0.8 higher than that of the polymer. By choice of a different solvent, the pore structure is determined. After polymerization the solvents are eliminated [1]. The products gained may be used as filters.

A bonding of nonwovens may be achieved by impregnating a substrate with at least 5 % of a catalyst, based on a peroxide, together with xylylene-bis-acrylamide, butandiol-bis-acrylate, etc., dissolved in a water/methanol mixture. After in situ polymerization, water and solvent are removed [2].

Nonwovens consisting of polyamide fibers may be impregnated by monomers with at least two double bonds, like methylene-bis-acrylamide or triethylene glycol diacrylate together with compounds with only one double bond like butyl-acrylate. The polymerization is started at – 30 °C by a photoinitiatior, like benzoin methyl ether, and radiation. Then the solvents are removed [4, 6]. The presence of (co)polymers helps in the formation of the bonding. A variation of the process is to polymerize in the presence of a vinyl chloride/methacrylate copolymers plastisized, e.g., by ethyl hexyl phthalate [5].

Solutions or emulsions of olefinic unsaturated monomers containing 0.1–30 % of a polymer with the same constitution are used for impregnation of nonwovens. The polymerization is started with light waves of a wavelength of 0.39 μm [9]. By means of water and an emulsifying agent, a powdered polymer, insoluble in water, styrene and methylmethacrylate, are transferred to a water-in-oil dispersion, polymerized and water is evaporated [7]. The wavelength of the light to start the polymerization should be 0.39 μm.

Monomers of polymers, able to form a film after polymerization, like methylmethacrylate, laurylmethacrylate and methacrylic acid, are dissolved in solvent/nonsolvent mixtures, like toluene/butan-2-one, polymerized and the solvent is removed by selective evaporation [3].

Coatings can be obtained with olefinic copolymers emulsified in water by means of an emulsifier. This is carried out by the addition of a filler and a catalyst on fabrics and polymerization on a calender during coating [8].

On a release paper a fine layer of frozen water crystals are produced. Then a nonwoven is posed onto the layer. After that, a dispersion of monomers, which may also contain some polymer, is added, cooled to get at least a partial crystallization and polymerized by electromagnetic radiation of a wavelength of 0.7 μm. After polymerization and drying a porous substrate with a velvet-like surface is produced [10].

An urethane acrylate oligomer is dissolved in dimethylacetamide and mixed with polyvinylpyrrolidone and a crosslinker [Irgacure® 184 of Ciba (CH)], applied onto a glass plate, the solvent partially evaporated and irradiated with UV rays then dipped in water then in ethanol to create a porous membrane [11].

10.6
Reactive Processes by Elimination of a Component

In a reaction mixture able to form a polyurethane, finely dispersed fillers like ammonium, sodium or potassium chloride are included, the polymerization is carried out and the fillers are leached [1].

In a reacting mass, materials, able to dissolve or evaporate, may also be included which can be removed after the polymerization by radiation or by electronic beams [2].

Monomers, like styrene, or prepolymers of polyurethanes or epoxy resins possibly in mixture with a polymer like PVC, are emulsified or dissolved. Finely ground calcium carbonate, starch, sodium chloride or sodium sulfide is incorporated into this emulsion or solution. With this mixture a film or a coating is formed and the monomers are polymerized. Then the salts are leached by means of water and starch with the help of an enzyme [3].

In an organic solvent the metal or ammonium salt of an organic acid is dissolved. At the boiling point of the solvent the salt should have a solubility of 60 g/100 ml. Monomers, catalysts etc. are added to the solution. After coating on a fabric, the polymerization is carried out. The coated fabric is immersed in water to leach the salts and to create microporosity [4].

References to Part 2

6
Creating Microporosity

1. Hollemann-Wiberg (1985) Lehrbuch der anorganischen Chemie, Berlin, p 784

7
Elimination of Solid Particles – Especially Leaching of Salts

Also of interest: 11.1 [3, 38], 12 [37], 16 [17], 16.2 [28], 19 [18], 27 [112]

1. DAS 1 129 921 (Jan Spirit, Olomouc, Vojtech Smejkal; CZ Prior. 10.9.56; 4.9.57/24.5.62)=GB 816 095=FR 1 182 092=DDR 21 102
2. RA 138 583 (IA Stern et al.; 19.12.59/8.6.61)
3. RA 145 539 (IA Stern et al.; 1.2.61/21.3.62)
4. J 49 126–801 (16.4.73/4.12.74; Shiga Akiresu)
5. DOS 1 569 282 (Porous Plastics; L David et al.; GB Prior. 25.11.63 and 21.10.64; 25.11.64/24.7.69)= BE 656 239=SZAS 15 180/64
6. OS 2 052 831 (3M; RL Elton; US Prior. 20.10.69; 19.10.70/29.4.71)
7. DBP 910 960 (C Freudenberg; CL Nottebohm; 1.10.48; 17.9.53/10.5.54)
8. US 2 819 981 (BF Goodrich; RE Schornstheiner et.al.; 23.2.55/14.1.58)
9. DAS 1 204 186 (H Becker; 13.2.61/4.11.65)
10. FP 1 131 453 (Göppinger Kaliko; D Prior. 8.2.54, 17.2.54, 4.10.54 and 16.11.54; 8.2.55/21.2.57)= DAS 1006 387; DAS 1 018 837 (Göppinger Kaliko; V Jilge; 4.10.54/ 7.11.57); SZ 314 298 (Kötitzer Ledertuch- und Wachstuchwerke; D Prior. 4.4.53; 23.3.54/31.7.56)=GB 757 814
11. DAS 1 138 534 (Pritchett and Goldand E.P.S. Comp.Ltd.; EMO Honey, CR Hardy, F Sharp; GB Prior. 17.2.58 and 26.1.59; 17.2.59/25.10.62)=BE 575 780=FR 1 220 322=GB 844 801=US 2 959 822
12. RA 165 994 (P.F. Sapilevskii; 5.1.63/March 65)
13. RA 144 149 (I.A. Stern; 18.5.61/6.2.62)
14. FR 1 375 715 (Walkerlan Ltd.; GB Prior. 11.4.62; 10.4.63/14.9.64)=BE 630 861=GB 1 055 872
15. FR 1 477 810 (Porous Plastics Ltd.; GB Prior. 1.4.65; 1.4.66/13.3.67)=NE 6 604 403=BE 678 894
16. GB 1 126 849 (W. R. Grace Co., US Prior. 19.5.66; 19.5.67/11.9.68)
17. DOS 2 835 607 (BF Goodrich; CD Segmer; 4.8.54/20.5.68)=FR 1 143 016=GB 782 781
18. NE 6 907 740 (General Electric Co., US Prior. 20.5.68; 20.5.69/24.11.69)
19. US 2 983 960 (Göppinger Kaliko; V Jilge; 4.10.55/16.5.61, cip 8.6.54; D Prior. 4.4.53)=SZ 314 298=GB 757 814; US 3 054 691 (US Rubber; HD Myers; 4.6.58/18.9.62); DAS 1 010 267 (The Chloride Electrical Storage Comp.; EM Honey et.al.; 6.7.53/28.11.57; GB Prior. 17.7.52+11.7.53); DAS 1 013 421 (The Chloride Electrical Storage Comp.; EM Honey et.al.; 7.7.53/6.8.57; GB Prior. 17.7.52+18.6.53)
20. DOS 1 444 154 (Göppinger Kaliko; S Dorogi; 21.12.63/25.3.71)
21. DAS 1 004 136 (DuPont; JA Piccard et al.; 18.7.53/14.3.57; US Prior. 29.7.52)=GB 737 137
22. US 3 882 054 (IPRC Corp.; R Hofstettler; 6.2.73/6.5.75)=US 73–330 102
23. JA 21 471/66 (Kobe Jushi Co. Ltd.; 15.7.60/15.12.66)

24. BE 610 266 (Reeves Corp.; CH Teague, HJ Strauss, DA Davis; US Prior. 17.11.60; 14.11.61/14.5.62)
25. FR 1 369 563 (Soc. Rhodiaceta; R Bolliand; 5.7.63/6.7.64)
26. US 3 054 691 (US Rubber; HD Myers, HT Nelson; 4.6.58/18.9.62)
27. DOS 1 569 282 (Porous Plastics, L David et al.; GB Prior. 25.11.63 and 21.10.64; 25.11.64/24.7.69) =BE 656 239=SZAS 15 180/64
28. US 3 169 885 (Interchemical Corp.; MM Golodner, GE Salrino; 15.3.63/15.2.65)
29. OS 1 619 246 (Göppinger Kaliko; S Dorogi; 12.12.63/29.4.71; addition to 1 444 154)
30. JA 29 453/64 (Toyo Spinning Co. Ltd.; 2.9.63/18.12.64)
31. OS 2 122 367 (Kureha Kagaku Kogyo KK; N Murayama, T Katto; JA Prior. 8.5.70=JA 38 613/70; 6.5.71/13.7.72)=FR 2 091 382
32. US 3 379 658 (CR Kemper; 25.5.64/23.4.68; cip 22.8.62)
33. JA 1 669/65 (Shibata Gomu Kogyo KK; 17.10.62/29.1.65)
34. GB 1 072 417 (C Freudenberg; D Prior. 4.10.63; 5.1.64/14.6.67)=SZ 420 598
35. OS 2 031 340 (Union Carbide Co.; WA Miller, RD Jenkinson; US Prior. 27.6.69; 25.6.70/25.3.71)
36. OS 2 052 831 (3M; RL. Elton; US Prior. 20.10.69; 19.10.70/29.4.71)
37. US 3 654 065 (Göppinger Kaliko; St Dorogi; 22.7.68/4.4.72)=NE 6 904 488=FR 1 602 488 (31.12.68/8.11.71)
38. JA 31 004/72 (Kuraray Co. Ltd.; 8.2.68/11.8.72)
39. NE 6 706 867 (AKU; 18.5.67/27.5.68)
40. FR 95 376 (Porous Plastics; GB Prior. 31.7.67; 25.7.68/11.9.70)=NE 6 810 718; addition to FR 1 418 420
41. US 3 496 042 (Porvair Ltd.; 30.3.66/17.2.70)
42. OS 1 469 266 (BASF; G Welzel; 22.7.65/12.12.68)
43. SZ 475 730 (Porous Plastics Ltd.; MW Dento; GB Prior. 15.6.65; 14.6.66/15.9.69)
44. OS 2 021 305 (Am Cyanamid Co.; VL Gallacher; 30.4.70/21.1.71; US Prior. 1.5.69)=BE 749 791
45. DAS 1 239 262 (BASF; G Welzel; 30.1.64/27.4.67)
46. OS 2 152 596 (F. Uhde GmbH; KD Hammer et al.; 22.10.71/26.4.73)
47. OS 2 331 896 (Kureha Kagaku Kogyo KK; T Nagamura, H Hagiwara; JA Prior. 22.6.72)=JA 61 790/72; (22.6.73/10.1.74)
48. FR 1 483 411 (Porous Plastics Ltd.; MW Denton; GB Prior. 15.6.65; 15.6.66/2.6.67)
49. US 3 718 561(EJ Jacog; 7.7.70/27.2.73; cip 4.12.67)
50. US 3 549 398 (Fiber Ind. Inc.; GA Watson 3.4.67/22.12 70)
51. GB 1 048 985 (ESB-Reeves Corp.; US Prior. 27.6.62; 26.6.63/23.11.66)
52. JA 6 246/64 (Toyo Spinning Co. Ltd.; 14.1.61/2.5.64)
53. JA 14 580/66 (T Ishizuka; 31.8.62/17.8.66)
54. RA 182 105 (IA Stern et al.; 31.8.62/Nov. 66)
55. JA 14 759/66 (Nippon Orimono Kako Co. Ltd.; 29.12.63/19.8.66)
56. JA 837/67 (T Ishizuka; 2.5.63/17.1.67)
57. JA 56 169 877 (Mitsubishi Rayon KK; 3.6.80/26.12.81)
58. JA 21 91 778 (Kokoku Chem. Ind. KK; 20.1.89/27.7.90)=JA 89–11084
59. JA 23 07 987 (Achilles Corp.; 18.5.89/21.12.90)=JA 89–125536
60. DDR 221 216 (VEB Vowetex; W Guenther, B Hellwich, 28.11.83/17.4.85)=DDR 83–257163
61. JA 82–66 186 (Toyo Cloth, 9.10.80/22.4.82)=JA 80–141288
62. JA 51 029 201 (Kyowa Leather Cloth; 31.8.74/12.3.76)=JA 74–100 124
63. US 4 237 083 (Evans Prod. Co.; J Young et al.; 13.2.79/2.12.80)=US 79–11 900
64. JA 56 011 931 (Kuraray KK; 10.7.79/5.2.81)=JA 79–87 856
65. DOS 2 258 527 (De Bell and Richardson Inc.; LA White, WH Holley; 29.11.72/7.6.73; US Prior. 30.11.71)=US 71 203 471
66. JA 74 008 507 (Toray Ind. Inc.; 26.2.70/26.2.74)=JA 70–25 039
67. JA 58 160 326 (Fujikura Rubber Works KK; 18.3.82/22.9.83)=JA 82–43 529
68. US 4 425 395 (Fujikura Rubber Works KK; I Negishi, K Hunyu; 30.3.82/10.1.84)=JA 81–65 564+JA 81–65 565+JA 81–65566
69. DOS 3 330 031 (Toyo Cloth Co. Ltd.; K Takashima et al.; 19.8.83/26.7.84) JA Prior. 24.1.83= JA 83–10 393
70. DOS 3 543 217 (Dai-Ichi Kogyo Seiyaku Co. Ltd.; T Tanaka et al.; 6.12.85/3.7.86); JA Prior. 7.12.84=JA 84–259 743

71. JA 62 062 991 (Asahi Chem. Ind. KK; 10.9.85/19.3.87)=JA 85–198 590
72. DAS 1 469 542 (Göppinger Kaliko und Kunstleder-Werke GmbH; St. Dorogi; 16.11.64/23.7.70)
73. JA 06 093 571 (Daiichi Lace KK; Hosokawa Micron KK; 26.9.91/5.4.94)=JA 91–0318 744
74. JA 061 92 433 (Kyoeisha Kagaku KK; Osaka Prefecture; Tomen KK; 24.12.92/12.7.94)

7.1
The Coagulation Process

The following references are also of relevance for this section: 15 [27], 17.1 [21, 22, 29], 19 [9, 21, 22, 26–30, 32, 33, 35–41, 43], 23 [1–3]; the coagulation of poly-γ-glutamic acid, see 18.3[18], 21 [7]

1. Sittig M (1969) Synthetic leather from petroleum. Noyes Development Corp., New Jersey, London
2. DDR 55 928 (Inst. für Textiltechnologie; G Schröder, D Scharch; 18.10.66/20.5.67)
3. Toxicology of DMF: see, for instance, Sax NI (1962) Dangerous properties of industrial materials. Reinhold, New York; Patty FA (1963) In: Fassett DW, Irish DD (eds) Industrial hygiene and toxicology, 2nd edn., vol II Interscience Publishers, New York; Gleason MN, Gosselin RE, Hodge HC, Smith RP (eds) (1969) Clinical toxicology of commercial products, 3rd edn. Williams & Wilkins Co, Baltimore
4. DRP 275 697 (M Wünschmann; 30.10.13/23.6.14)
5. JA 39 328/72 (Mitsubishi Rayon Co. Ltd.; 17.9.68/4.10.72)
6. JA 17 315/63(Teijin Co. Ltd.; 6.4.63/9.7.68; JA 16 282/68)
7. JA 44 026/72 (Toray Ind. Inc.; 31.5.69/7.11.72)
8. US 3 520 874 (Fuji Shashin Film KK; W Neno, H Kawaguchi; JA Prior. 7.10.66=JA 66/41 096; 6.10.67/21.7.70)=FR 1 540 625
9. OS 1 669 615 (AB Tudor; EG Sundberg et al.; SW Prior. 1.9.66; 24.8.67/22.10.70)=NE 6 712 022=BE 703 285=GB 1 183 470
10. JA 73–00 534 (Mitsubishi Rayon; 3.6.70/9.1.73)
11. JA 50–019 904 (Asahi Chem. Ind.; 30.6.73/3.3.75)
12. JA 50–025 892 (Teijin Cordley Ltd.; 13.7.73/18.3.75)
13. JA 19 910/66 (Toyo Spinning Co.; 29.10.62/18.11.66)
14. e.g. microcapsules for carbonless copying papers (see Becker/Braun Kunststoffhandbuch, vol. 7, G Oertel Polyurethane. München, 1993, p 634)
15. Hirose S, Shimizu A (1979) Struktur und Leistungsfähigkeit poröser Membranen aus PVC. Kobunshi Ronbunshu 35:435
16. Koehnen D (1977) Phasentrennungsvermögen bei der Bildung asymmetrischer Membranen. J Appl Polym Sci 21(1):199
17. Fourier St (1989) Untersuchungen zur Herstellung von PUR-Lösungen für den Koagulationsprozess. Coating Feb, pp 34–38
18. Fourier St, Stephan W, Masczyk D, Reich G Stand der Technik von PUR-Koagulationsverfahren zur Herstellung hochwertiger Poromeriks. Leder Schuhe Lederwaren 1980/1, pp 9–12; Hebestreit G (1995) Die Poromerikentwicklung im Forschungsinstitut für Leder- und Kunstledertechnologie Freiberg. Das Leder 88; Stoll M, Mädler A (1995) Leder und synthetische Schaftmaterialien – eine vergleichende Bewertung. Das Leder 292
19. JA 79–126 268 (Kuraray; 24.3.78/1.10.79)=JA 78–34 571
20. JA 81–20 045 (Sumitomo Chemical; 26.7.79/25.2.81)=JA 79–95 826
21. DDR 147 121 (Forschungsinstitut für die Leder- und Kunstledertechnologie; D Mascyk et al.; 5.11.79/18.3.81)=DDR 79–216 671 (WP)
22. DDR 147 122 (S Fourier et al.; 29.10.79/18.3.81)=DDR 79–216 521 (WP)
23. JA 63 282 378 (Unitika KK; 11.5.87/18.11.88)=JA 87–113 842
24. JA 61 034 287 (Toray Ind. Inc.; 25.7.84/18.2.86)=JA 84–153 108; JA 61 034 286 (Toray Ind. Inc.; 23.7.84/18.2.86)=JA 84–151 337
25. DDR 117 051 (Forschungsinstitut für die Leder- und Kunstledertechnologie; P Hertel; 19.11.74/20.12.75)=DDR 182 430 (WP)
26. DOS 2 706 522 (Bayer AG; U Reinehr et al.; 16.2.77/17.8.78)

7.2
Coagulation of Polyamide and Other Polymers

1. JA 18 619/68 (Kogyo Gijutsu-in; 4.9.65/12.8.68)
2. JA 73 11 924 (Nippon Cloth Ind. Co. Ltd.; 26.7.69/17.4.73)
3. DAS 1 242 552 (NV Lederfabriek; L Mombers, PJ de Nijs; NE Prior. 26.4.63; 27.4.64/22.6.67); DOS 2 418 945 (NV Lederfabriek; L Mombers, WLR. Mombers; NE Prior. 22.6.73; 19.4.74/30.1.75)
4. JA 17 845/63 (Shigeo; 28.12.60/11.9.63); GB 957 377 (Toyo Cloth +Toyo Rayon; T Aoki; 16.10.62/6.5.64); DOS 1 444 169 (Toyo Cloth Comp.; A Tomoo; 16.10.62/2.1.69)
5. GB 1 091 935 (Toyo Gomu KK; JA Prior. 21.3.64; 22.3.65/22.11.67)
6. FR 1 485 472 (Toyo Rubber; JA Prior. 6.7.65; 5.7.66/16.6.67)=JA 40 740/65
7. US 3 275 468 (Toyo Cloth Co.; T Aoki; 4.9.62/27.9.66)
8. JA 6248/64 (Fujikura Rubber Works Ltd.; 20.7.62/2.5.64)
9. JA 12 298/65 (Kanegafuchi Spinning Co. Ltd.; 10.12.62/ 17.6.65)
10. JA 16 641 (Katakura Ind. Co. Ltd., 25.10.62/13.8.64)
11. BE 616 904 (Kunstzijdespinnerij Nyma N. V.; NE Prior. 28.4.61; 28.4.62/26.10.62)
12. JA 8 436/65 (Kyowa Leather Co. Ltd.; 27.10.60/30.4.65)
13. JA 26 891/63 (Nikko Kasei Kogyo Co. Ltd.; 12.4.62/ 27.12.63)
14. JA 3 143/64 (Toyo Cloth Co. Ltd. 28.6.60/26.3.64)
15. JA 18 236; JA 18 237; JA 18 238/65 (Toyo Cloth Co. Ltd. 17.7.62 and 23.10.62 and 18.5.63/17.8.65)
16. JA 17 849/63 (Toyo Cloth Co. Ltd.; 1.2.62/11.9.63)
17. JA 48 085 703 (Toyota Central Res. and Dev.; 15.2.72/ 13.11.73)
18. JA 3 144/64 (J Kanebo; 8.8.61/26.3.64)
19. OS 1 619 208 (Courtaulds Ltd.; ME Baguley; GB Prior. 19.5.65; 20.5.66/17.9.70) addition to OS 1 469 519=FR 1 401 832=GB 1 145 856=OE 262 225=BE 681 258
20. OS 2 124 491 (S Bocciardo et C. SpA; G Albanesi, DN Bocciardo; IT Prior. 27.5.70; 18.5.71/9.12.71)
21. BE 634 592 (Grace; RF Stierli; US Prior. 6.7.62; 5.7.63/4.11.63)
22. JA 16 641/64 (Katakura Ind. Co. Ltd.; 25.10.62/13.8.64)
23. JA 12 298/65 (Kanegafuchi Spinning Co. Ltd.; 10.12.62/ 17.6.65)
24. JA 26 891/63 (Nikko Kasei Kogyo Co. Ltd.; 12.4.62/27.12.63)
25. JA 26 889/63 (K Nagao; 9.3.62/27.12.63)
26. BE 650 763 (Courtaulds Ltd.; GB Prior. 18.7.63 and 14.1.63; 20.7.64/16.11.64)=NE 6 408 154
27. DOS 1 469 532 (Fujikura Rubber Works Ltd.; S Nishino; JA Prior. 21.11.63; 20.11.64/12.12.68)= JA 13 155/66
28. FR 1 578 254 (NV Lederfabriek; L Mombers; NE Prior. 14.7.67; 15.7.68/14.8.69)=NE 6 709 853
29. JA 73 5 014 (Kuraray Co. Ltd.; 26.4.69/13.2.73)
30. JA 73 5 009 (Nippon Ikoru Kagaku Kogyo KK; 10.2.69/13.2.73)
31. JA 48 42 051 (Unitika Co. Ltd.; 27.9.71/19.6.73)
32. JA 8 436/65 (Kyowa Leather Co. Ltd.; 27.10.60/30.4.65)
33. OS 1 444 164 (DuPont; EL Yuan; US Prior. 31.10.61; 11.10.62/6.11.69)=BE 624 250
34. US 3 645 668 (Kuraray Co. Ltd.; K Nagoshi, H Hayanami; 29.9.69/29.2.72; cip 11.7.66)= JA 7 014 590
35. GB 1 105 032 (Polymer Corp; US Prior. 2.4.64; 2.4. 65/6.3.68)
36. JA 47 085/72 (Nihon Matai Co. Ltd.; 6.6.70/28.11.72)
37. JA 48 61 602 (Toyota Chem. Res. Dev. Lab. Inc.; 6.12.71/29.8.73)
38. JA 10 744/63 (Kokoku Chem. Ind. Co. Ltd.; 10.11.60/29.6.63)
39. JA 3 143/64 (Toyo Cloth Co. Ltd.; 28.6.60/26.3.64)
40. JA 24 799/68 (Katakura Kogyo Co. Ltd.; 12.2.63/25.10.63); JA 8 240/66 (Katakura Ind. Co. Ltd.; 2.10.62/28.4.65)
41. JA 18 397/69 (Taira Okuda; 5.7.67/12.8.69)
42. RA 144 148 (AD Zaionckovskii et al.; 2.2.61/6.2.62)
43. GB 1 068 781 (Kurashiki Rayon KK; JA Prior. 16.4.63 (=JA 19 494/63); 16.4.64/17.5.67)= FR 1 389 341

44. JA 2 344/70 (Kureha Seni Co. Ltd.; 27.1.65/26.1.70)
45. JA 50 018 796 (Unitika KK; 22.6.73/27.2.75)
46. RA 685 739 (Ivan Chem. Tech. Inst.; 17.4.78/25.9.79)
47. DOS 2 822 265 (Millipore Corp.; DJD Grandine; 22.5.78/7.12.78; US Prior. 25.5.77)=US 77–800575
48. US 4 537 817 (Soc. Resp Limit. Styled; J-L Guillaume; 6.12.83/27.8.85; FR Prior. 8.12.82)=FR 82–20 594

8
Coagulation of Polyurethanes

References also covering the subjects of this chapter: 9.3 [28], 15 [27], 18.2 [19, 28], 19 [27, 29, 30, 35, 39, 40, 44], 21.1 [6–8, 14, 15], 21.1 [31, 32], 23.1 [3, 8, 11], 22 [14, 15, 21, 35, 40, 46, 48, 58, 73, 74, 76, 78, 79], 25.4 [15, 33], 27 [8, 16, 53, 110];
coagulation of a polyurethane, modified with γ-methyl-glutamate derivative see 18.3 [40–45]

1. DBP 888 766 (Farbenfabriken Bayer; W Brenschede; 22.5.51/20.11.51)=GB 722 723=FR 1 050 942=US 2 755 266
2. Pepper KW (1966) J Soc Leather Trades Chem 4
3. Riess W (1974) J Coated Fabr 47
4. Zorn B (1971) Das Leder 147
5. Nakayo (1966) Jap Chemical Quarterly 2, 4:56
6. Nihon Keizai, dated 14.1.76
7. Daily News Record, 12.9.75
8. Chemiefasern Nov, 1976
9. Yatake S Japan Textile News, Feb, 1976
10. Hayashi T (1974) Chem Economy and Engin Review 6(1):26
11. Fukushima O, Kogame K (1976) Melliand Tex Ber 8:673
12. Japan Textile News Jan, 1971
13. Zorn B (1974) Textil-Praxis-International 29:1706, 1711, 1712
14. Anon (1975) Modern Plastics International 12:12
15. Zorn B (1984) Poröse PUR-Filme und -Beschichtungen. J Coated Fabr 13:166
16. Kellert HJ, Reich G (1978) Aufbau von Poromeriks als Schichtenverbund und deformations-mechanisches Verhalten der Aufbaukomponenten. Leder, Schuhe, Lederw 13(6):270
17. Anon (1982) Permair – Kombination von Leder und Synthetik Schuh-Technik Dec, pp 1076–1079
18. DE 2 350 205 [Mitsubishi Paper Mills Ltd.; M Amano et al.; 5.10.73/18.4.74; JA Prior. 11.10.72 (101 741–72)]
19. JA 51 62 154 (Toyota Jidosha KK; 12.12.91/29.6.93)=JA 91–351 597
20. JA 51 86 631 (Asahi Chem. Ind. Co. Ltd.; 14.1.92/27.7.93)=JA 92–23 438
21. JA 41 85 777 (Kuraray Co. Ltd.; 16.11.90/2.7.92)=JA 90–312 408
22. JA 41 94 086 (Achilles Corp.; 28.11.90/14.7.92)=JA 90–327 052
23. JA 41 94 085 (Achilles Corp.; 27.11.90/14.7.92)=JA 90–324 493
24. EP 507748 (R Besana; 5.4.91/7.10.92)=EP 91–227=US 5 413 846
25. DOS 3 143 064 (Hornschuch; K Falk, K Haeger; 30.10.81/19.5.83)
26. JA 78–46 881 (Kanebo; 26.7.68/16.12.78)=JA 68–53 202
27. JA 2 139 483 (Kanebo KK; 22.6.88/29.5.90)=JA 88–153 840
28. JA 77–41 201 (Lonseal Corp.; 20.3.77)
29. JA 540 055 703 (Kyowa Rubber Ind. KK; 8.10.77/4.5.79); JA 540 055 704 (Kyowa Rubber Ind. KK; 11.10.77/4.5.79)
30. JA 55 098 971 (Sehren KK; 23.1.79/28.7.80)=JA 79–6 189
31. JA 55 128 078 (Suzutora Seisen KK; 26.3.79/3.10.80)=JA 79–35 272
32. W.4 (1983) Die Herstellung von Polyurethan-Kunstledern nach dem Koagulationsverfahren. Coating 16:90, 128
33. Fukushima O, Saito Y (1982) Die chemische Struktur und die Koagulation von Polyure-thanen. Kobunshi Ronbunshu 39:535, 543
34. JA 60 162 877 (Kokoku Chem. Ind. KK; 30.1.84/24.8.85)=JA 84–14 883

35. JA 60 144 318 (Sanyo Chem. Ind. Ltd.; 7.1.84/30.7.85)=JA 84–1 148
36. US 2 973 333 (DuPont; M Katz et al.; 1.9.54/28.2.61)

8.1
Coagulation of Polyurethanes Without Additives

Also covering the subject of this section, see 18.4 [40–43], 19 [26], 22 [48], 27 [60]

1. OS 1 619 106 (Kurashiki Rayon Co.; Nakajo S et al.; 9.3.67/10.9.70, JA Prior. 12.3.66 + 22.3.66)= FR 1 514 549=US 3 481 765
 2. JA 07 598/71 (Mitsubishi Rayon Co.; 14.6.66/25.2 71)
 3. OS 1 923 611 (Kurashiki Rayon Co.; O Fukushima et al.; 8.5.69/5.2.70, JA Prior. 8.5.68+16.5.68+ 1.8.68+27.5.68+8.6.68)
 4. OS 1 940 772 (Kalle AG; H Mahl et al.; 11.8.69/25.2 71)
 5. JA 73–20 018 (Toray Ind.; 3.7.69/18.6.73)
 6. JA 44–026/72 (Toray Ind.; 31.5.69/7.11.72)
 7. GB 1 203 709 (Kalle AG; 19.12.68/3.9.70; D Prior. 21.12.67)=NE 6 818 423
 8. JA 73–4 623 (Toray Ind.; 21.8.69/9.2.73)
 9. OS 2 244 520 (Kabushiki Kaisha Kobunshi Oyo Kenkyusho, T Takahashi et al.; 11.9.72/7.6.73; JA Prior. 27.1 71)
10. OS 1904 278 (Kalle AG; D Beissel et al.; 29.1.69/17.12.70)=BE 744 958
11. JA 73–43599 (Daiichi Lace Co.; 20.3.70/19.12.73)
12. OS 1 953 626 (Kalle AG; K Andrä et al.; 24.10.69/6.5.71)
13. US 3 515 573 (BF Goodrich; AB Japs et al.; 28.3.69/2.6.70; cip 18.2.66)
14. JA 73–00 927 (Mitsubishi Rayon; 12.3.70/12.1.73)
15. JA 73–11 824 (Kuraray Co.; 9.4.69/16.4.73)
16. JA 73–18 942 (Kyowa Leather Cloth; 9.10.69/9.6.73)
17. JA 73–20284 (Kanebo Co.; 7.3.70/20.6.73)
18. GB 1 172 325 (Kurashiki Rayon; 5.2.68/26.11.69; JA Prior. 10.2.67)
19. DDR 101 713 (Forschungsinstitut für Leder- und Kunstledertechnologie; R Steinhardt et al.; 8.1.73/12.11.73)
20. JA 06 264 371 (94 264 371) (Teijin Cordley Ltd.; N Ookawa, Y Suzuki, S Yamauchi; 10.3.93/ 20.9.94); WO 9 420 665 (10.3.93/1.4.93)=JA 93–96 359+JA 93–74 995=EP 640715=JA 65 17 893
21. JA 07 207 052 (94–4071) (Mitsubishi Kagaku KK; S Uchida, T Oohori, K Mori; 19.1.94/8.8.95)
22. JA 07 216 752 (95,216,752) (Unitika Ltd.; T Furuta, K Kamemaru, M Shinomya; 31.1.94/15.8.95); JA 07 216 757 (94–28 868) (Unitika Ltd.; T Furuta, K Kamemaru, H Tsujimura; 31.1.94/15.8.95)
23. JA 50 71 078 (Daiichi Lace KK; 3.9.91/23.3.93)
24. JA 216 977 (Kokoku Chem. Ind. KK; 23.12.88/29.6.90)=JA 88–325 358
25. JA 21 75 736 (Nankai Gum KK; 4.4.88/9.7.90)=JA 88–82 764
26. JA 11 68 975 (Kanebo KK; 24.12.87/4.7.89)=JA 87–327 946
27. JA 85–239 573 (Kokoku Chem. Ind.; 10.5.84/28.11.85)=JA 84–93 795; JA 85–252 780 (Kokoku Chem. Ind.; 23.5.84/13.12.85)=JA 84–104 160
28. JA 51 067 702 (Fuji Spinning; 11.6.76)
29. JA 53 096 302 (Kyowa Rubber Ind. KK; 2.2.77/23.8.78)
30. JA 56 085 479 (Daiichi Lace KK; 7.12.79/11.7.81)=JA 79–159 692
31. JA 57 112 470 (Achilles KK; 29.12.80/13.7.82)=JA 80–187 642
32. JA 61 201 083 (Kanebo KK. 26.2.85/5.9.86)=JA 85–40 801
33. JA 61 225 376 (Kokoku Chem. Ind. KK; 29.3.85/7.10.86)=JA 85–66 077

8.2
Coagulation of Polyurethanes and Additional Small Sized Solids

Also of interest: 17.1[21,22,29], 18.2[28,67], 19[29,32–35], 23.1[2], 23.2[2]

1. DAS 1 068 660 (DuPont; EA Rodman; 30.9.55/12.11.59; US Prior. 1.10.54)=US 2 723 935= FR 1 139612
 2. OS 1 419 150 (DuPont; JL Hollowell; 30.6.60/12.12.68; US Prior. 24.8.59)=BE 591 648=GB 914 713

3. JA 73–23886 (Sanyo Chem. Ind.; 3.12.70/17.7.73)
4. FR 1 411 110 (Soc.Rhodiaceta; 27.5.64/9.8.65)
5. OS 1 469 562 (DuPont; RVC Einstman; 5.3.65/5.12.68; US Prior. 6.3.64)=CH 434 182
6. OS 1 419 149 (DuPont; JL Hollowell; 17.11.59/12.12.68)
7. OS 1 619 301 (Toyo Rubber Ind.; K Fukada et al.; 20.5.67/22.10.70; JA Prior. 20.5.66)= FR 1 522 692
8. OS 1 938 990 (Kalle AG; H Porrmann et al.; 31.7.64/18.2.71)
9. OS 1 704 777 (Kalle AG; D Kaempgen et al.; 30.11.67/27.5.71)
10. OS 1 619 287 (Svit. NP; Z Hrabal et al.; 21.5.66/24.9.70; CZ Prior. 14.7.65)=FR 1 486 730
11. OS 1 719 375 (Kurashiki Rayon; O Fukushima et al.; 7.2.68/2.9.71; JA Prior. 10.2.67)= GB 1 172 325
12. DAS 1 146 473 (DuPont; JL Hollowell; 3.7.59/4.4.63; US Prior. 3.7.58+3.2.59)=BE 596 948= FR 1 233 735+addition 76 988=GB 914 712 + addition to 919 500
13. JA 17 233/66 (Toyo Rubber Ind.; 11.11.63/30.9.66)
14. FR 1 414 241 (Toyo Rubber Ind.; 6.11.64/6.9.65; JA Prior. 27.12.63)
15. JA 21 877/69 (Toyo Rubber Ind.; 18.2.63/18.9.69)
16. FR 1 444 201 (Soc. Rhodiaceta; 20.5.65/23.5.66)
17. BE 687 006 (Soc. Rhodiaceta; 19.9.66/16.3.67; FR Prior. 17.9.65+28.12.65)=NE 6 612 755= FR 1 456 929+addition 90 012
18. JA 73–20281 (Daiichi Lace KK; 29.9.69/20.6.73)
19. JA 73–33361 (Toray Ind.; 21.11.66/13.10.73)
20. JA 07 597/71 (Mitsubishi Rayon; 2.6.66/25.2.71)
21. JA 47 20 244 (Toray Ind.; 19.2.71/28.9.72)
22. JA 04 667/68 (Toyo Rubber Kogyo; 15.3.65/20.2.68)
23. JA 49 080 205 (Tsunoda Kagaku; 7.12.72/2.8.74)
24. OS 1 469 559 (DuPont; WF Manwaring; 5.3.65/28.11.68; US Prior. 27.3.64)=BE 660734
25. OS 1 469 560 (DuPont; AW Batersnan; 5.3.65/2.1.69; US Prior. cip 27.3.64)=US 3 520 765= BE 660 733
26. OS 1 469 561 (DuPont; AV Patsis; 5.3.65/5.12.68; US Prior. 27.3.64)=BE 660 735
27. NE 6 713 073 (Goodrich; 26.9.67/4.4.68; US Prior. 3.10.66)=BE 704 350
28. DDR 60 281 (Ceskoslov. Zavody Gumarenski a Plastikavske; C Holcik; 13.4.67/26.2.68; CZ Prior. 13.4.66)
29. JA 948/69 (Kurashiki Rayon; 3.12.64/17.1.69)
30. FR 2 013 330 (Genset Corp.; WL Wang; 6.6.69/3.7.40; US Prior. 6.6.68)=BE 734 178=NE 6 908 650=OS 1 928 600
31. GB 1 277 438 (Tenneco Chemicals; 6.6.69/14.6.72; US Prior. 6.6.68)=US 3 634 184
32. OS 1 560 880 (Kurashiki Rayon; E Morita et al.; 17.10.66/1.10.70; JA Prior. 15.10.65)=GB 1 160 237
33. JA 73–29 513 (Toray Ind.; 27.11.69/11.9.73)
34. DDR 67 691 (Deutsches Lederinstitut; G Hebestreit et al.; 15.5.68/5.7.69)
35. JA 43 475 (Banto Chotai Rubber; 25.1.68/23.12.71)
36. OS 1 922 701 (Statni Vyzkumny Ustav Eozedelny; Z Hrabal. et al.; 3.5.69/13.11.69; CZ Prior. 6.5.68)
37. FR 1 495 017 (Goodrich; WT Murphy; 16.8.66/7.8.67; US Prior. 18.8.65)=NE 5 611 626=BE 685 532
38. BE 691 588 (DuPont; J Hochberg; 21.12.66/21.6.67; US Prior. 22.12.65)
39. BE 707 932 (Bayer AG, B Zorn et al.; 13.12.67/16.4.68; D Prior. 13.12.66)=NE 6 716 891
40. OS 1 960 992 (Genset Corp.; WL Wang; 4.12.69/25.6.70; US Prior. 4.12.68)
41. US 3 701 681 (WT Murphy, AB Japs; 16.2.71/31.10 72)
42. JA 43 046/72 (Toray Ind; 29.1.69/31.10.72)
43. FR 1 485 472 (Toyo Rubber Ind.; 9.7.66/16.6.67; JA Prior. 6.7.65)
44. FR 2 203 901 (Toray Ind.,; K Okazaki et al.; 24.10.72/17.5.74)=US 3 841 897=JA 49–066 803 (JA Prior. 26.10.72)
45. FR 1 531 923 (Toyo Rubber Ind.; K Fukada et al.; 19.7.67/23.7.68; JA Prior. 22.7.66)=OS 1 619 302=DDR 66 161
46. GB 1 179 321 (Toyo Rubber Ind.; 5.1.67/28.1.70; JA Prior. 10.1.66)
47. OS 1 619 303 (Toyo Rayon KK; T Shinohara et al.; 18.9.67/16.4.70; JA Prior. 19.9.66+4.3.67)=GB 1 165 228

48. OS 1 769 808 (Toyo Rubber Ind.; KI Fujita et al.; 17.7.68/2.9.71; JA Prior. 18.7.67)
49. JA 5 634/72 (Kuraray Co.; 3.2.65/17.2.72)
50. JA 02 592/71 (Toyo Spinning Co.; 9.4.65/22.1.71)
51. OS 1 769 277 (Bayer AG; B Zorn et al.; 30.4.68/18.11.71)
52. DAS 1 270 276 (Bayer AG; B Zorn et al.; 22.6.66/12.8.68)=BE 700 304=NE 6 708 653
53. DAS 1 178 586 (Farbenfabriken Bayer; D Dieterich et al.; 5.12.62/24.9.64)
54. DAS 1 179 363 (Farbenfabriken Bayer; D Dieterich et al.; 28.2.63/8.10.64)
55. DAS 1 184 946 (Farbenfabriken Bayer; D Dieterich et al.; 26.10.62/7.1.65)
56. OS 2 345 256 (Bayer AG; A Reischl; 7.9.73/20.3.75)
57. DAS 2 311 418 (Geoscience Instr. Corp.; E Jensen; 8.3.73/12.12.74; US Prior. 23.6.72)
58. DE 2 558 350 [Kuraray; T Yamasaki et al.; 23.12.75/8.7.76; JA Prior. 25.12.74 (4 082-75)]=GB
    1 496 369=US 4 053 546
59. EP 310 037 (Kuraray Co Ltd.; T Nishimura; 28.9.88/5.4.89, JA Prior. 28.9.87)=JA 87-244 817
    and JA 87-244 818
60. DE 1 619 641 (Toyo Rubber Ind. Co.; I Minobe, T Suzuki; 5.8.66/24.9.70)
61. JA 07 54 277 (95 54 277) (Unitika Ltd.; T Furuta, K Kamemaru, H Tsujimura; 10.8.93/25.2.95)
62. JA 7 11 580 (95 11 580) (Soko Seiren KK; O Takamura, H Hamai. O Hiroshi, N Ogawa,Y Kama-
    hata; 24.6.93/13.6.95)
63. JA 06 346 379 (94 346 379) (Achilles Corp.; K Oosawa; 2.6.93/20.12.94)
64. JA 07 97 461 (95 97 461) (Asahi Chem. Ind.; M Fukui. K Maki, S Okajima; 12.5.93/11.4.95)
65. JA 07 166 479 (95,166,479) (Unitika Ltd.; 15.12.93/27.6.95)
66. JA 07150 478 (Kuraray Co.; M Nakano, T Yamazaki, T Akazawa; 1.12.93/13.6.95)
66a. JA 07 145 569 (Kuraray Co.; S Nakanashi,Y Yoshida; 24.11.93/6.6.95)
67. JA 06 330 473 (94, 330, 473) (Toray Ind.; T Oonu, M Kunieda, H Nakamura; 21.5.93/29.1.94)
68. Dubjaga EG, Simonovskij FI (1993) Modifizierung der Struktur und der Eigenschaften von
    mikroporösen Folien auf Basis von Polyetherurethanen. Plast. Massy 5:13
69. JA 060 01 875 (Dai Nippon Ink & Chem KK; 17.6.92/11.1.94)
70. RU 2 005 748 (Med. Polymer Res. Inst.; LP Gaidarova, LK Tsivinskaya; 9.6.92/15.11.94)
71. JA 50 71 080 (Dainichiseika Color & Chem. Mfg.; 6.9.91/23.3.93)
72. JA 79-147 901 (Teijin; 8.5.78/19.11.79)=JA 78-53 786
73. JA 08 253 609 (Asahi Chemical Ind.; S Okajima, A Takeuchi; 14.3.95/1.10.96)=JA 96-253 609
74. JA 09 316 784 (Unitika Ltd.; K Kamemaru, S Soejima; 29.5.96/9.12.97)=JA 96-134 676
75. SU 1509382 (Polimersintez Combi.; 22.1.87/23.9.89)=SU 87-184924
76. JA 82 91 473 (Asahi Chemical Ind.; H Yamazaki, N Kuramoto; 20.4.95/5.11.96)=JA 95-95 453
77. JA 61 160 487 (Kokoku Chem. Ind. KK; 28.12.84/21.7.86)=JA 84-279 944
78. DOS 2 456 071 (Interbrinderea de Piele Sintetitca Bucure; M Bordei; 12.8.76)
79. DOS 2 756 671 (Hoechst AG; W Busch, A Holst, W Fischer; 19.12.77/29.6.78)
80. JA 54 070 403 (Ichikawa Keori KK; 11.11.77/6.6.79)
81. JA 54 080 363 (Kuraray KK; 9.12.77/27.6.79)
82. JA 54 126 268 (Kuraray KK; 24.3.78/1.10.79)
83. JA 54 107 502 (Kuraray KK; 7.2.78/23.9.79)=JA 78-13 067
84. DOS 2 951 307 (Akzo GmbH; K Gerlach et al.; 20.12.79/2.7.81)
85. JA 56 000 382 (Kunimine Kogyo KK; 13.6.79/6.1.81)=JA 79-74 269
86. JA 56 159 370 (Teijin KK; 14.5.80/8.12.81)=JA 80-62 692
87. DOS 3 143 064 (K Hornschuch AG; K-H Falk, K Häger; 3.10.81/19.5.83)
88. JA 01 033 284 (T Yamada; 30.7.87/3.2.89)=JA 87-191 189
89. JA 57 149 567 (Toyo Cloth KK; 7.3.81/16.9.82)=JA 81-32 838
90. CAN 1 034 323 (Porvair Ltd.; B Hays, D Price, G Kleinerman, JA Macpherson; 20.3.74/
    25.10.78)=DOS 2 512 058=GB 74-12 378; DOS 2 413 461 (Porvair Ltd.; D Price et al.; 20.3.74/
    23.10.75)=BE 812 592
91. GB 1 414 961 (Brit. Millerain; CA Redfarn et al.; 13.7.71/19.11.75)=GB 71-32 882
92. JA 74 008 841 (Kokoku Chem. Ind. Co.; 27.1.70/28.2.74)=JA 70-6 835
93. JA 73 000 177 (Toray Ind. Inc.; 13.3.70/6.1.73)=JA 70-20 833
94. JA 9 111 670 (Unitika Ltd.; K Kenichi, S Mamoru; 13.10.95/28.4.97)=JA 95-265 164

8.3
Coagulation of Polyurethanes – Addition of Other Polymers

See also 18.1 [28, 67]
 1. DAS 1 110 607 (DuPont; EK Holden; 28.1.58/13.7.61; US Prior. 28.1.57)=US 3 000 757
 2. DAS 1 444 167 (DuPont; DG Hulfslander et al.; 13.8.63/17.10.68; US Prior. 13.8.62)=BE 636 018
 3. OS 1 444 163 (DuPont; EK Holden; 21.2.62/10.10.68; US Prior. 21.2.61)
 4. OS 1 469 550 (Kurashiki Rayon; O Fukushima et al.; 16.11.64/19.12.68; JA Prior. 15.11.63 +
    11.12.63 + 29.1.64)=FR 1 420 623=BE 655 812=NE 6 413 333
 5. BE 660 730 (DuPont; RV Einstman; 5.3.65/6.9.65; US Prior. 6.3.64)=NE 6 502 848
 6. BE 691 588 (DuPont; J Hochberg 21.12.66/21.6.67; US Prior. 22.12.65)
 7. OS 2 056 052 (General Tire; EC Brown; 13.11.70/27.5.71; US Prior. 14.11.69)
 8. BE 707 932 (Bayer AG; B Zorn et al.; 13.12.67/16.4.68; D Prior. 13.12.66)=NE 6 716 891
 9. JA 75–004 036 (Toray Ind.; 22.5.69/13.2.75)
10. JA 18 427/72 (Toray Ind.; 12.12.68/27.5.72)
11. OS 1 779 417 (Toyo Rubber Ind.; K Fukada et al.; 9.8.68/7.10.71; JA Prior. 9.8.67)
12. JA 13 638/71 (Kuraray; 26.1.67/12.4.71)
13. JA 30 596/71 (Toyo Rubber Ind.; 10.8.67/6.9.71)
14. JA 17 463/72 (Toyo Cloth Co.; 2.7.68/22.5.72)=US 3 676 206
15. JA 73–41 265 (Toray Ind.; 12.6.69/5.12.73)
16. JA 02 761/73 (Toray Ind.; 19.3.69/26.1.73)
17. OS 2 101 160 (Statni Vyzkumny Ustav Kozedny; E Mück et al.; 12.1.71/ 23.9.71; CZ Prior.5.3.70)
18. OS 2 030 703 (DuPont; WA Hare et al.; 22.6.70/14.1.71; US Prior. 23.6.69)
19. OS 1 936 073 (Glanzstoff AG; G Seibert et al.; 16.7.69/25.2.71)
20. DDR 60 281 (Ceskoslov. Zavody Gumarenske & Plastikarske; G Holcik; 3.4.67/20.2.68; CZ
    Prior. 13.4.67)
21. JA 18 236–38/65 (Toyo Cloth Co.; 17.7.62 +23.10.62+18.5.63/17.8.65)
22. GB 1 091 935 (Toyo Gomu KK; 22.3.65/22.11.67; JA Prior. 21.3.64)=OS 1 469 576
23. OS 1 619 305 (Toyo Rubber Ind.; 11.11.67/10.9.70; JA Prior. 12.11.66)=GB 1 179 756=FR 1 557
    046=DDR 66 163
24. JA 01 036/72 (Toyo Rubber Chem Ind.; 5.12.67/12.1.72)
25. OS 1 694 252 (Bayer AG; H Träubel et al.; 8.3.68/12.8.71)
26. JA 949/69 (Kurashiki Rayon; 4.12.64/17.1.69)
27. JA 57 038 833 (Dainippon Ink Chem. KK; 19.8.80/3.3.82)
28. JA 54 127 499 (Toyobo KK; 27.3.78/3.10.79 (35 783))
29. DOS 1 669 829 (Ceskoslov. Zavody Gumarenske & Plastikarske; F Hadobas, C Holcik, Z Janek;
    20.2.67/23.10.69; CZ Prior. 17.3.66)
30. US 3 551 830 (DuPont; GR Hodge, AV Patsis; 22.10.68/15.2.72)=GB 1 242 640
31. JA 58 098 480 (Komatsu Seiren KK; 3.12.81/11.5.83)=JA 81–193 739
32. JA 59 059 735 (Mitsubishi Rayon KK; 30.9.82/5.4.84)=JA 82–171 184
33. DOS 3 200 942 (H von Blücher et al.; 14.1.82/21.7.83)
34. US 4 707 400 (Burlington Ind. Inc.; DR Towery; 3.9.86/17.11.87)=US 86–903 130; WO 88–01 570
    (Burlington Ind. Inc.; DR Towery, BR. Hill et al.; 3.9.86/17.11.87; US Prior. 3.9.86=US 86–903
    130 and 9.9.86=US 86–905 135)
35. JA 63 075 189 (Sehren KK; 12.9.86/5.4.88)=JA 86–214 160
36. JA 49 068 074 (Kyowa Leather Cloth Corp.; 7.11.72/2.7.74)=JA 72–110 807
37. FR 2 211 560 (Toray Ind. Inc.; 21.12.73/19.7.74)=JA 72–128114 (JA Prior. 22.12.72)+JA 73–20 708
    (Prior. 22.2.73)
38. Enomoto M, Muraoka Y, Suehiro K (1997) Effect of coating polymer composition and co-
    agulation structure in waterproofed moisture-permeable fabrics prepared by wet coagula-
    tion process on wearing comfort and end-use properties. J Jap Res Assoc for Text End
    Uses 38:41

8.4
Coagulation of Polyurethanes – Addition of Nonsolvents

1. JA 17 430/68 (Kyowa Leather Co.; 7.9.63/23.7.68)
2. DAS 1 110 607 (DuPont; EK Holden; 28.1.58/13.7.61; US Prior. 28.1.57)=US 3 000 757
3. US 3 516 883 (DuPont; LR Harper; 1.2.67/23.6.70)
4. OS 1 803 021 (Toyo Rayon; K Mitusakawa; 14.10.64/22.5.69; JA Prior. 13.10,67)
5. OS 2 061 151 (General Tire; JF Barnes et al.; 11.12.70/16.6.71; US Prior. 11.12.69)
6. OS 2 025 616 (Bayer AG; H Conrad et al.; 26.5.70/9.12.71)
7. DDR 89 719 (Deutsches Lederinstitut; E Döring et al.; 11.12.70/5.5.72)
8. JA 34 922/72 (Toray Ind.; 17.4.68/2.9.72)
9. FR 1 412 310 (Goodrich; WT Murphy; 23.10.64/16.8.65; US Prior.31.10.63)
10. OS 1 469 534 (C Freudenberg; T Schachowskoy et al.; 9.3.65/23.1.69)
11. OS 1 469 535 (C Freudenberg; J Knoke; 8.4.65/23.1.69)
12. DAS 1 444 165 (DuPont; EK Holden; 13.12.62/17.10.68; US Prior. 5.1.62)=GB 981 642=FR 1 355 577=BE 626 816
13. JA 73-27 441 (Toyo Spinning Co.; 22.10.64/22.8.73)
14. DAS 1 444 164 (DuPont; EL Yuan; 30.10.62/30.4.63; US Prior. 31.10.61)=BE 624 250
15. OS 1 469 563 (DuPont; RV Einstman; 27.7.65/6.11.69; US Prior. 31.8.64)=US 3 492 154=BE 667 624=NE 6 511 220
16. US 3 536 639 (C Freudenberg; T Schachowskoy; 21.4.66/27.10.70; DB Prior. 24.4.65)=OS 1 263 292
17. US 3 532 529 (Suehiro Seni Kogyo KK; H Endo et al.; 29.11.67/6.10.70, JA Prior. 27.12.66)= GB 1 178 240
18. JA 32 597 (Mitsubishi Rayon; 2.11.67/22.9.71)
19. OS 1 619 641 (Toyo Rubber Ind.; I Minobe; 5.8.66/24.9.70)
20. JA 31 003/72 (Toyo Rubber Ind.; 26.1.68/11.8.72)
21. NE 6 708 454 (Glanzstoff AG; 16.6.67/19.12.67; D Prior. 18.6.66)
22. FR 1 517 287 (Ceskoslov. Zavody Gumarenke & Plastikarske; K Koutny et al.; 30.3.67/15.3.68; CZ Prior. 25.5.66)
23. FR 1 495 017 (Goodrich; WT Murphy; 16.8.66/7.8.67; US Prior. 18.8.65)=BE 685 532=NE 6 611 626
24. JA 73-4622 (Kyowa Leather Cloth; 7.7.69/9.2.73)
25. OS 1 815 043 (Glanzstoff AG; H Schulze et al.; 17.12.68/25.6.70)=BE 739 674
26. JA 1 034/72 (Mitsubishi Rayon; 22.12.66/17.1.71)
27. JA 49/062 603 (Toyo Cloth Co.; 23.10.72/18.6.74,
28. US 3 748 287 (DuPont; KH Lee; 1.9.71/24.7.73)
29. JA 02 594/71 (Nippon Cloth Ind.; 7.5.66/22.1.71)
30. GB 1 162 851 (Ichikawa Woolen Textile Co.; T Hamano et al.; 3.1.67/27.8.69)
31. DDR 99 811 (Deutsches Lederinstitut; HJ Kellert et al.; 21.3.72/20.8.73)
32. US 3 536 553 (DuPont; FJ Farrell et al.; 19.12.66/27.10.70)
33. FR 90 012/1 456 929 (Soc. Rhodiaceta; 28.12.65/29.9.67)
34. JA 73-5 893 (Kyowa Leather Cloth; 15.12.70/21.2.73)
35. JA 73-24 721 (Kao Soap Co.; 12.6.70/24.7.73)
36. JA 73-05 893 (Kyowa Leather Cloth; 15.12.70/ 21.2.73)
37. J A 74-045 748 (Teijin; 22.7.70/5.12.74)
38. OS 1 901 950 (Kurashiki Rayon; O Fukushima et al.; 13.1.69/19.2.70; JA Prior. 17.1.68)
39. JA 01 168 975 (Kanebo KK; 24.12.87/4.7.89)=JA 87-327 946
40. JA 63 182 475 (Unitika KK; 21.1.87/27.7.88)=JA 87-13 144

8.5
Coagulation of Polyurethanes – Addition of Emulsifying or Polar Auxiliaries

1. FR 1 455 522 (Kurashiki Rayon; 13.11.65/5.9.66; JA Prior. 14.11.64)
2. FR 1 466 907 (Kurashiki Rayon; 2.2.66/12.12.66; JA Prior. 3.2.65+2.8.65)
3. JA 73-4939 (Kurashiki Rayon; 25.11.65/13.2.73+JA 73-4941)
4. JA 39 635/70 (Kurashiki Rayon; 25.11.65/12.12.70)

5. JA 03 052/64 (Kuraray; 25.11.65/15.2.73)
6. JA 02 593/71 (Kuraray; 4.3.66/22.1.71)
7. FR 1 524 724 (Kurashiki Rayon; 26.5.67/10.5.68; JA Prior. 27.5.66+9.6.66)
8. JA 490 224/73 (Kuraray; 22.6.72/27.2.74)
9. OS 1 903 402 (Kurashiki Rayon; 17.1.69/2.10.69; JA Prior. 17.1.68)
10. OS 2 343 295 (Kuraray; O Fukushima; 28.8.75/14.3.74; JA Prior. 30.8.72)
11. OS 1 809 574 (Kanegafuchi Boseki KK; T Shikada; 18.11.68/18.12.69; JA Prior. 18.11.67)
12. OS 2 414 251 (Kuraray; H Shimamura et al.; 26.3.74/17.10.74; JA Prior. 31.3.73)
13. OS 1 694 127 (Farbenfabriken Bayer; H Träubel et al.; 10.1.67/2.3.72)=BE 709 183
14. DAS 1 240 653 (C Freudenberg; T Schachowskoy; 24.4.65/18.5.67)
15. OS 1 544 912 (C Freudenberg; T Schachowskoy; 2.9.65/31.7.69)
16. DAS 1 263 292 (C Freudenberg; T Schachowskoy; 24.4.65/14.3.68)
17. JA 17 462/72 (Toray Ind.; 3.6.68/22.5.72)
18. JA 72–46 890 (Toray Ind.; 6.8.69/27.11.72)
19. JA 39 331/72 (Toray Ind.; 9.10.68/4.10.72)
20. FR 1 495 017 (Goodrich; 16.8.66/7.8.67)=NE 6 611 626
21. US 3 642 966 (Goodrich; RT Morrissey et al.; 24.11.69/15.2.72)
22. JA 02 597/71 (Daiichi Lace KK; 31.1.67/22.1.71)
23. JA 73–19 381 (Dai Nippon Ink and Chem. Co; 14.8.70/13.6.73)
24. JA 73–19 923 (Kokoku Chem. Ind. Co.; 3.6.69/18.6.73)
25. JA 57 112 371 (Achilles KK; 6.12.80/13.7.82)
26. DAS 1 694 291 (BF Goodrich; WT Murphy; 18.8.66/26.8.71; US Prior. 18.5.65)
27. JA 7 039 638 (Kanegafuchi Spinning Co.; 8.7.66/12.12.70)
28. DE 2 948 892 (Toray Ind. Inc.; 5.12.79/26.6.80; JA Prior. 6.12.78); DE 2 948 892 (Toray Ind. Inc.; 6.12.78/26.6.80; JA Prior. 6.12.80)
29. US 5 387 437 (China Textile Res. Inst.; S Yao; 26.11.93/7.2.95)
30. RU 202 990 (Film Materials Synt. Leather Res. Inst.; LV Kuzina, LA Skripo, IA Otikova; 4.6.91/15.11.94)
31. JA 08 060 556 (Achilles Corp. S Arai, S Oikawa; 24.8.94/5.3.96)
32. JA 78–109 901 (Honey Chemical Industry; 4.2.77/26.9.78)=JA 77–10 719
33. JA 82–112 471 (Achilles; 6.12.80/13.7.82)=JA 79–74 269
34. DDR 134 968 (Forschungsinstitut für Leder und Kunstleder; G Reich, W Stephan, S Fourier, E Steinert; 4.4.79)
35. DOS 2 908 314 (Forschungsinstitut für Leder- und Kunstledertechnologie; G Reich et al.; 3.3.79/27.9.79; DDR Prior. 17.3.78)=DDR 204 255 (WP)
36. JA 58 087 373 (Kuraray KK; 16.11.81/25.5.83)=JA 81–184 229
37. JA 59137577 (Mitsubishi Rayon KK; 17.1.83/7.8.84)=JA 83–6425
38. JA 59228086 (Kokoku Chem. Ind. KK; 3.6.83/21.12.84)=JA 83–99965
39. JA 01239177 (Kuraray KK; 16.3.88/25.9.89)=JA 88–64172
40. JA 01 033 283 (Kuraray KK; 23.7.87/3.2.89)=JA 87–184 957
40a. JA 01 029 439 (Kuraray KK; 23.7.87/31.1.89)=JA 87–184 956
41. JA 61 186 570 (Kuraray KK; 11.2.85/20.8.86)=JA 85–24 199
42. JA 60 141 734 (Kuraray KK; 28.12.83/26.7.85)=JA 83–251 969
43. JA 61 201 084 (Kanebo KK; 25.2.85/5.9.86)=JA 85–37 178
44. JA 61 034 289 (Kuraray KK; 19.7.84/18.2.86)=JA 84–150 635
45. JA 61 266 679 (Kanebo KK; 15.5.85/26.11.86)=JA 85–104 737
46. JA 49 059 167 (Kuraray Co. Ltd.; 7.10.72/8.6.74)=JA 72–100 821
47. JA 51 030 852 (Toray Ind. KK; 11.9.74/16.3.76)=JA 74–103 803
48. Chu C, Mao Z, Yan H (1996) Influence of the temperature of condensation bath on microstructure and properties of the PU asymmetric membrane. J Coated Fabr 26:137; Chu C, Mao Z, Yan H (1995) An investigation on the formation mechanism of the porous structure of the PU films. J Coated Fabr 24:298

8.6
Coagulation of Polyurethanes – Addition of Leachable Solids

See also 7 [68, 69], 16.2 [28]

1. JA 18 229/70 (Toyo Rubber Ind.; 13.5.64/23.6.70)
2. JA 33 795/70 (Hitachi Kasei KK; 27.10.65/30.10.70)
3. JA 33 796/70 (Kanegafuchi Spinning; 13.11.65/30.10.70)
4. BE 695 136 (Porous Plastics; 7.3.66/7.9.67; GB Prior.7.3.66 + 11.1.67)
5. BE 698 306 (Porous Plastics; 10.5.67/10.11.67; GB Prior. 7.3.66)=NE 6 703 587
6. OS 1 753 668 (Porvair Ltd.; CR Cunningham et al.; 10.1.68/23.12.71; GB Prior. 11.1.67)=BE 709 258=NE 6 800 390
7. GB 1 220 218 (Porvair Ltd.; CR. Cunningham et al.; 11.1.67/20.1.71)
8. FR 1 565 893 (Kanegafuchi Boseki KK; 16.5.68/2.5.69; JA Prior. 17.5.67)=NE 6 806 928
9. OS 1 694 181 (Farbenfabriken Bayer; H Träubel et al.; 9.8.67/26.8.71)=BE 719 078
10. JA 24 800/68 (Toyo Rubber Ind.; 18.12.63/25.10.68)
11. OS 1 817 211 (Nippon Cloth Ind.; T Maeda et al.; 27.12.68/12.11.70)
12. OS 2 123 198 (Akzo GmbH; G Seibert et al.; 15.11.71/23.11.72)
13. OS 2 060 616 (Porvair Ltd.; F Ch Loew et al.; 9.12.70/16.6.71; GB Prior. 9.12.69 + 28.4.70)
14. OS 2 117 350 (Kanegafuchi Boseki KK; T Shikada; 8.4.71/28.10.71; JA Prior. 13.4.70)
15. OS 2 161 445 (Porvair Ltd.; EA Warwicker et al.; 10.12.71/29.6.72; GB Prior. 11.12.70)
16. OS 2 226 879 (Porvair Ltd.; EA Warwicker et al.; 2.6.72/28.12.72; GB Prior. 2.6.71)
17. OS 2 435 880 (Porvair Ltd.; AW Pearman et al.; 25.7.74/6.2.75; GB Prior. 25.7.73)
18. JA 73–28 044 (Daiichi Chem.Ind.; 29.8.70/29.8.73)
19. OS 2 010 332 (Chem Fabrik Stockhausen & Cie.; R Peppmöller; 5.3.70/16.9.71)
20. OS 1 922 303 (Porvair Ltd.; GR Hull; 2.5.69/13.11.69; GB Prior. 1.5.68)
21. JA 18 791/70 (Kanegafuchi Spinning; 1.12.66/27.6.70)
22. JA 73–22 547 (Kanebo Co.; 6.7.70/14.7.73)
23. OS 2 035 975 (Kanegafuchi Boseki KK; S Oohara et al.; 20.7.70/4.2.71; JA Prior. 19.7.69)
24. FR 1 565 943 (Algemene Kunstzijde Unie N.V.; 17.5.68/2.5.69; NE Prior. 18.5.67)
25. JA 43 045/72 (Kuraray Co.; 28.12.68/31.10.72)
26. JA 44 025/72 (Kyowa Leather Cloth Co.; 23.5.68/7.11.72)
27. JA 73–11 928 (Kuraray Co.; 13.12.68/17.4.73)
28. JA 73–29 513 (Toray Ind.; 27.11.69/11.9.73)
29. JA 73–5 889 (Japan Vilene; 16.9.70/21.2.73)
30. JA 39 333/73 (Kakuda Kagaku KK; 13.11.68/4.10.72)
31. JA 73–28 043 (Toray Ind.; 14.8.70/29.8.73)
32. JA 73–36 832 (Toray Ind.; 29.3.68/16.9.72)
33. JA 47 648/64 (Teijin Co.; 21.8.64/16.7.68)
34. OS 1 785 080 (Toray Ind.; T Shinohara et al.; 8.8.68/27.5.71; JA Prior. 10.8.67+17.4.68)=US 3 565 670
35. JA 21 275/72 (Toray Ind.; 10.7.68/15.6.72)
36. JA 44 601/72 (Toray Ind.; 31.5.69/10.11.72)
37. JA 49/054 501 (Nippon Cloth Ind.; 30.9.72/27.5.74)
38. JA 73–27 443 (Kuraray; 29.7.65/22.8.73)
39. JA 37 199/71 (Toray Ind.; 20.1.68/1.11.71)
40. OS 2 016 522 (Kurashiki Rayon; O Fukushima et al.; 2.4.70/15.10.70; JA Prior. 2.4.69 + 24.4.69 + 26.4.69)
41. JA 44 606/72 (Toray Ind.; 14.7.69/10.11.72)
42. JA 39 324/72 (Kuraray; 17.7.68/4.10.72)
43. JA 00 924/73 (Toray Ind.; 5.2.70/12.1.73)
44. JA 00 925/73 (Toray Ind.; 21.2.70/12.1.73)
45. JA 040 921/74 (Kuraray; 15.10.70/6.11.74)
46. US 4 632 860 (JP Antonio et al.; US Prior. 2.3.84/30.12.86)
47. JA 06 240 583 (94,240,583) (Teijin Cordley Ltd.; H Kimura, Y Suzuki; 16.2.93/30.4.94); JA 06 248 578 (94,248,578) (Teijin Cordley Ltd; H Kimura, T Sakamoto; 13.2.93/6.9.94)

48. US 4 803 116 [Toray Ind.Inc.; J Amano, M Shimada, K Takano, S Tohyama; 30.10.86/7.2.89; JA Prior. 11.10.85 (60–242960), 4.5.86 (61–45328) and 19.5.86 (61–112773)]=EP 225 060
49. US 4 429 000 (Toray Ind. Inc.; Y Naka, K Kawakami; 24.7.81/31.1.84; cip 11.12.79)
50. JA 78–46 881 (Kanebo; 26.7.68/16.12.78)=JA 68–53 202
51. DOS 2 546 414 (3M; RC Edberg; US Prior. 15.10.74; 14.10.75/29.4.76)=US 74–514808
52. JA 6 476/66 (Toyo Cloth Co.; 5.11.63/9.4.66)
53. JA 54 026 305 (Asahi Chemical Ind. KK; 28.7.77/27.2.79); JA 54 026 304 (Asahi Chemical Ind. KK; 27.7.77/27.2.79)
54. JA 54 002 302 (Dainichi Seika Kogyo KK; 3.6.77/9.1.79)
55. JA 54 015 096 (Dynic Corp.; 7.7.77/3.2.79)
56. JA 54 020 107 (Kanebo KK; 12.7.77/15.2.79)
57. JA 54 084 002 (Mitsubishi Rayon KK; 14.12.77/4.7.79)
58. JA 01 092 482 (Kanebo KK; 30.9.87/11.4.89)=JA 87–246 750
59. US 4 560 611 (Toray Ind. Inc.; Y Naka, K Kawakami; 28.10.83/24.12.85; div. 24.7.81; cip 11.12.79)
60. JA 83–18 484 (Kuraray; 21.7.81/3.2.83)=JA 81–114798
61. GB 2 114 585 (Porvair; D Price, EA Warwicker; 1.12.81/24.8.83)=GB 81–036141
62. JA 81–382 (Kunimine Kogyo; 13.6.79/5.1.81)=JA 79–74 269

8.7
Coagulation of Polyurethanes – Addition of Blowing Agents

1. JA 74–024 645 (Kohkoku Chem. Ind.; 30.12.70/25.6.74)
2. JA 74–024 646 (Kohkoku Chem. Ind.; 5.12.70/25.6.74)
3. JA 72–8 943 (Kuraray; 17.2.67/15.3.72)

8.8
Coagulation of Polyurethanes by Reacting Prepolymers of Polyurethanes

1. OS 1 569 081 (Kalle AG; H Porrmann et al.; 4.4.64/14.8.69)=SZ 446 264; see also FR 1 439 127+ GB 1 104 174
2. JA 8151/67 (Toyo Cloth Co.; 13.11.63/4.4.67)
3. JA 4721 461 (Honey Chem.Co.; 24.2.71/4.10.72)
4. FR 1 486 730 (Svit, NP; Z Hrabal et al.; 13.7.66/30.6.67)
5. JA 9154/71 (Mitsubishi Rayon; 8.3.66/8.3.71)
6. FR 1 373 717 (Soc Rhodiaceta; 23.7.63/24.8.64)
7. SU 231516 (Fil Mat. Leather; 5.8.66/7.11.80)=SU 66–95 821
8. JA 08 003 876 (Seiko Kasei KK; M Enomoto, K Ogasawara; 9.1.96)

9.1
Spraying of Polymer Solutions

See also 11.1 [59]
1. US 2 999 788 (DuPont; PW Morgan; 22.1.59/12.9.61)
2. US 3 364 063 (Kendall Co.; D Satas; 20.7.64/16.1.68)
3. DAS 1 225 380 (Bayer AG; A Reischl, B Zorn; 4.6.64/22.9.66)=BE 664 870=NE 6 507 007
4. DDR 34 258 (VEB Kunstblume Sebnitz; E Pilz, E Augst; 27.12.63/28.12.64)
5. FR 1 501 187 (Bayer AG; D Prior. 5.11.65; 4.11.66/10.11.67)=NE 6 615 459=DAS 1 267 841
6. FR 1 530 717 (Uniroyal Inc.; RN Steel, PV Butsatz, RT Nojiri; UR Prior. 31.5.66; 31.5.67/28.6.68)=NE 6 707 452
7. US 3 496 056 (Uniroyal Inc.; RN Steel et al.; 6.2.69/17.2.70 div. from Appl. of Prior. 31.5.66, see also FR 1 530 717
8. US 2 950 752, (Am. Viscose Corp. 24.12.53); US 2 988 469, (Am. Viscose Corp. 22.12.59)
9. GB 1 196 090 (Uniroyal Inc.; US Prior. 18.5.67; 19.4.68/24.6.70)=FR 1 580 935=NE 6 806 362 =US 3 537 947

10. DBP 830 585 (Alkor; K Lissmann, L Wolf; 9.12.44/16.10.52)
11. FR 1 298 959 (Kendall Co.; D Satas; US Prior. 23.5.60; 23.5.61/12.6.62)
12. JA 27 273/67 (Toyo Rubber Ind. Co. Ltd.; 13.7.64/23.12.67)
13. GB 1 081 406 (C Freudenberg; H Fabricius et al.; D Prior. 25.10.63; 28.10.64/31.8.67)
14. JA 13 156/66 (Teijin Co.Ltd.; 27.11.63/25.7.66)
15. Satas D (1965) Porous sprayed sheets and coatings. Ind Eng Chemistry 57:38
16. JA 2 233/68 (Teijin Co. Ltd.; 24.2.65/26.1.68)
17. JA 74–004 069 (Yuasa Battery Co. Ltd.; 30.6.66/30.1.74)
18. SZ 417 085 (C Freudenberg; E Demme et al.; 30.10.63/31.1.67)
19. DOS 4 425 793 (M Keppeler; 21.7.94/1.2.96); DOS 4 425 792 (M Keppeler; 21.7.94/1.2.96)
20. DOS 2 621 141 (C Freudenberg; K Schmidt, H Hoffmann; 6.4.78)
21. JA 53 075 304 (Honey Chem. Ind.; 6.2.74/4.7.78)=JA 74–156 444+JA 74–14 246

9.2
Selective Evaporation Process

See also 8.4 [40], 29.1 [13]; selective evaporation with poly-γ-methyl glutamate together with polyurethane, see 18.3 [16]

1. BE 652 899 (Feldmühle; OE Prior. 13.9.63; 10.9.64/4.1.65)=OE 246 093
2. One drop of a solvent is dropped on a paper at 20°C and 66% air humidity. The paper becomes transparent. The time required to evaporate the drop is measured. (The paper becomes intransparent again.) The same is done with diethyl ether as a comparison. Then the time required to evaporate the solvent to be tested is divided by the time the diethyl ether needs to evaporate. The evaporation number is the quotient of this division. The evaporation number is a figure without a dimension (as e.g. second etc.) with diethyl ether as evaporation number as 1. It is measured according to DIN 53 170
3. DBP 1 694 059 (Bayer AG; A Reischl, H Träubel, B Zorn; 3.1.66/3.6.71)=BE 692 116=FR 1 510 261
4. A survey of solvent parameters of polymers and solvents can be found in: Bandrup J, Immergut EE (1966) Polymer handbook. vol I. New York, p 341 etc.; solvents and nonsolvents for different polymers are named in Bandrup J, Immergut EH (1966) Polymer handbook, 3rd ed. pp VII/391
5. OS 1 694 179 (Bayer AG; A Reischl et al.; 9.8.67/26.8.71)=FR 1 578 378=NE 6 811 167=BE 719 174
6. IUP 15=DIN 53 333, see Das Leder (1961) 86
7. OS 1 694 205 (Bayer AG; H Träubel, B Zorn; 12.10 67/4.11.71)=BE 706 920=NE 6 715 596, addition to OS 1 694 085
8. Dyck M, Hoyes F (1964) Löslichkeits- und Wasserstoffbrückenparameter. Farben und Lack 70:522
9. JA 6 149/68 (Yuasa Battery Co.; 22.7.63/6.3.68)
10. JA 24 548/71 (Yuasa Battery Co.; 29.3.66/14.7.71)
11. BE 600 256 (DuPont; DT Bottorf, JL Hecht, VE James; US Prior. 7.2.60; 15.2.61/16.8.61)
12. GB 1 051 834 (Polymer Corp; GB Prior. 22.8.62; 16.8.63/21.12.66)
13. RA 226 835 (NA Abaturova, IN Vlodavets; 6.12.66/March 69)
14. NE 6 715 596 (Bayer AG; D Prior. 22.11.66; 16.11.67/24.5.68)
15. OS 1 619 213 (Ceskoslov. Zavody Gumarenski & Plastikavske; K Hlustik; CZ Prior. 18.7.66; 27.6.67/10.9.70)=DDR 63 360=FR 1 531 873
16. JA 24 547/71 (Yuasa Battery Co. Ltd.; 25.3.66/14.7.71)
17. US 3 546 001 (Immont Corp.; Ch Giannone et al.; 11.2.69/8.12.70; cip. 16.5.66)
18. US 3 403 046 (Interchemical Corp.; FH Schwacke, Ch Giannone; 8.10.65/24.9.68)
19. OS 2 136 558 (Goodrich; WTh Murphy; 22.7.71/3.2.72; US Prior. 24.7.70)
20. OS 2 523 740 (Stamicarbon B. V.; AJ Pennings et al.; NE Prior. 30.5.74; 28.5.75/11.12.75)
21. JA 73–36 940 (Yuasa Battery Co. Ltd.; 29.12.70/8.11.71)
22. Träubel H (1991) Homogene und heterogene Raumkörper – Leder und seine mikroporösen Substitute; ein Vergleich. Das Leder 42:109
23. DE 3 507 467 (Bayer; K Nachtkamp et al.; 2.3.85/4.9.86); DE 3 521 762 (Bayer; 19.6.85/2.1.87)
24. JA 59 179 636 [Tejin KK; 31.3.83/12.10.84 (53 741)]

25. GB 1 135 463 (Interchemical Corp.; 9.5.66/4.12.68; US Prior. 16.6.65)=FR 1 514 955=SZ 479 655 =DE 1 619 255

26. WO 94 20 665 (Tejin Ltd.; K Okawa, K Sasaki, Y Suzuki; 10.3.93/15.9.94)=EP 640 715=JA 065 17 893; JA 062 64 369 (Teijin Cordley Ltd.; 10.3.93/20.9.94); JA 062 64 370 (Teijin Cordley Ltd.; 10.3.93/20.9.94); JA 062 64 371 (Teijin Cordley Ltd.; 10.3.93/20.9.94)

27. JA 79–157 803 (Fujikura Rubber Works; 1.6.78/13.12.79)=JA 78–66 140

28. JA 53 023 101 (Teijin KK; 7.9.76/27.3.78)=JA 76–106 228

9.3
Selective Evaporation Process of Dispersed Polymers

See also 18.2 [29, 52], possibly also 7 [72]

1. SZ 1 546/66 (Chemgene Corp.; R Smith-Johannsen; US Prior. 3.2.65; 3.2.66/15.10.69)=NE 6 601 383

2. FR 1 547 182 (Gurit AG; CH Prior. 28.12.66; 14.12.67/22.11.68)=NE 6 716 617

3. JA 49 111 970 (Nikko Physiochem.Shik; 26.2.73/24.10.74)

4. OS 1 809 610 (BASF; CH Krauch, A Sanner; 19.11.68/11.6.70)

5. JA 49 111 970 (Nikko Physiochem. Shik; 26.2.73/24.10.74)

6. JA 5117586 (Mitsubishi Kasei Corp.; 24.10.91/14.5.93)

7. DAS 1 045 359 (Göppinger Kaliko; W Gräbner; 22.6.56/4.12.58)

8. FR 1 554 754 (Teijin Ltd.; JA Prior. 7.11.66; 7.11.67/24.1.69)

9. NE 6 815 306 (Bayer; K König, H Träubel, A Reischl, B Zorn; D Prior. 25.10.67; 25.10.68/ 29.4.69)=OS 1 694 213

10. DAS 1 020 300 (HC Bick, RS Horn, RF Patt; US Prior. 21.11.51; 18.11.52/5.12.57)=FR 1 070 592 =GB 734 791=US 2 746 941

11. OS 2 004 276 (Teijin Ltd.; K Kigane et al.; JA Prior 31.1.69 (=JA 7 088–69); 30.1.70/24.9.70)= FR 2 033 850

12. JA 49 682/72 (Teijin Ltd.; 25.12.69/13.12.72)

13. OS 2 063 949 (PPG Industries Inc.; JA Steiner; US Prior. 22.6.70; 28.12.70/30.12.71)

14. NE 6 516 286 (USM Corp., US Prior. 18.12.64; 14.12.65/20.6.66)=FR 1 462 597

15. OE 213 367 (Göppinger Kaliko; 22.2.60/15.7.60)

16. JA 19 623/72 (Teijin Ltd.; 14.10.68/5.6.72)

17. JA 43 047/72 (Teijin Ltd.; 31.1.69/31.10.72)

18. OS 2 020 153 (Teijin Ltd.; K Kigane et al.; JA Prior. 2.4.69; 24.4.70/17.12.70)

19. JA 16 764/63 (Nippon Mikusani Kogyo KK; 21.8.61/3.9.63)

20. JA 6 196/66 (Nankai Gum Co.Ltd.; 12.2.62/1.4.66)

21. RA 164 579 (IV Plotnikov et al.; 15.1.62/Jan.65)

22. FR 1 418 697 (IBM; P Chebiniak, RT Wiley; US Prior 30.12.63; 24.12.64/11.10.65)

23. OS 1 469 597 (USM; CG. Newton; US Prior. 18.12.64; 18.12.65/2.1.69)

24. JA 49 067 952 (Nikko Shikiryo Kogyo KK; 4.11.72/2.7.74)

25. JA 49 130 476 (Teijin; 17.4.73/13.12.74)

26. JA 06 294 077 (94,394,077) (Achilles Corp.; K Oosawa, K Mitsumura, K Sugaya; 10.2.93/21.10.94)

27. JA 07145570 (Unitika Ltd.; T Furuta, K Kamemaru; 22.11.93/6.6.95)

28. JA 4202859 (Achilles Corp.; 29.11.90/23.7.93)

29. DOS 3 836 030 (Bayer AG; W Thoma, R Langel, W Schröer; 22.10.88/3.5.90); EP 238 991 (Bayer AG; W Thoma, R Langel et al.; 26.3.86/30.9.87); DOS 3 507 467 (Bayer AG; K Nachtkamp, W Thoma et al.; 2.3.85/4.9.86)

30. FR 2 384 058 (Adidas; H Dassler; 17.3.77/13.10.78)=DOS 2 711 579

31. JA 51 024 666 (Bando Chem. Ind. Ltd.; 23.8.74/28.2.76)

32. FR 1 488 995 (Soc. Anon. Peltex; 13.9.65/12.6.67)=NE 6 612 797=BE 686 221

33. JA 74 039 172 (Toyo Rubber Ind. Co.; 22.5.70/23.10.74)=JA 70–44 137

34. FR 1 337 536 (R. T. Vanderbilt Co. Inc.; RR Waterman et al.; 11.2.63/20.4.64; US Prior. 6.12.62)= US 62–242 653

35. JA 60 017 181 (Teijin KK; 4.7.83/29.1.85)=JA 83–120 316); JA 60 021 982 (Teijin KK; 4.2.83/ 8.7.83)=JA 83–129 557

36. DOS 3 327 862 (Dainichiseika Color & Chem. Mfg. Co. Ltd., Ukima Colour & Chem.; K Kuri-yama et al.; 2.8.83/14.2.85)
37. JA 63 286 466 (Dainichiseika Color Chem.; 20.5.87/24.11.88)=JA 87–121 089
38. JA 01 138 263 (Dainichiseika Color Chem.; 26.11.87/31.5.89)=JA 87–295 963
39. JA 5117586 (Mitsubishi Kasei Corp.; 24.10.91/14.5.93)

10
Reactive Processes

1. ADIPRENE is a trademark of DuPont (NCO-Prepolymer); MOCA=Methylene-bis-*o*-chloro-aniline=3,3′-dichloro-4,4′-diaminodiphenylmethane
2. BE Appl. P 663 102 (Wyandotte Chem. Corp. 27.4.64); DAS 1 097 678 (DuPont; US Prior. 30.6.53; 28.6.54/19.1.61)
3. Dieterich D, Reiff H (1972) Angew Makromolekulare Chemie 26:85
4. Will G, DAS 1 444 172 (4.5.62/2.10.69)
5. Will G, DAS 1 150 524 (30.7.58/20.6.63 not for man-made leather)
6. GB 890 228 (ICI; E Knowles, WF Smith; 9.3.60/28.2 62)
7. DOS 1 795 631 (G Will; 21.3.61/9.11.72) addition to 1 420 851
8. JA 10 356/66 (Toyo Spinning Co. Ltd.; 28.8.62/4.6.68)
10. DAS 1 069 562 (Göppinger Kaliko; W Gräbner; 17.9.58/26.11.59) addition to DBP 1 043 275
11. FR 1 184 642 (Göppinger Kaliko; D Prior. 18.4.57; 30.7.57/9.2.59), see also DAS 1 074 001
12. OS 1 619 222 (Dunlop; MW Higgs, DI Clarke; GB Prior. 6.4.66; 6.4.67/3.9.70)=FR 1 518 134=NE 6 704 883
13. GB 1 103 600 (General Tire; US Prior. 20.1.64; 14.1.65/21.2.68)=FR 1 424 061=US 3 311 527
14. GB 1 099 236 (General Tire; US Prior. 20.1.64; 14.1.65/17.1.68)=FR 1 426 443=US 3 328 225
15. GB 1 141 718 (Goodyear Tire and Rubber Co.; US Prior. 3.11.65; 6.10.66/21.1.69)
16. US 3 734 894 (General Tire; AF Finelli, JC West; 1.7.71/22.5.73; cip 27.8.69 and 25.6.65)
17. BE 720 615 (Bayer; D Prior. 6.9.67; 6.9.68/17.2.64)=DOS 1 621 910
18. US 3 542 617 (Fiber Industries Inc.; GA Watson; 8.6.67/24.11.70)
19. GB 1 210 547 (Goodyear Tire; US Prior. 16.6.67; 23.5.68/28.10.70)=FR 1 571 632
20. BE 651 461 (US Rubber; GL Barnes; US Prior. 6.8.63; 6.8.64/1.11.64)=NE 6 408 850
21. GB 1 228 254 (Textron Inc.; US Prior. 1.9.67; 13.8.68/15.4.71)
22. US 3 539 424 (Wharton Ind. Inc.; I Tashlick; 10.4.69/10.11.70; cip 9.5.68)
23. US 2 973 284 (BF Goodrich Co.; ST Semegen; 30.4.57/28.2.61)
24. Bartl H, Bonin Wv (1962) Über die Polymerisation in umgekehrter Emulsion. Makromole-kulare Chemie 57:74
25. Logemann H, Bartl H, Nogradi J, Glosauer O, Kopet K (1961) Allgemeines zur Polymerisation von Vinyl- und Divinylverbindungen in heterogener Phase. In: Houben-Weyl, Methoden der organischen Chemie, vol XIV/1 Thieme, Stuttgart, pp 24–126, 840
26. OS 2 053 705 (Continental; F Koch, K Witt; 2.11.70/10.5.72)
27. OS 2 316 454 (Ugine Kuhlmann; AP Strassel; FR Prior. 5.4.72; 3.4.73/11.10.75) addition to OS 1 570 524

10.1
Polyaddition of Emulsified Oligomers and Monomers with Nonsolvents in the Inner Phase

See also 18.3 [30], reactive coating with a water-in-oil emulsion containing γ-glutamate modified urethane

1. DAS 1 248 926 (Continental; P Aufleger; 14.12.64/31.8.57)
2. US 3 551 364 (USM Corp.; JJ McGarr; 23.1.69/29.12.70; cip 21.10.65)=FR 1 497 310=GB 1 162 215 =SZ 484 229=DOS 1 719 256
3. GB 1 150 995 (Bayer; H Träubel, W Klebert, K Breer; D Prior. 26.10.66 and 11.5.67; 25.10.67/ 7.5.69)=BE 705 612=NE 6 714 307
4. OS 1 694 081 (Bayer; H Träubel, W Klebert; 26.10.66/20.8.70)

5. BE 695 850 (Continental; D Prior. 11.6.66; 21.3.67/1.9.67)
6. US 3 625 871 (Bayer; H Träubel, K König, W Heydkamp, K Breer; D Prior. 12.5.67; 7.5.68/ 7.12.71)=DAS 1 694 152=BE 715 003
7. US 3 539 388 (USM Corp.; Shu Tung Tu; 17.4.67/10.11.70)
8. US 3 539 389 (USM Corp.; Shu Tung Tu; 19.5.67/10.11.70)
9. OS 1 807 124 (USM Corp.; JJ McGarr; US Prior. 14.11.67; 5.11.68/12.6.69)
10. NE 6 905 148 (Bayer; H Träubel, B Zorn, K König, W Heydkamp; D Prior. 2.4.68; 2.4.69/ 6.10.69)=DBP 1 769 089=BE 730 943
11. NE 6 907 643 (USM Corp.; US Prior. 17.5.68; 19.5.69/19.11.69)
12. FR 2 073 681 (USM Corp.; StI Hayes; US Prior. 12.12.69; 11.12.70/1.10.71)
13. OS 2 011 842 (USM Corp.; Shu Tung Tu, JJ McGarr; US Prior. 16.10.69; 12.3.70/27.5.71)
14. OS 2 129 706 (Kinyosha Co.Ltd.; MM Maeda et al.; JA Prior. 15.6.70 (51 351/70); 15.6.71/ 30.12.71)
15. OS 2 117 395 (USM Corp.; StI Hayes; US Prior. 8.4.70; 8.4.71/28.10.71)=GB 1 343 332
16. OS 2 330 601 (USM Corp.; JJ McGarr; US Prior. 16.6.72; 7.8.72 and 9.5.73; 15.6.73/3.1.74)
17. OS 2 527 283 (USM Corp.; JTh Day, JG Hollick; US Prior. 21.6.74; 19.6.75/2.1.76)
18. UCECOAT® 2000 of UCB (Belgium)
19. IMPRAPERM® 43153 of Bayer AG Leverkusen (Germany)
20. FR 1 542 644 (3M; 11.8.67/18.10.68; US Prior. 15.8.66+28.6.67)=US 3 595 732=DE 1 694 641
21. WO 9502009 A1 (Dow Chem.Co.; RL Tabor; 7.7.93/19.1.95); US 5 478 867 (Dow Chem.Co.; RL Tabor; 7.7.93/26.12.95)
22. DOS 2 448 133 (Bayer AG; H Träubel; 22.4.76)
23. US 3 178 310 (DuPont; RL Berger, MA Youker; 1.3.62/13.4.65)=US 62–176 818

10.2
Polyaddition of Dissolved Polymers

1. BE 725 052 (Bayer; D Prior. 7.12.67; 6.12.68/16.5.69)=DAS 1 694 230
2. Träubel H (1974) Das Leder 25:162
3. Träubel H (1976) Das Leder 27:1
4. OS 2 329 299 (Bayer; H Träubel et al.; 6.6.73/2.1.75)
5. Trademark of ICAP (Italy)
6. Trademark of Bayer AG (Germany)
7. Trademark of Ugine-Kuhlmann (France)
8. Trade mark of SNPE (France)
9. BE 719 272 (Bayer; D Prior. 9.8.67 and 7.12.67, 9.8.68/16.1.69)=DAS 1 694 180 and DAS 1 694 229 and DAS 1 694 231
10. OS 2 034 558 (Bayer; H Träubel, K König; 11.7.70/ 3.2.72)
11. OS 2 041 710 (C. Freudenberg; T v Schachowskoy et al.; 22.8.70/24.2.72)
12. OS 2 123 962 (Bayer; H Hetzel et al.; 14.5.71/23.11.72)
13. US 3 849 528 (Polysar Ltd.; AE Smith; CA Prior. 16.2.73; 8.1.73/19.11.74)
14. OS 2 526 843 (Union Carbide; AP Jones et al.; US Prior. 18.6.74; 16.6.75/2.1.76)=US 74–480 567
15. JA 07 041 731 (Asahi Kasei Kogyo KK; 26.7.93/10.2.95)
16. US 5 478 867 (The Dow Chemical Comp; RL Tabor; cip 7.6.93; 29.6.94/20.12.95)

10.3
Bonding of Nonwovens by Reactive Processes

1. DAS 1 098 909 (Goodrich; US Prior. 29.12.54; 28.12.55/9.2.61)=GB 804 669=FR 1 145 244
2. JA 9 197/66 (Toyo Rubber Kogyo Co.Ltd.; 11.2.63/16.5.66)
3. BE 668 641 (Bayer; W Klebert, W Wunder, W v Langenthal; D Prior. 27.8.64 and 8.7.65; 23.8.65/ 16.12.65) and NE 6 511 172
4. FR 1 462 456 (DuPont; EH Pagliaro; US Prior. 6.11.64; 5.11.65/7.11.66)
5. US 2 973 284 (Goodrich; ST Sennegen; 30.4.57/28.2.61)

6. JA 24 680/67 (Toyo Rayon Co.Ltd.; 21.5.64/27.11.67)
7. US 3 255 061 (US Rubber; ID Dobbs; 20.4.62/7.6.66)
8. JA 21 519/69 (Kurashiki Boseki KK; 12.11.65/13.9.69)
9. US 3 102 835 (Allen Ind.Inc.; MM White 25.4.60/3.9.63)
10. JA 25 392/69 (Kurashiki Spinning; 31.3.65/25.10.69)
11. GB 1 148 873 (Lantor Ltd.; BF Jones; 9.3.65/6.4.69)=NE 6 603 785
12. OS 1 964 064 (Continental Gummi-Werke; P Aufleger; 22.12.69/24.6.71)
13. OS 2 053 468 (Ugine-Kuhlmann; PGL Arbaud; FR Prior. 30.10.69; 30.10.70/13.5.71)
14. OS 2 034 537 (Bayer; H Träubel et al.; 11.7.70/3.2.72)

10.4
Polymerization of Emulsified Polymers

1. OS 1 806 652 (BASF; CH Krauch, A Sanner; 2.11.68/27.5.70)
2. FR 1 457 808 (Continental Gummi Werke AG; H Kaste; D Prior. 21.11.64, 25.11.64, 5.12.64 and 22.11.65/26.9.66)=DAS 1 229 285=NE 6 515 069
3. OS 2 253 851 [VEB Chemiefaserkombinat Schwarza; HE Seyfarth, E Meusel; DDR Prior. 24.12.71 (DDR Wirtschaftspatent 160 135) 3.11.72/5.7.73]
4. OS 1 495 227 (Chem. Fabrik Kalk GmbH; R Schell; 27.3.62/12.12.68) (see also D Pat Appl. W 30 983/39c)
5. OS 1 469 519 (Courtaulds Ltd.; ME Baguley; GB Prior. 18.7.63 and 14.1.64; 17.7.64/6.11.69)= BE 650 763=GB 1 087 867=NE 6 408 154=OE 249 625=US 3 377 190
6. OS 2 521 156 [VEB Chemiefaserkombinat Schwarza; HE Seyfarth et al.; DDR Prior. 10.7.74 (WP 179 830); 13.5.75/29.1.76]
7. BE 726 364 (BASF; 31.12.68/29.5.69)
8. NL 93 00 243 (Universiteit Twente, 8.2.93/1.9.94)
9. DOS 1 669 779 (Continental Gummi Werke AG; G Will, H Sieber et al.; 20.4.66/7.10.71)
10. DOS 1 469 521 (Continental Gummi Werke AG; R Fischer et al.; 4.11.65/12.12.68)
11. DOS 1 544 690 (Continnental Gummi Werke AG; H Hildebrandt et al.; 15.12.64/31.7.69)
12. DOS 1 619 310 (G Will; 11.1.66/10.9.70)
13. DAS 1 220 606 (G Will; 2.11.61/7.7.66) DAS 1 240 665 (G Will; 23.5.62/18.5.67) addition to DOS 1 220 606; DOS 1 469 590 (18.11.65/12.3.70); addition to DOS 1 444 172 (see also 4.3 [4]); DOS 1469 585 (G Will; 2.1.65/12.3.70) addition to DOS 1 444 172; DOS 1 619 311 (G Will; 6.6.66/ 10.9.70); DOS 1 569 540 (G Will; 14.11.63/30.4.70)
14. Will G (1964) Poröse Kunststoffe aus Wasser in Öl Emulsionen. Kunststoffe 54:513

10.5
Polymerization Reaction of Dissolved Polymers

1. US 3 322 695 (Dow; T Alfrey, WG Lloyd; 7.1.66/30.5.67; cip 22.1.63)
2. OS 1 594 938 (BASF; H Reinhard et al.; 4.8.67/18.9.69)
3. FR 1 518 715 (Pittsburgh Plate Glass; JA Steiner; US Prior. 1.2.66; 21.12.66/29.3.68)
4. DDR 69 213 (BASF; CH Krauch, A Sanner; 30.7.68/5.10.69)
5. US 3 640 753 (BASF; CH. Krauch, A Sanner; D Prior. 2.11.68; 31.10.69/8.2.72)=OS 1 806 652
6. OS 1 806 290 (BASF; H Bittermann, CH Krauch; 31.10.68/6.5.70)
7. OS 1 808 391 (G. Will; 12.11.68/27.5.70)
8. OS 1 619 230 (Dunlop AG; R Hank, H Wolf; 1.8.67/4.2.71); cip OS 1 289 656=BE 718 848=GB 1 214 940
9. SZ 485 067 (BASF; A Sanner, CH Krauch; D Prior. 21.9.68; 11.9.69/13.5.70)=OS 1 795 378
10. OS 2.202 568 (BASF; G Bleckmann et al.; 20.1.72/9.8.73)
11. EP 803 533 (Dai Nippon Ink and Chemicals Inc.; Kawamura Institute of Chemical Research; H Zhang, T Anazawa, Y Watanabe, M Miyajima; 17.3.97/29.10.97)=JA 97–62 904

10.6
Reactive Processes by Elimination of a Component

1. US 3 882 054 (IRPC Corp., F Hostettler; 6.2.73/6.5.75)
2. DDR 104 534 (Akademie der Wiss.; R Barthel et al.; 13.9.71/12.3.74)
3. US 3 062 760 (Electrical Storage Battery Co.; WF Dermody, TH Meltzner; 6.10.59/6.11.62)
4. JA 23 856/64 (Toyo Rubber Ind.Co.Ltd.; 2.8.61/24.10.64)

Part 3

Porosity by Other Means

CHAPTER 11

Blowing and Foaming

Blowing is the liberation of gas which may create pores. Pores can only be obtained if the structure of the product wherein the gas is developed is strong enough to keep at least a part of the gas inside. To obtain porous products the gas creates pores in the blowing process. The gas is mostly liberated in the polymer during solidification to a film, coating or impregnation. Gas can be generated by adding a blowing agent like azoisobutyric acid dinitrile or azodicarbonamide (see Chap. 2.1) to the polymer(solution) prior to the solidifying step. By raising the temperature, the blowing agent liberates nitrogen gas [3]. Another product that liberates gas by raising the temperature is ammonium bicarbonate, which liberates carbon dioxide [8].

Another method is to add small sized polymer particles to a polymer solution or dispersion. The particles contain a swelling agent, which after liberation by a thermal shock is a gas, i.e. a blowing agent. An example is a vinylidene chloride–acrylonitrile–divinylbenzene copolymer containing isobutane or pentane [9].

Another method exists in the manufacturing of polyurethane-foams: A molar excess of isocyanate, compared to the hydroxyl functional polyether or polyester also present, and added water, liberates carbon dioxide [4] (see Eq. 8.1) which – during the polyaddition reaction and solidification of the polyurethane – blows up the foam.

In the conventional methods blowing was achieved by the evaporation of fluorochlorohydrocarbons to develop small closed pores with excellent thermal insulation properties. Today, volatile alkanes of a low boiling point are used together with the carbon dioxide liberation of the isocyanate/water reaction [11].

Fluorochlorohydrocarbons seemed to be suited very well due to their non-inflammability. They were used worldwide in the production of insulating foams or foams for mattresses, etc. Due to the impact on ozone destruction in the stratosphere, their use today is banned.

Polyurethane foam (Fig. 11-1), rubber sponge [7] (Fig. 11-2), or a man-made PVC-leather [5] (Fig. 11-3) show different pore structures. All these materials are produced by a blowing reaction.

Materials with a foamed polymer layer between a water vapor permeable top layer and an adhesive coat are, for example, suited for the production of bags [13].

Another method to produce foams is to whip air in a polymer dispersion [2]. In this case the foam is produced by mechanical means. This process is, however, described in more detail in Chap. 11.1.

Fig. 11-1. Pore structure of a polyurethane foam. The rigid polyurethane foams are used mostly for insulating products, like in refrigerators. The best insulation property of the foam is achieved if the foam contains a gas of low thermal conductibility. Therefore the pores must be closed to keep the insulating gas inside

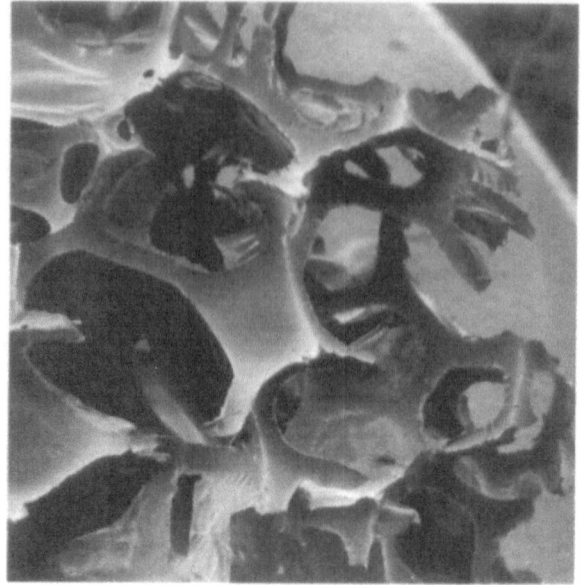

200 µm

Fig. 11-2. Pore structure of a rubber sponge

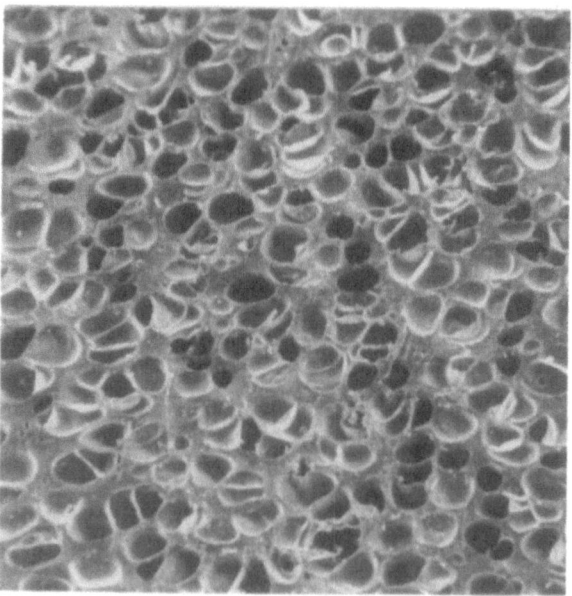

200 µm

Fig. 11-3. Pore structure of
a man-made PVC leather.
This is a detailed photo
of an artificial PVC leather
substitute to demonstrate
the closed pores of the
intermediate coat

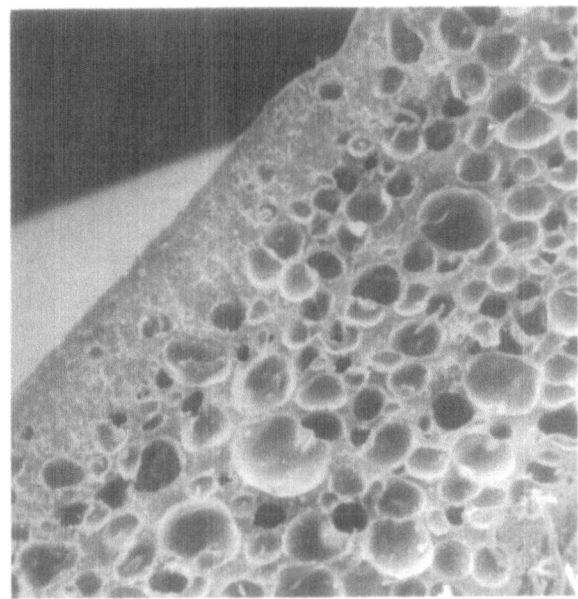

200 µm

Another method described to produce pores, open to water vapor transport, is
to press air through a foamed PVC-leather and a freshly prepared polyurethane top
layer. The air does not allow the polyurethane to close its surface during drying [1].

Solid carbon dioxide can be mixed with PVC. By heating to a temperature of
180 °C under pressure the evaporating carbon dioxide also creates pores [6].

Expanded polytetrafluoroethylene can be adhered to a textile by an adhesive
containing carbon or metal particles to avoid electrostatic charges on the surface
of the material [10]. A polyurethane can be used as a top layer and in a blown
stage as an adhesive layer for man-made leather [12].

Blowing creates middle sized pores (0.1–1 mm in diameter); i.e. the pores are
visible to the naked eye. The pores are mostly closed and have no connection
with each other – a necessity for the transport of water vapor through the
polymer. Therefore a water vapor permeability is not necessarily the result of a
blowing action. It is often claimed that pores can be opened by the addition of
tensioactive substances. There are some processes described which claim to
achieve water vapor permeability, but there are no processes known being used
technically. In most cases the surface of a layer produced by a blowing process is
totally closed and not porous. Therefore this process – according to the opinion
of the author – is not suited for water vapor permeable materials. Blowing is
mostly suited for 100 % systems i.e. systems containing no or only small quantities of solvents. Such kind of systems consist e.g. of a thermoplastic polyurethane
composition with thermally decomposable blowing agents. The polyurethane
composition is coextruded and the resulting cellular layer may be laminated to
a fibrous substrate [14].

11.1
Production of Foams

In this chapter methods to transfer polymers into a foam using dissolved or dispersed products are discussed [70]. In the production of mattresses, cushions or parts for the production of shoes, foams are used. These foams may be produced in different manners [67].

In the Dunlop process a sodium silicofluoride solution is mixed into a rubber latex mixture which can be vulcanized. During the vulcanizing step, hydrogen fluoride is eliminated which blows up the solidifying rubber.

In the Talalay process [66] the rubber latex after mechanical whipping is brought into a mold. Then the mold is frozen under vacuum at a temperature of below −30 °C. The latex totally solidifies at that temperature. Then carbon dioxide is brought in to get an acidic pH. The latex coagulates. Even under vulcanizing conditions at 110 °C the latex keeps its structure (see also Sect. 17.1, bonding of nonwovens).

Applications of Water-Based Dispersions. A blowing agent, boiling between −46 and +104 °C, is added to a water-based dispersion of a polyurethane or a polyacrylate. The mixture is coated onto leather and by heating a foam is produced [18, 19]. A heat-sealable polyacrylate, polyurethane or nitrocellulose dispersion is additionally applied [62].

A dispersion consisting of 50–83 parts of cellulose fibers, 0.2–1 part of an emulsifier, like sodium lauryl sulfonate, 5–45 parts of polyvinylacetate and 0.3–5 parts of sodium alginate or sodium polyacrylic acid is mechanically whipped to form a foam. This foam is applied, dried at 100–200 °C and embossed [46]. In a similar way, a dispersion of a copolymerisate of ethyl acrylate, acrylonitrile, acrylamide and methoxymethyl methacrylate is whipped to obtain a foam. The increase in volume should be 15–400 %. The foam is used to get a porous coating [48]. An acrylate dispersion is used similarly [54].

A water-based emulsion is dispersed into a plastisol of PVC and a plasticizer. The mixture is poured onto a film which is heated at a temperature above 100 °C to get a foam structure [50]. A textile coated with PVC containing a foaming agent is covered by a release paper, embossed at 180–250 °C and at the same time the blowing agent is activated to develop a foam structure of the PVC coating [53].

A water-based dispersion of a polymer is mixed with a thickening agent and air is whipped in or blown into the dispersion [56]. The foamed structure may be stabilized by the addition of polymeric hollow spheres containing air or a gas or a blowing agent [63].

Mixtures of foaming and vulcanization agents, which are calcium carbonate, silica and a rubber latex are vulcanized and cast to a porous shoe material [43].

A releasing surface is treated with a solution of a polyurethane. Then a whipped water-based dispersion of a polyurethane or a styrene–acrylonitrile copolymer is applied and dried at 110 °C and the textile is then laminated [55].

A water-based dispersion of a polymer, able to crystallize, is dispersed into a tacky one, being in the continuous phase. After being cast into a form and frozen, the resulting porous materials may be used for medical or biotechnological applications or in the manufacture of shoes [67].

Textiles are coated with polyurethanes able to crosslink by heat [21]. Components able to produce a polyurethane with a foam structure [26] and possibly in the presence of a thickening agent are applied to a nonwoven [22, 24, 28]. An application in an indirect process is also possible [27]. Often a isocyanate prepolymer, able to form a foam, is used as the layer between the top coat and the textile [29 – 31, 33, 35 – 37].

Fiber fleeces consisting of polyamide and polyester are impregnated with a mixture of components, able to form a polyurethane foam [49]. If the reaction is carried out in appropriate molds, finished shoe parts may be produced in this way.

Applications of Solutions. A polyurethane solution is treated with a water-based solution or dispersion of silica or bentonite and a water-based solution of a blowing agent, like ammonium bicarbonate or a petrol fraction of a boiling range of 80 – 200 °C, and applied as a coating product for textile substrates [20]. The foam may be stabilized by the addition of an organosiloxane [69].

A solution of a vinyl chloride copolymer and a blowing agent combined with a powder able to be sintered is coated onto a textile substrate and heated to obtain a porous coating [42].

Coatings may also be carried out with a mixture of an isocyanate prepolymer and a hydroxy-group-containing vinyl polymer in a water-miscible solvent. The mixture is foamed by -exposure to vapor [58].

A polyurethane dissolved in THF or DMF is mixed with 100 – 200 parts of ammonium bicarbonate (for every 100 parts of PUR dissolved) in a film cast and heated to 100 – 150 °C [40].

Isocyanate prepolymers mixed with methylene chloride or fluorochlorohydrocarbons and carbonates of (cyclo)aliphatic diamines are claimed to give compounds which are stable on storage. Textile substrates can be coated with these mixtures. If the substrate is heated to the boiling temperature of the solvents and the decomposition temperature of the carbonate, the polyurethane develops a foam structure [32].

Bonding of nonwovens can be carried out with these mixtures, e.g. in butan-2-one [17]. Addition of acrylates or leather dust to these mixtures is possible [13]. Nonwovens of collagen fibers may also be bonded by this method [14]. Mixtures of a polyurethane with a vinyl polymer containing hydroxy groups are coated on a textile substrate then foamed by means of a blowing agent [60].

Nonwovens may be bonded by a mixture of latex and isocyanate prepolymers [34] and fully reacted under pressure [15].

A nonwoven is treated with water, then coated with an isocyanate group containing urethane, sprayed with water and heated to 100 – 130 °C to start the foam reaction [59].

Application of Solid Materials. Systems containing no solvents, do not need heat to evaporate water or solvent. Their viscosity – even if the polymer is not applied in its high molecular form but from its components which react during or after application – is so high that these systems cannot be sprayed or applied by a doctor's knife in thin layers. Solid systems therefore are often applied by *extruding*. Another way is the application of oligomeric components which are able to react to the high polymeric form after or during application. Examples for so called 100 % systems are:

- Melted polyethylene, containing a blowing agent like pentane, carbon dioxide etc., is extruded through a nozzle [45] or sprayed and foamed [2]. Calendering of rubber, PVC, vulcanizing and blowing agent, like dinitrosopentamethylene-tetramine, onto a textile substrate also gives a porous coating [39]. Mixtures of PVC, polystyrene or polybutadiene, bicarbonate and oxalic acid after application and heat treatment result in porous coatings [41].
- Ethylene propylene copolymerisates, mixed with ethylene vinylacetate copolymers and a blowing agent are applied to a cotton fabric and blown to get a foam structure. The foam is finally covered by a polyamide top [57]. Mixtures of polyethylene, PVC and a blowing agent like azodicarbonamide can be used to produce water vapor permeable man-made leather [61]. Shavings of chrome leather, blowing agent and ethylene vinylacetate copolymer are used to produce – after foaming and embossing – an artificial leather [64].
- 100 parts of a styrene–methylmethacrylate–butadiene polymer are treated with a solvent of a boiling temperature of 30 – 90 °C, like n-pentane. The polymer is treated in such a way that it absorbs around 5 – 10 % of the solvent. The mass is extruded to form a porous film [47].
- A porous PVC-layer can be laminated onto a semi-dried polyurethane which has been applied by the indirect process onto a release paper to give a man-made material [38, 65].
- TDI, polyethylene glycol adipate and azo-dicarbonamide are mixed with titanium dioxide and calendered at 100 – 120 °C to a film with a thickness of 0.2 mm and laminated onto a textile. By heating this up to 200 °C a foam is developed which raises the film to a thickness of 0.8 mm [52]. Fabrics or knitted textiles are treated with a mixture able to build a polyurethane foam. They are then foamed and pressed to increase the density by 0.2 – 0.5 [25].
- Besides coatings [1, 3, 11, 68], impregnations or bondings of textile substrates can also be carried out by polyurethane foams [9, 44]. The water, used to react with the isocyanate groups, can be added in the form of sodium tetraborate decahydrate [10].
- Compressed layers consisting of PVC, polystyrene, polyolefin or polyurethane [7,16] may be used as water vapor permeable coatings of nonwovens [4, 5]. It is possible to bond these compressed layers additionally with an isocyanate prepolymer [12]. Foamed polyurethane layers can be adhered to ABS-copolymers, PVC, polyamide etc. by high frequency. The final products can be used as materials for upholstery [6].
- Polyurethane foam laminates can be transformed to finished shoe uppers by high frequency molding [8].

11.2
Treatment of Foams

A foam is not necessarily permeable to water vapor. Most foams have closed pores. Foamed layers are often used as intermediate coats (see Fig. 5-4) because they are able to eliminate irregularities of a substrate and offer a softer touch to the product.

The following examples discuss foamed coats: A method which has been well used for a long time is to laminate a foam consisting of polyurethane [22, 23], rubber sponge [9] or PVC [10, 13] on one side with a thin film for the later top layer and on the other with a textile substrate. Blowing, as explained previously, results in rather thick pores with closed walls. Therefore, laminates produced in such a manner, are normally not water vapor permeable.

In most cases this foam is reduced in thickness by pressing [14 – 17, 19, 20, 25, 26, 43 – 45, 52 – 55, 62]. By this pressing the existing water vapor permeability will be reduced further [30]. The compressed foam may be impregnated by an isocyanate prepolymer [56]. This pressing operation is usually carried out at temperatures above the melting range of the foam [27, 28, 47]. Almost every bonding of a foam layer with a textile substrate is combined with a pressing ope-ration [21, 31, 36, 49].

The lamination of foam layers to textile substrates may be carried out by high frequency [57].

Foams are almost always laminated by *flame lamination* [46, 49, 59] (see Fig. 11-4). Textiles having a foam interlayer are good thermoinsulating materials which are used to produce cheap apparel to be used at cold temperatures. Micro-porous films may also be adhered to textiles by a foam system [63].

Decreasing the porosity by pressing is a current method for improving the surface structure and to increase the physical properties of the foams [5]. This pressing operation may be done under friction [41]. Impregnating the open pores of a polyurethane foam layer with a polyisocyanate, bonding to a textile and pressing the article also results in an artificial leather [8].

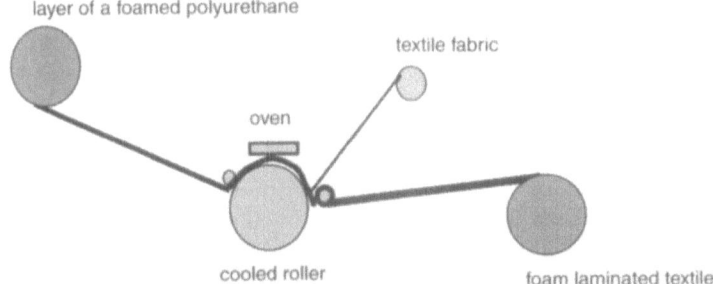

Fig. 11-4. Flame lamination. In the flame lamination process a foam is heated by an oven or an open fire. Thereby the foam decomposes and melts. The melting side of the foam is laminated to a substrate which adheres extremely well to it afterwards

In the *explosion process,* a foam containing a blowing agent is compressed. Then – very rapidly – the material is decompressed. Air and the blowing agent, which are in the pores, evaporate in an instant whereby the walls of the pores are destroyed. In a foam, the pores of which had been opened by the explosion-process, a second, hydrophilic foam is produced [7]. This material can be laminated to produce a man-made leather.

Fibers are put on the surface of a foam [32, 33], bonded by a latex and pressed. The foam itself is further bonded by a solution of an elastomer [29].

Flocks of a PUR foam are bonded by means of a polyurethane, possibly able to foam onto a textile [40].

Filling of pores may be achieved, for instance, by applying an isocyanate prepolymer to a foam layer. The prepolymer partially penetrates the foam [6]. Another method is to impregnate a foam layer with a PVC plastisol [18] or a polyurethane, dissolved in butan-2-one [37].

PVC, polyurethane or polyacrylate foams with hydrophilic properties are laminated onto a textile. The final product is able to absorb water and is water vapor permeable. Textiles treated in this manner may be used in the ocean rescue service [60].

Opening of pores: There has been a lot of research published in the literature regarding the opening of pores or attempts to obtain pores with a consistently small pore size [4]. One such treatment of a polyurethane foam is by a water/DMF mixture [2]. Then a swelling agent is used which is removed afterwards by heat and vacuum [34], or by the application of a needling process [24, 61]. Another mechanical process to increase porosity is to buff the surface [48].

A further way to create an open pore structure is to achieve a partial hydrolysis [12] of the polymer material. Usually, afterwards, an impregnation with a dispersion or a solution must also be done to obtain a smooth surface [45].

Dissolving low molecular parts of a polymer and leaching them from the foam also increases the porosity [3]. Foams consisting of polymer mixtures may also be increased in porosity if the foam is treated with a solvent which dissolves one of these polymers [38]. Rapid evaporation of the solvents which are in the walls of a foam create permeability of the foam [11].

Synthetic suede is obtained if the top layer of a PVC foam is torn off. The artificial PVC leather has a solid top and a foamy inner layer (see Fig. 11-3). By this method a suede-like surface is obtained [1]. A PVC foam is directed around a heated roller. By this action the surface melts and forms an even and uniform structure. Then this closed surface layer is torn off. The result is a suede-like surface [51].

Open pores of polyurethane foams are impregnated with solutions or dispersions of polyurethanes or rubber heated and pressed to get smooth materials [39]. High frequency may also be applied [35]. If the foam is put into the mold of a shoe upper, a formed shoe upper may be achieved by applying high frequency [58].

CHAPTER 12

Microporosity by Controlled Melting Processes

Glass filter frits are known to be produced by a controlled melting process. The melting is carried out in such a way that when a finely divided powder is heated up only the surface of the particles is melted. The particles flow together and adhere by their walls – they sinter. After cooling down, a porous membrane is created. If the particles are heated further, a homogeneous film is created.

Thermoplastic powdered products are applied to a releasing surface which is then heated to a temperature near the melting range of the polymer. After controlled, partial melting, a substrate is laminated to it and after cooling the releasing surface is stripped off [19, 20]. It is rather difficult to get the powdered polymers in an appropriate form onto a surface. The problem is to achieve an equal thickness of the powder. Therefore spraying, doctor's knife application or spreading evenly in an electrostatic field may be applied [26, 31, 36]. The heat transfer can be done by gas heaters [34] or high frequency [24]. If the sintering is done in the presence of water or a blowing agent porous articles can be produced [29]. To apply the powder a rotating screen printing system can also be used [32].

Due to the difficulty in applying polymer powders in a uniform way, a suspension of these kinds of powders, e.g. in a nonsolvent, are used. Thermoplastic polymers such as PVC, polyacrylate, polyester and polyurethane with a particle size of more than 0.1 mm are dispersed in a nonsolvent, applied to a substrate and sintered by heating at 150–180 °C [9]. Polyurethane particles with a size of 0.005–150 µm are sintered to microporous films [7].

Since 1955 a way to produce polyurethane-dispersions has been known [3, 4]. The polyurethanes are polyadded directly in water [2]. As in the production of paper the polyurethane particles can be separated from water by using a Fourdrinier wire. The resulting film has no tear strength at all because the polymer particles do not stick together. By a thermal action afterwards, microporous films are produced whereby the microporosity depends on the intensity and duration of the thermal action [1, 9].

Sedimenting polyurethane dispersions with polyurethanes [11, 13] of a particle size of more than 5 µm which are redispersable can be sintered after separation of the particles at 80–220 °C [5]. The dispersions may also contain thermoplastic polymers of a small particle size [6]. To improve the physical properties the polyurethanes may contain N-methlylol groups able to crosslink the polymers [6, 8]. A mixture of 50 % polyurethane and 50 % vinylidene chloride–acrylonitrile copolymers can be sintered and possibly crosslinked by an isocyanate [12].

Instead of a polyaddition in water, polyurethanes may be produced in a non-solvent like xylene. At the beginning the polyurethane precursors are soluble in the nonsolvent. By an ongoing polyaddition reaction the solution becomes turbid and the polyurethane is dispersed in it. By the action of water vapor xylene is evaporated. From the resulting dispersion, films may be produced by drying at 60 – 80 °C [14].

Other polymers, like polystyrene, PVC or polymethylmethacrylate, can be treated with two different nonsolvents, which in mixture are able to dissolve the polymer. Shaped parts consisting of these polymers are treated with nonsolvents and heated to a temperature near the melting range. There parts are then put together. They are melted together by this action [10].

Shrinkable PVC fibers and other fibers containing plasticizers are melted on a substrate to form porous films [23]. Dispersions of a vinyl chloride polymer in a nonsolvent like glycerol are formed to a film and sintered. Then the nonsolvent is removed. The resulting material may be used in the production of filters [27].

Nonwovens consisting of thermoplastic fibers are treated by heat and pressure to create a porous surface on the top [21, 25]. Often the surface is not exposed directly to the heated rollers but instead it is protected by a release foil during the action of heat and pressure. The resulting surface will be more uniform or shaped according to the structure of the releasing sheet [22].

Nonwovens consisting of polyamide-6, -6,6 or -12 or polyethylene, which are impregnated by a polyamide, polystyrene or polyurethane are treated with a

Fig. 12-1. Photo of a film produced by sintering. The formally discrete polyurethane particles are partially melted together

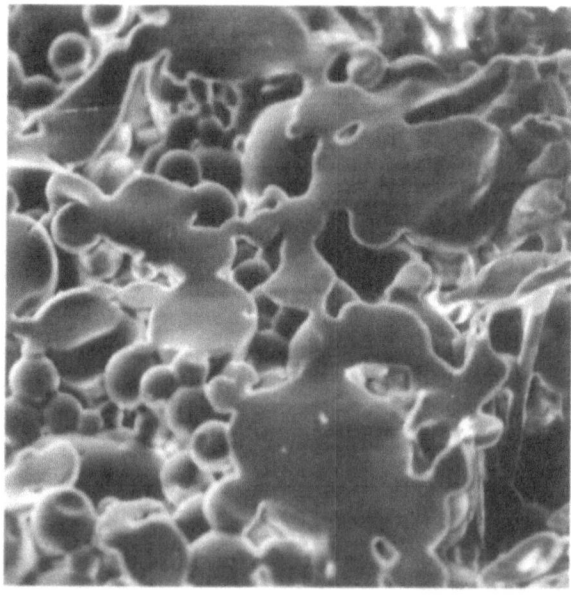

50 µm

solvent like DMF or dimethyl sulfoxide (DMSO), heated and melted. On the surface pores of a size of up to 30 µm are created [28].

Polyethylene powder is put on the surface of a porous substrate and calendered to a porous coating [6]. Amorphous polyamide powder can also be coated on a substrate and treated with a hot formaldehyde solution [18].

Thermoplastic powders of PVC, polyethylene or polyamide are treated with 5–20% of a liquid like DMF whose boiling point is between 5 to 20 °C higher than the melting range of the polymer. Preferably the liquid should also contain potassium hydrogen carbonate. Then a film is produced. The film is heated and the rest of the liquid and, if present, the potassium hydrogen carbonate is washed out [37].

Nonwovens are coated with a binder containing powdered polymeric binding agents. Preferably the nonwoven is heated prior to the coating step. Then the coated nonwoven is directed between heated rollers to melt the binding agent [15]. The coating may be protected against the rollers by a release paper during the heating action [16]. Copolymers containing 90–97% vinyl chloride and 3–10% vinyl acetate are mixed with a plasticizer and a water-repellent agent. A fabric is coated with this mixture. Under pressure and the specified temperature the coating is sintered together [17]. Polytetrafluoroethylene (PTFE) may be applied to nonwovens, other textile substrates or glass-fiber fleeces and sintered [33]. A nonwoven is sprayed by a PVC dispersion containing a plasticizer. Then a sintering action can be carried out to create a porous coating [30]. These films may be used as coating materials etc. In Fig. 12-1 a film is shown consisting of polyurethane particles which have been partially melted together.

Sheets with a good internal adhesion of its layers are manufactured by melt molding of a polyurethane and a thermoplastic ethylene–ethylacrylate copolymer. Then the sheet is laminated to a polyethylene nonwoven. The resulting product may be used as disposable diapers [39, 40]. More or less analogous is a lamination of a moisture permeable elastomer onto a nonwoven of polypropylene and an ethylene–α-olefin copolymer [41].

The production of microporous films and coatings by sintering is not technically used for textile coating. Although melting looks ecologically suited, due to the critical factors, however, such as thickness and evenness of the films and the difficult handling of the non-sintered, meta-stable polymer films the process is not feasible. The grade of microporosity depends on an extreme exact temperature management. If the intensity and duration of the heat action is not controlled extremely well, the grade of porosity and the physical properties cannot be managed easily enough. On the other hand sintering is often used to laminate foils on textile substrates or – in shoe production – to adhere parts together. In the production of automotives thermal bonding of parts is often also used [38].

CHAPTER 13

Perforation

Perforation by stitching or cutting can also result in a porous surface. This perforation should be carried out during a stressing action of a coated substrate [14, 15]. Apparently it is almost impossible to produce fine, invisible pores by such a standard perforation technique.

If the perforation is carried out by electrical sparks [2, 13, 17] which penetrate the surface, water vapor permeability can be created. Optimally in such a case the coating should contain particles of a size of 10 – 50 μm of a conductive substance like Mg, Al, Sb etc. salt or oxide. Suitable polymers are polyethylene, polyurethane, PVC or a vulcanized rubber [12].

Films consisting of microspheres containing polyester–polyurethane become porous by electronic beams. In contrast, foils consisting of polyester, polyamide or polypropylene become cracked during such a treatment [16].

Fig. 13-1. Stereoscan photo of an article whereby its porosity has been created by a cutting cylinder

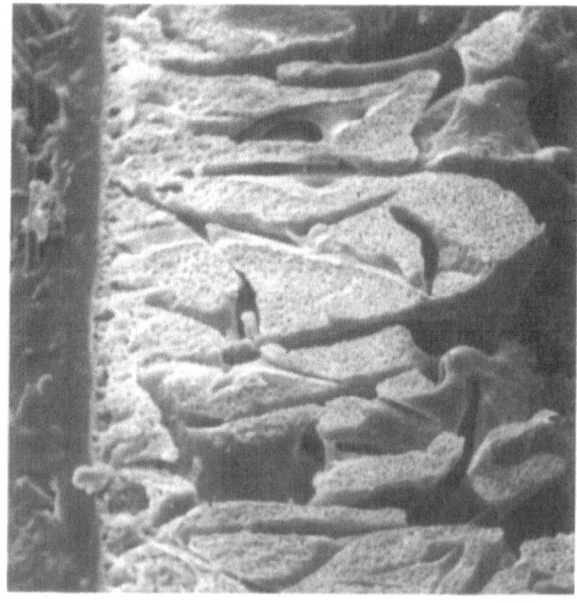

100 μm

Mechanical cutting of surface coatings has been applied to obtain a porous surface (see Fig. 13-1) [3].

The perforation can be carried out using needles [5], beams of high energy [11 – 13] or by a fiber application in a strong electrostatic field [10] (see chapter 15). For instance, a polypropylene film can be needled onto a nonwoven and then adhered by a thermal action to it [7]. Alternatively, a polyurethane film vulcanized by sulfur can be needled onto a nonwoven [8]. A resin layer put on a nonwoven can be needled with needles at a temperature of 150 – 250 °C and 200 – 500 stitches per square centimeter [18]. A multilayer compound is produced by needling a coated textile onto a nonwoven [9].

A perforation by a combined chemical and mechanical treatment is the following: A polyurethane film is laminated to one side of a bonded polyester nonwoven. Then the composite material is crumpled mechanically in an aqueous sodium hydroxide solution to obtain a leather-like sheet with natural appearance and softness [19].

Hydrophobic films of polypropylene become porous by treatment with a raising gig. After that they are treated with a surface active substance. These films can be used as filters or as wound dressings [1]. PVC is the polymer mostly used in the perforation technique. BALAN® [4], a trademark of Balamundi (The Netherlands), was produced by the perforation technique. The photograph in Fig. 13-1 is of a BALAN sample.

CHAPTER 14

Discontinuous Coatings – e.g. by Printing

Usually in textile coating an outer layer consisting of a hard polymer is produced during the coating process by a doctor's knife application of a polymer solution or application. This process creates a homogeneous film. Poromeric surfaces are usually (top)coated by a finishing step, in particular to improve their abrasion resistance and to modify their color.

With printing techniques the polymer is applied in little discrete spots. Printing is also a method to create a superficial coat onto substrates. If the printing is carried out in such a way that fine spots of the polymer are created onto the substrate, the resulting coat is also water vapor permeable .

In addition a thin printed layer applied onto a fabric in such a way that the textile is covered partially results in a porous coating [2, 5]. Even if a second layer is put on such a discontinuous layer the article remains air and water vapor permeable [3]. A coating on a raised fabric, where the coating layer partially penetrates the fabric, also results in permeable products [4, 7].

Fabrics with raised patterns are coated in such a way that only the raised parts are covered [8, 14]. A similar method is to coat a fabric then emboss it and remove the upper parts by buffing [9, 10]. If the solution applied as a coating also contains a blowing agent then, after coating and heating, the resulting foamed parts of the coating are buffed [12]. So a certain air permeability can be achieved by applying 0.12 – 1-cm coating layers consisting of PVC or polyurethane onto substrates [1] by a finishing operation and using a printing machine (see Chap. 22). The printing is carried out discontinuously [6].

A solution of a thermoplastic polyurethane is applied in a discontinuous manner onto the surface of a fabric or a nonwoven. After drying the coating is calendered by a hot roller to homogenize the surface [13].

Fibers consisting of different polymers, like methoxymethylated polyamide-6 and polypropylene or polyethylene are treated with a solvent for one of the polymers. As solvent a calcium chloride/methanol mixture has been described. Polyamide is dissolved in such a solvent mixture. A nonwoven is printed by such a mixture and the solvent is evaporated to form a discontinuous coating [11].

CHAPTER 15

Flocking

Velvet-like structures can be achieved by flocking. Thereby metal, wood or textile substrates are coated with an adhesive onto which the fibers are applied. The treatment of the surface of a substrate by fibers is carried out in an electrostatic field. Mostly small holes in the adhesive coat on the surface of the substrate result, thereby increasing its porosity. Flocking is carried out in the following way: a fabric, cardboard or leather is coated with an adhesive coat. Usually viscose fibers are either sprayed by air or in an electrostatic field [22] onto the coated surface. [9, 28]. Flocking [1] does not create eo ipso water vapor permeability.

Fibers with less than 10 den and at maximum 1.5 mm length [13] are drizzled onto the substrate. The energy with which the fiber touches the surface determines how deep the fiber penetrates the adhesive coat. Short fibers result (< 0.5 mm length) in a suede-like surface and longer fibers (ca. 1.5 mm length) in a plush-like surface. The fibers should be equally orientated vertically to the surface of the substrate. Therefore, nowadays, an electrostatic field is mostly used. Only fibers touching or, even better, penetrating the adhesive coat to a certain extent are fixed to the substrate. The surplus fibers are collected and may be reused. Figure 15-1 shows a schematic view of a flocking unit.

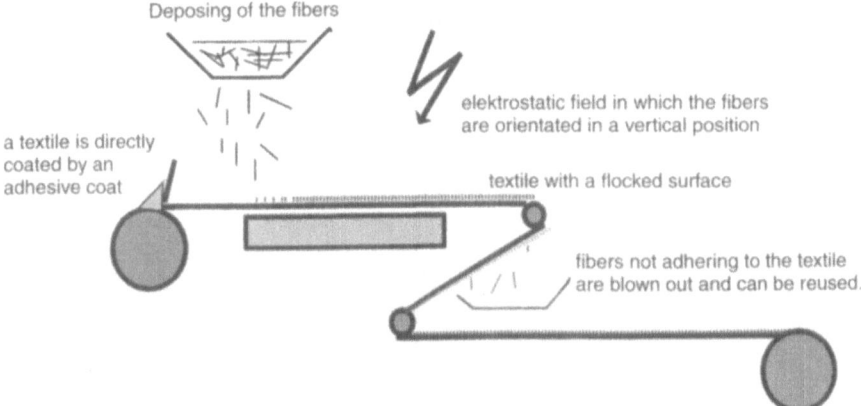

Fig. 15-1. Schematic view of a flocking unit

Generally, textile substrates like nonwovens, fabrics or knitted textiles are coated with an adhesive coat based on poly(meth)acrylate, rubber, soft PVC [21] or polyurethane [19, 30]. Cellulose, leather [15], collagen or wool fibers are used [3, 5–9, 11]. Textile substrates may also be flocked on both surfaces [12].

An indirect way of flocking is the following method: A release paper is coated with a water-based adhesive and then flocked. This layer is then transferred to a textile substrate by polyurethane. Then the material is torn off and the first water-based adhesive coat is washed off [2, 4].

Besides textile substrates, polyurethane foam [14] can also be treated with flock fibers to get a suede-like structure or polyurethane foils, which afterwards may be transferred to split leather [16, 18].

To increase the water vapor permeability of a material, for instance, of PVC, a 2–15% water-containing PVC plastisol is coated onto a release paper. Then hydrophilic fibers are applied by flocking. After that, another water-containing PVC plastisol layer may follow as well as a further flocking. Finally, textile is laminated, dried and peeled off the release paper. By this flocking with the hydrophilic fibers the water vapor permeability may be increased remarkably [17].

If the flocking is carried out on a microporous substrate and an adhesive coat consisting of a hydrophilic polyurethane, a suede-like product with an excellent water vapor permeability is obtained (Fig. 15-2 [26]). A rayon fabric having a flock fiber surface is transferred onto a layer of a polyurethane, dissolved in DMF, coagulated and dried to a thick microporous product [27].

Fig. 15-2. Fibers flocked onto a microporous surface which has been obtained by the selective evaporation process of reacting polymers

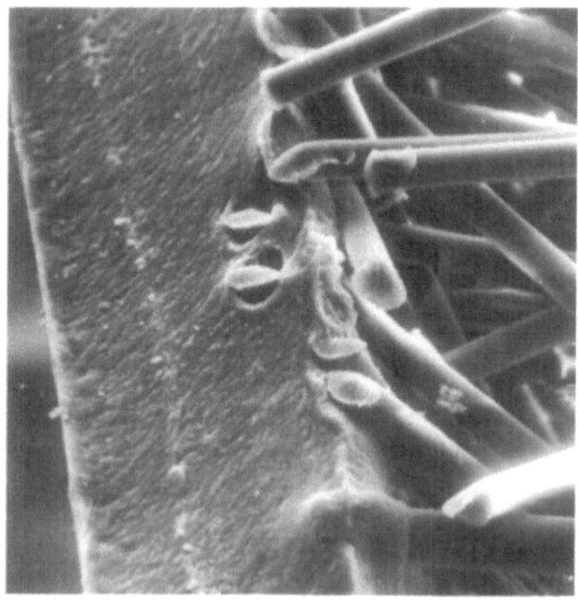

50 µm

A release paper is treated with a dispersion of a polyurethane containing a blowing agent or is mechanically whipped to a foam and flocked with fibers [20].

The tops of the fibers flocked onto a textile may be dissolved by an adequate solvent [10] or by the action of a solvent and by brushing defibrillated [25].

Flocking in an intermediate coat may improve the grain break of the surface layer: A textile substrate with a porous layer consisting of PVC is flocked with fibers and coated with a solution of a polyurethane in isopropanol/toluene [23]. A coat consisting of a water-based polyurethane dispersion is applied onto a release paper, 0.2 – 1-mm fibers of 1 – 2 dtex are flocked in, then a textile is laminated and torn off the release paper [24].

Prior to a (re)tanning operation leather fibres are flocked onto a textile substrate [29].

CHAPTER 16

Crystallization of Homopolymers

Homopolymers consisting of crystalline and possibly amorphous parts can be transferred to microporous products. A homopolymer, like polyethylene for instance, is treated with a swelling agent like perchloroethylene. Then the foil is stretched and the swelling agent removed [3, 20]. Homopolymers like polypropylene or polytetrafluoroethylene (PTFE) (9.3 [11]) are in a crystalline state at room temperature. If films are stretched a kind of mini-crack occurs, which remain as micropores. Polymers with a partial crystalline structure such as polyolefins, polyacetates or polyamides become microporous by stretching [21].

The same technique can be used to produce microporous fibers. Prior to spinning, calcium carbonate is incorporated into polypropylene. During the spinning, the fibers are stretched to five times their original length [17]. The microporosity of the fibers may be further increased by leaching of the calcium carbonate with a methanol solution of hydrochloric acid.

The stretching of homopolymers can be carried out by stretching of films in a biaxial way. Gore-Tex® (trademark of W.L. Gore and Ass.) is a stretched, microporous polytetrafluoroethylene (PTFE) (Fig. 16-1) [1, 2].

Naphtha and PTFE are extruded to foils. Naphtha is evaporated at 100 °C and the foils are additionally heated at < 350 °C. These foils, consisting of microporous PTFE, can be laminated with adhesives, based on hydrophilic polyurethanes (21.2 [13], [4 – 7, 11]), onto textile substrates. A unit to do this on a technical scale has been described [8]. As hydrophilic polyurethanes, isocyanate prepolymers, able to react with water or crosslinkers based on diamine carbamate, are also suited [9, 10].

Microporous products produced by stretching can be used in the production of shoes [15, 16]. The microporous PTFE films are usually laminated on a textile by means of a hydrophilic polyurethane. Additionally, the microporous PTFE film may also be filled by a hydrophilic polyurethane [23].

Films consisting of polyethylene are stretched in one direction. Then a liquid filling agent is applied to the surface of the stretched film. The filler should enter the pores; then the film is stretched again in a second direction and the liquid filler is hardened by heat. The product is suited for the production of protective clothing or as a separating agent for batteries [18]. Polyethylene containing other polymers, like PTFE or polyamide-6, where the melting range should be lower than that of the polyethylene, is extruded to a film and stretched to become microporous [19].

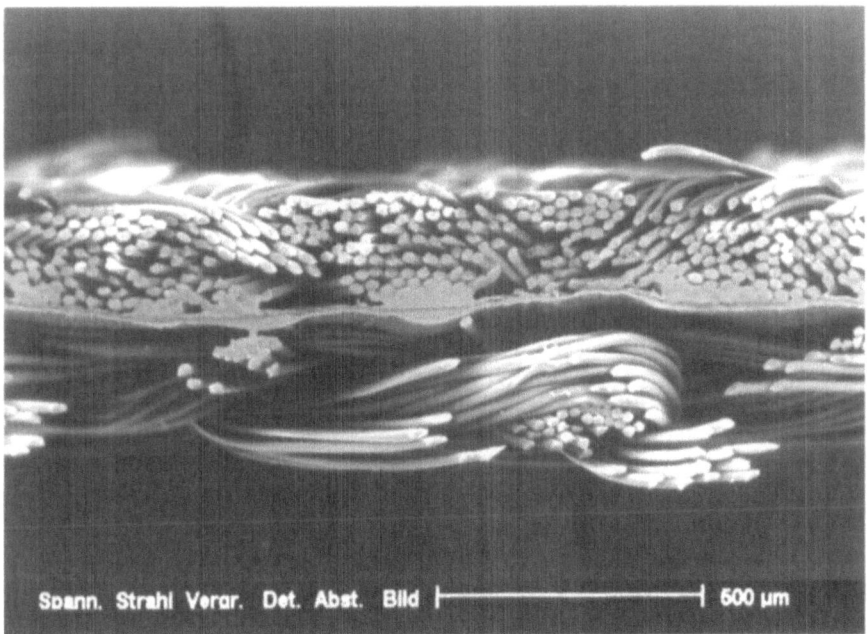

Spann. Strahl Verar. Det. Abst. Bild ├─────────────┤ 500 µm

Fig. 16-1. Gore-Tex. The microporous layer is embedded between two textile layers

Isotactic polypropylene or a statistically polymerized C_4-C_{10} α-olefin can be formed to a film, etc. by slush molding. The resulting products can be used as indoor panels for automotives [14].

Stretched microporous polyethylene films are printed with an isocyanate prepolymer, dissolved in butan-2-one so that then the solution enters the pores of the stretched polymer. After evaporation of the solvent, the resulting microporous film can be adhered to fabrics, which can be used in the production of gloves or wound dressings [12].

Water-repellent or hydrophobic stretched, microporous polymers are produced with fluorine-containing products like FREON® (DuPont) [13].

Nonwovens may be impregnated with a dissolved thermoplastic material like chlorinated polyethylene or vinylidene chloride–acrylonitrile copolymers. After the evaporation of the solvent the material is stretched to become microporous [22].

16.1
Mixtures of Incompatible Polymers

As homopolymers with crystalline segments mixtures of – preferably – incompatible polymers can also be used to create microporosity by stretching of films produced from them [1]. The only requirement is that the polymers are miscible at an adequate temperature. Polymers which are soluble with each other at higher temperatures and become no more miscible at stretching temperature are well suited. By stretching these then immiscible polymers, the interfacial areas being extended by the stretching operation are separated from each other and form micropores.

Examples of incompatible polymers are:

- polyurethane (100 parts) with polyethylene (10–60 parts) [7],
- polyamide-6 and polypropylene [10], and
- polyethylene and PVC [11]

Microporous panels are produced from solutions or plastisols of polyamide and PVC or epoxy resin and polysulfide, which are stretched [3]. These mixtures may also be applied to fabrics [2]. Microporous films are produced from plasticizer-containing vinyl polymers together with polyethylene or polypropylene [4]. PVC plastisol or polyurethane is mixed with a non-compatible polymer like polyethylene of a particle size of 8–420 μm. Then a film is produced. The film is heated, the incorporated polymer is melted and after cooling and stretching pores are produced [5].

A film is produced consisting of two incompatible polymers where the melting range thereof differs by at least 20°C [6]. The film is heated to a temperature between the melting range of the two polymers. The films are stretched in two directions to become porous [8].

Polyethylene (high-pressure type: HDPE) (60 parts), PVC (40 parts) and an inorganic filler (5–30 parts), like carbon black or chalk, are mixed, formed to a film and stretched [9]. Instead of a filler fibers can be incorporated in two incompatible polymers [8].

In alcohol a mixture of polyurethane, PTFE and a dispersing agent like polyaminoamide are ground to form a compound. The compound, which creates its microporosity by the incompatibility of the polymers, is used for coatings [12].

Polyurethanes containing 0.1–50 parts of PTFE are used for microporous coatings with high abrasion resistance [14].

Porous membranes consisting of PTFE and ethylenemethacrylic acid can be used for water tanks releasing their content slowly [13].

Two polymers insoluble in each other are mixed at 250 °C and extruded through a slit shaped nozzle to form a film. The polymers may be polyamide-6 and polypropylene. The film is stretched and then fibrillated to form a nonwoven [15].

16.2
Incorporation of Solid Compounds

Films containing inhomogeneous particles included in them also have a certain water vapor permeability. Inhomogeneous particles may consist of inorganic crystalline materials, polymeric gel particles, crystalline polymers or crystalline parts of the film-forming polymer itself [14, 16, 23] or as small cut fibers. It is usually necessary to create or increase this porosity by stretching, tumbling etc. [11].

Applications with Solutions or Dispersions. In a solution or a dispersion of a polymer an insoluble powdered polymer is mixed. The insoluble powdered polymer, which has a particle size of 0.05–0.1 mm, is stirred into the solution or the dispersion of a polymer. After the coating of a substrate the material is heated to a temperature which is able to melt the film-building polymer as well as the insoluble one. As the film-building polymer, polyamide or polyurethane is used and as an insoluble product polyurethane or polyacrylate in a powdered form [7]. It is also possible to produce a porous polyurethane film by including 1–30 parts of hollow spheres consisting of glass. The glass particles should have a particle size of less than 150 μm [2].

A releasing surface is coated with a polyurethane or polyacrylate solution prior to an electrostatic application of flock fibers of 5–20 mm length and adhered under pressure [26, 27].

A raised fabric is treated with an adhesive coat consisting of a solution of polystyrene in butyl acetate. Then a powder consisting of leather, asbestos or cellulose is applied and the material is plated. A leather-like sheet is produced in this way which is claimed to be water vapor permeable [18, 19]. It is also possible to apply the powder in the polymer solution [9].

A polyurethane solution containing small sized fibers is applied onto a textile coated with a PVC foam. The resulting material looks like man-made suede [20]. Oligourethanes containing some hydroxy groups, polyisocyanates and a catalyst are dissolved in an organic solvent. Then an insoluble inorganic powder is mixed in. The mixture is used to coat a nonwoven [24]. The same can be done with a solution of a polyurethane and trimethylammonium alkyl(meth)acrylic acid in mixture with an inorganic powder [8]. Crosslinked polyurethane coatings containing inorganic whiskers like TiN, BN or CB can be used as windbreakers [5].

Polyurethanes possibly containing some PVC are mixed with hydrophilic, fibrous or powdered materials. The resulting mixture, which may also contain some blowing agent, is dissolved or melted to form a film. After drying the film is stretched [3].

A PVC solution in dimethylformamide is mixed with powdered cellulose and water and a film is produced. Then the film is heated and gelated to a microporous product [21].

A latex or a solution of a vinyl chloride–vinylidene chloride copolymer is mixed with a powder of a polymer such as PVC, a plasticizer, a hydrophilic product like an ethoxylated fatty alcohol and a solvent like methylcyclohexanone, etc. The mixture is coated onto a cotton fabric and heated to 145 °C [17].

A chloroprene–methylmethacrylate copolymer latex is sprayed onto a textile substrate, then wood, leather or textile dust is applied. This action may be repeated several times [25].

100 parts of water-insoluble, powdered polymer, like polyamide-6, are dispersed in 3–300 parts of a binder, like a polyurethane dispersion. This mixture is applied on supports like a nonwoven consisting of polytetraethylene glycol terephthalate, and dried to a microporous coat [31].

Applications in the Solid State. A polymer like PVC or polyurethane is mixed with small sized particles consisting of PVC, vulcanized rubber, sawdust or carbon, then melted to a film and at the same time laminated to a textile substrate. The material is stretched then in biaxial directions to create a micro-porosity [10].

15–50% of an inorganic filler like calcium sulfate with a particle size of 1–100 μm is mixed with 15–85% of a polyolefin like polyethylene or -propylene. The mixture is calandered and 1 to 5 times of its length stretched [1].

A polyolefin like polyethylene or polypropylene with a melting index below 10 is mixed with 50–90% calcium sulfate semihydrate then transferred to a film which by flexing becomes microporous [22]. Polyethylene–polypropylene block polymers and calcium carbonate may be transferred to a microporous film which may be used in ultrafiltration or as a membrane for reverse osmosis [29].

A PVC plastisol consisting of 30–70 parts of PVC and a plasticizer is mixed with 1–5 parts of polypropylene fibers with a size of 50–200 μm. A coating is made out of this and applied onto the textile substrate then the gelation is carried out. Possibly an intermediate coat consisting of the same plastisol with some blowing agent, like azodicarbonamide therein, can be applied additionally [12].

A rubber, which can be vulcanized, may be mixed with small sized solids as asbestos or powdered cotton to create microporosity [13]. Elastomers like acrylic craft polymers on polyurethanes are mixed with crystalline polymers such as polyamide or polyester [15], powdered glass fiber, magnesium sulfate or dia-tomaceous earth, then heated and stretched. The included powders should not exceed a particle size of 10 μm [30].

Thermoplastic elastomers are mixed with 60–150% sodium bicarbonate and 3–15% carbonic acid, like oleic or palmitic acid, prior to a calendering action. The resulting mixture can be leached [28].

Polyester, polyamide, polyolefin or vinyl polymers are treated prior or after the polymerization with 1–25% of finely dispersed materials like chalk, glass or mica and transferred to a film which may be stretched at 80 °C. It is possible to write on the surface of such a film [4].

CHAPTER 17

Precipitation of Polymers in Water-Based Dispersions

The precipitation of dispersed monomer, oligomeric and polymeric compounds in water is easily achieved by changing the pH value of the dispersion or solution or by increasing the quantity of ions by adding a neutral salt.

The best known example of such a precipitation for a non-expert is the coagulation of milk. Milk is a rather complicated mixture of fatty and proteinic substances in water. Milk separates after the addition of citric acid or vinegar into a clear solution and curd.

Polymers are mostly dispersible in water due to a cationic or anionic charge or by means of emulsifiers often bearing such charges. The addition of an acid to a negatively charged emulsifier or polymer or of an alkaline substance to a cationic charged substance results in a coagulation of the dispersed system. The addition of salts may also create a coagulation.

This effect is used broadly in the production process of polymers. Polymers polymerized in a water-based emulsion are very often treated with acid to precipitate them into the solid state [1, 5, 6].

Another method is to mix a coagulant into the water-based dispersion, which precipitates with increased temperature. Such coagulants are, for example, substances with a good solubility at room temperature and a poor one at a higher temperature (ca. 60 °C). By increasing the temperature they coagulate and coprecipitate the dispersed polymer. This technique is broadly used in the binding of nonwovens (see Chap. 21). This method may also be used to create microporous coatings [8].

A polymer dispersion mixed with a polysiloxane is coated onto split leather. By raising the temperature the polymer precipitates by coagulation [3].

Salts or electrolytic compounds able to build an insoluble salt with the ionic group of the polymer are especially effective coagulants. A polymer bearing carboxylic groups like a sodium polyacrylate solution may be coagulated easily with an aqueous calcium chloride or aluminum sulfate solution. A dispersion containing a polymer consisting of 37 % styrene, 60 % butadiene and 3 % acrylic acid, mixed with a sodium polyacrylate solution may be used to impregnate a nonwoven. Afterwards this is dipped into a bath of calcium chloride or aluminum sulfate and coagulated therein [2] (for other examples, see Chap. 21).

A coating of a cationic polyurethane dispersion, then top-coated by an anionic polyurethane, results in a water vapor permeable coating [7]. For a coating, an anionic polyurethane mixed with a Neoprene® (trademark of DuPont) latex which coagulates can be used [9].

A fabric is put into a mold and warmed up. The fabric is dipped into a PVC emulsion containing polyvinyl methyl ether or polysiloxane as a coagulant. The PVC emulsion coagulates to a porous coating [4].

17.1
Bonding of Nonwovens

Most nonwovens (see Chap. 21) need to be bonded or filled with a polymeric substance. Woven or knitted textiles differ in their elasticity or tensile strength if they are stretched in the working direction – the direction during the manufacturing process – or against the working direction. This effect is an anisotropy which nonwovens also have (Fig. 21-5). Bonding reduces this effect or equalizes it totally. Bonding has another positive effect on textile substrates. If bonded textiles are knitted they break in a round smooth manner, a manner which is preferred.

The polymeric bonding material should ideally be positioned at the cross points of the fibers (see Figs. 17-1 and 17-2) to get a good isotropic behavior of the processed textile.

The bonding agent is normally coagulated in the nonwoven. Alkoxylated amines with an inverse solubility (good solubility at room temperature and insoluble at higher temperature) (see Chap. 17) are responsible for such a thermal coagulation [1]. Besides a difference in pH or an increase in temperature [2, 15], even with an open fire [11], freezing [5, 13] or high frequency [8, 18] may also induce coagulation.

After impregnation of a nonwoven by a polymer latex, exposure to vapors of organic C_1–C_{12} amines also causes coagulation [10]. Sodium silicofluoride (see Sect. 11.1) [6, 12], methyl polysiloxane [14] solutions of electrolytes or organic solvents [7] may also be used to coagulate a latex. N,N-Disubstituted alkyl sulfimides are known as coagulants which can be degraded biologically [17]. Mixtures of alkoxylated amines with 1,1,2,2-tetrafluoroethane are able to coagulate a polyethyl acrylate dispersion [23].

As will be shown later (see Chap. 21) nonwovens normally contain fibers with a thermal contractility. The nonwovens are heated up to induce the thermal contractility and to increase their density. The thermal coagulation may, therefore, be combined with such a shrinking process [9, 16].

Water-based dispersions of polystyrene containing a volatile organic solvent with a boiling point of 40–95 °C, like isopropyl acetate, can be used to impregnate a nonwoven. The polymer dispersion coagulates by immersing the nonwoven in a water bath of 50–70 °C [3]. Mixtures of a Neoprene® latex and a polyurethane dispersion may be used to bond a nonwoven to produce a material suited for the production of shoe uppers [20].

Emulsions of natural rubber, polyacrylates, polychloroprene or copolymerisates of butadiene, styrene and acrylonitrile and/or (meth)acrylic acid can be easily coagulated by the addition of 0.2–8 % of a thermal sensibilizing agent like polyvinylalkyl ether or polysiloxane–polyalkylene oxide block polymer [8].

Polymer particles which are redispersible are dispersed in water. The redispersability depends on an emulsifying agent which loses its dispersing ability by

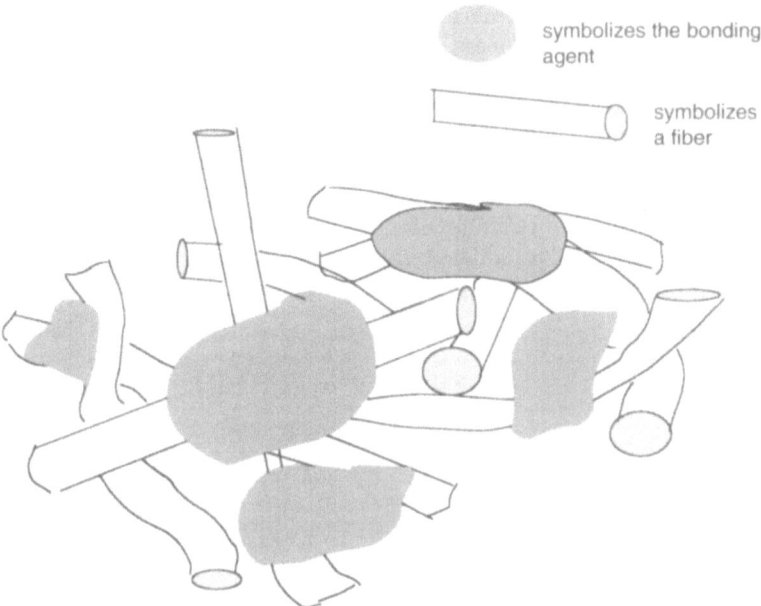

symbolizes the bonding agent

symbolizes a fiber

The fibers in a nonwoven are then bonded in an ideal way when the bonding agent is deposited in the cross-points of the fibers.

Fig. 17-1. Bonded fibers in a nonwoven (model)

Fig. 17-2. Bonded fibers in a nonwoven (photo)

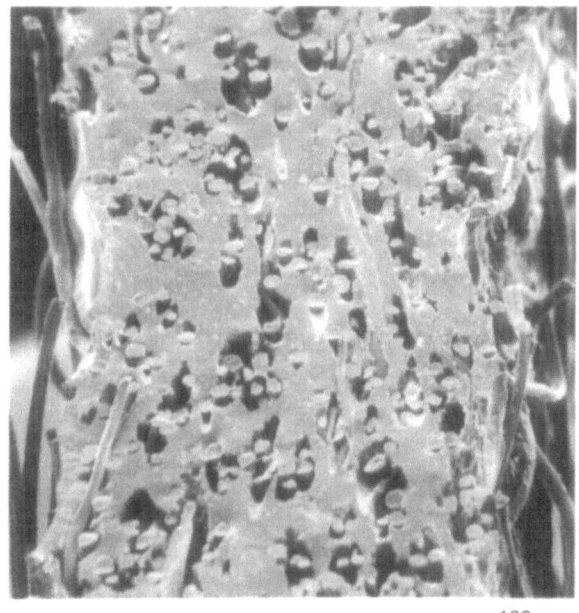

100 µm

a treatment with UV rays. Monomers which can be polymerized by the action of light can also be used in this process [19].

In an inverse process, a microporous polyurethane layer is produced (see Chap. 10). During this reaction an unbonded nonwoven is laminated onto it. After termination of the polyaddition, the reverse side with the unbonded nonwoven is bonded by an acrylonitrile–butadiene copolymer latex containing 10–100 % of a polyurethane dispersion [4].

Nonwovens with a certain combination of fibers like polyamide and rayon bonded by a coagulated polyurethane [21, 22] or nonwovens consisting of polyethylene and polyamide-6 fibers [29] are known to look especially like genuine leather.

CHAPTER 18

Hydrophilic Polymers

In nature hydrophilic products like protein, keratin or wool are responsible for water vapor permeability. The easiest way to obtain water vapor permeability for a chemist should be the of use a hydrophilic polymer. There are a lot of existing polymeric materials which are hydrophilic, e.g. materials of natural, semi-natural or of synthetic origin. Examples of those hydrophilic materials are proteins, cellulose, polyethylene glycol ethers, polyamides, polyacrylic amides, polyurethanes with polyethylene glycol ether soft segments, ethoxylated graft polymers, etc. Polymers which have a nonhydrophilic structure may be hydrophilized by mixing them physically or reacting them chemically with hydrophilic products or components. For example, ethoxy or polyacrylamide groups increase the hydrophilic properties of a polymer.

Under ordinary conditions apparel and shoes are exposed to rain or water. If an article consists only of a hydrophilic polymer – whose water vapor transport occurs by ab- and desorption of water – this article would swell by the water-(vapor) adsorption. Due to swelling the polymer is softened and loses its physical properties such as rub and bonding resistance. Therefore shoes or apparel exposed to water would change their surface and would not be resistant to wearing conditions. It is therefore not possible to just take any hydrophilic polymer to get water vapor transmission. A lot of experimental work was necessary to create good solutions for water vapor permeable materials with the appropriate wet properties.

As previously stated, hydrophilic polymers swell when they get wet. An outside coating of a textile worn in rainy weather would swell with every drop touching the surface of the coating (see Fig. 18-1). The customer would never accept a shoe or apparel with such wearing properties.

Hydrophilic polymers can be water vapor permeable due to ionic groups such as acid or – better yet – being in the form of a salt like a sulfonate, ammonium or carboxylate group or by hydrophilic elements like ethoxy ethers or amide segments [2].

An example of a polymer being hydrophilic due to ionic groups is polyacrylic acid, which by the addition of ammonium or sodium hydroxide becomes extremely well soluble in water. When this polymer is slightly crosslinked it builds a pH of more than 6 with water, gels or becomes so viscous that it can be used as a thickening agent.

Polyacrylamide has hydrophilic segments and – due to a partial hydrolysis to carboxylate – ionic groups. Polyacrylamide is used as an absorber of water in diapers.

Fig. 18-1. A coating consisting of a hydrophilic polymer with localized swelled spots by droplets of water

An example of a polymer being hydrophilic due to hydrophilic segments is carboxymethyl cellulose and polyvinyl alcohol. These two polymers are used as thickening agents in ice cream as an example.

Polyacrylamide, often described as a "superabsorber" (for water), is able to absorb water by more than 100 % of its own weight without getting wet. Polyacrylamide is, for example, today broadly used for hygiene articles. In the literature this polymer has almost not been mentioned as a polymer or an additive for water vapor permeable materials [5, 9]. Perhaps its future for this purpose is to come.

Hydrophilic properties – due to ionic groups – also depend on the cations used. The capacity of water absorbency of polyacrylamide, for example, is higher if sodium ions are used instead of potassium ions. Both salts, the sodium as well as the potassium salt bind more than three times the quantity of water than the free acid itself [6].

The water (vapor) transport through hydrophilic membranes functions by ad- and desorption of the water – depending on the difference of the water vapor pressure on the two sides of the membrane. Model experiments show that this phenomenon is a rather complicated process.

Water absorbed in hydrogels of polyacrylamide, polyvinyl alcohol or dextrin results in two different forms as opposed to normal water: in a melting and a non-melting form. The non-melting form of water does not melt at a sharp melting point: "non-melting" water has a melting range. The explanation for this could be that water is partially bound in a polar form; this water would melt at a lower melting temperature than free water. The other part of the water is not bound as fixed as the first one: this water is in the outer sphere of the ionic structures. This water is in a less organized structure – its crystalline state is not formed as well. Therefore its melting point is slightly below 0°C. The rest of the water melts at 0°C due to the fact that it is present in the polymer in the free form [7].

A way to improve a hydrophilic material in its wet properties is to crosslink the polymer. In leather finishing crosslinking is the state-of-the-art today. Otherwise the necessary wearing behavior in wet conditions could never be achieved [1].

Another possible way to avoid this localized swelling is to wear the hydrophilic layer under a water-resistant textile, or – even better – between two textile layers as a *liner*. Sympatex® (trademark of Akzo Nobel) is such an article. Under a water impermeable outer layer and a lining textile is a hydrophilic polyester film. Gore-Tex® (trademark of Gore) has more or less the same structure with the only difference being that instead of a hydrophilic polyester film a microporous polyfluoroethylene film (see Chap. 16) is used [3]. This type of liner guarantees a high water vapor permeability, good insulation against heat and cold and a wind-protection effect. The wearing comfort of apparel having these liners is very good. This technique has a broad use in the manufacturing of sports and functional clothing.

A situation where the swelling caused by water plays a minor role is the use of hydrophilic polyurethanes as adhesives for poromeric films (see 16 [3]). Adhesives usually never come into close contact with liquid water.

Laminates only able to transport humidity in one direction can be achieved by coating a highly permeable film of polyvinyl alcohol by using almost nonpermeable phenoxy resin crosslinked by a tris(isocyanatotolylurethane) of trimethylolpropane [4].

Hydrophilic coatings for heat exchangers are produced by a coagulated mixture of aluminum hydroxide with a copolymer of hydroxyethyl acrylate, acrylamide and 2-methylpropene sulfonic acid. When heated at 250°C these kind of coatings on aluminum panels give a high corrosion resistance [8].

Water-based dispersions of polyesters with hydrophilic sulfonate or carboxylate groups are water vapor permeable [10].

Fabrics, foils or fibers consisting of polyamide are treated at a high temperature for 24 h with potassium hydroxide containing some propanol then, after cooling, washed and heated under acidic conditions to become a significantly better hydrophilic product. The product is used for diaphragms [11].

The water vapor permeability of a PVA membrane depends on its degree of crosslinking [17].

A dissolved polyamide whose nitrogen atoms are substituted by methylol or methylol ether groups are used as water vapor permeable coatings [14].

Nonwovens become reversible water absorbent if their bonding agent is a graft polymer of ethylene oxide on acrylic acid [12]. A nonwoven containing hygroscopic fibers is treated with humidity and bonded [18]. Nonwovens produced by spinning endless fibers are bonded with a latex containing a swelling agent like carboxymethyl cellulose [13].

Coatings with crosslinked polyvinyl alcohol (PVA) are water vapor permeable. The degree of formylation or acetylation of the PVA is important for the water vapor permeability [22]. A degree of acetylation of 60–65 mol% is advantageous [20, 21]. Polyvinyl alcohol, together with an acetylating agent, may also be used as a coating product. The product acetalyzes at a high temperature [15].

Hydrophobic copolymers based on butadiene, isoprene or chloroprene containing 10–50% of a hydrophilic monomer like acrylamide, maleic anhydride or acrylonitrile, may be used as bonding agents for nonwovens [13].

Vinyl ether polymerisates which can be crosslinked are able to absorb water vapor reversibly [16].

Polymerisates with at least 5% of an ester with an ethylenic unsaturated carbonic acid and a C_4–C_8 alcohol ad- and desorb water vapor easily. At higher temperatures these esters decompose and then the olefin is split off, resulting in a foaming action [19, 13]. Vinylphosphoric acid, vinylsulfonic acid or p-styrenesulfonic acid may also be copolymerized [19].

One or more breathable layers of Sympatex® (trademark of Akzo Nobel) are bonded with a reinforcing fabric, e.g. polyamide Charmeuse. The segments of the material are cut and sewn together. Then the seams are sealed with an adhesive tape. The fabric is prior to lamination impregnated by immersing in a water-based emulsion of a polyester polyurethane [22].

18.1
Incorporation of Hydrophilic Materials

There has been a lot of research done to obtain a compact, homogeneous impregnation, coating or film permeable to water vapor but impermeable to liquid water by mixing a hydrophobic polymer with a hydrophilic ingredient. This is more difficult than it seems to be prima facie due to the fact that it is almost impossible to get the hydrophilic component into a film or coating so that it is able to show its hydrophilic property. Normally this ingredient is so embedded by the hydrophobic film-forming material that it never has a chance to come into contact with water or its vapor.

In many publications where an increase in water vapor permeability by embedding hydrophilic ingredients is claimed, the amount of increase is not shown clearly or only shows if these materials are stretched in a manner that the embedded hydrophilic components lose their close contact to the film-forming non-hydrophilic polymer. Therefore these processes are discussed without an evaluation from the author's side about the effectiveness of the methods used.

Polymers with a low water vapor permeability like PVC certainly become a little more porous if 5–20% of high molecular substances like starch, cellulose ethers or esters, casein [3], gelatin or leather dust [29], polyvinyl alcohol, alkoxy-(meth)acrylate [28], collagen [48, 50] or polyamide are mixed in. Besides these hydrophilic ingredients, capillary active substances and blowing agents are also added [1]. Wool fibers may act as capillary active substance [2, 7]. Active carbon [4] in moss rubber or diatomaceous earth in PVC [5] or finely powdered silica in polyurethane coatings [6] are also claimed to raise the water vapor permeability.

A release paper is coated with a finish solution and dried. Then a water-based polyurethane dispersion containing 10–100% of a hygroscopic material such as silicium dioxide powder gel is applied followed by a lamination of a cotton fabric. After stripping the release paper a man-made leather is obtained [27].

Applications Together with the Film-Forming Materials. In many cases the hydrophilic ingredients are active due not to their hydrophilic property, but to an nonhomogeneous film forming because of their presence. In many cases added hydrophilic ingredients build channels through the film because their size corresponds almost to the thickness of the film or the mixing in was not done accordingly. Polyethylene, for instance, is claimed to become permeable to water vapor by the addition of 10 parts (per 90 parts of polymer) of a *breathability provider* such as ultrafinely powdered keratin of feathers or hair. The powder should have a particle size of 0.5–10 μm [49]. PVC with fibers of a length of 3–5 mm is transferred to a film which may be laminated onto a textile substrate to form a man-made suede [10].

A binder, a ferromagnetic material and finely ground, wet, powdered leather is applied onto porous substrates. Then an electromagnetic field or impulse is applied to the material. The result is a man-made leather of an extremely high stability in the wet stage [12].

100 parts of a polymer, like a vinyl polymer or plasticized PVC, together with 5–80 parts of a multivalent metal salt of Pb, Zn, Ti, Cd, Al, Ba or Mg of a *N*-substituted aminocarbonic acid, like *N*-fat-sarcosin or *N*-fat-β-iminodipropionic acid may be transferred to water vapor permeable films or laminates [9].

100 parts of synthetic or natural rubber become permeable to water vapor if they contain more than 10 parts of glue and/or gelatin and more than 10 parts of a hydrophilic, inorganic filler [11].

Pastes, dispersions or emulsions of elastomers like PVC, polyurethane or poly(meth)acrylate can be used for porous coatings or films if they contain 5–60% (based on polymer) of leather dust of a particle size of 0.0001–0.2 mm [5]. Adhesive coats containing leather fibers for microporous topcoats are also water vapor permeable [55].

Membranes consisting of PVDC, PVC, polyvinyl acetate, polyolefin, polyacrolein, polyvinyl ether, polyamide, cellulose ether, ester, polyester, protein, natural or synthetic rubber containing 25–200% of a material which increases the water vapor permeability, like cellulose or bentonite, become permeable to water vapor [17]. Polymers like PVC, polyethylene or natural rubber containing a plasticizer are treated with water-soluble starch. The starch should have a particle size of 100–200 mesh. The particle size should be similar to the diameter of the film to be formed. The starch particles create channels through the film. These channels guarantee the microporosity [22]. PVC, plasticizer, pigments, stabilizing agents, a hydrophilic filler like the alkali(earth) salt of polyacrylic acid and an azeotropic organic solvent mix are added to a paste. The paste is used for a coating. The solvent mix, which is able to swell the filler, is evaporated; the resulting coating is porous [37].

Permeable foils for food packaging consist of Li, Na or K salts of ethylene-(meth)acrylic acid copolymers. The foils are permeable to air and water vapor [32].

Films or articles consisting of polyester or polyether polyurethane are water absorbing if they contain materials like collagen or acrylic polymers [47].

Graft polymers of diacetone(meth)acrylamides on polyurethane can be used for water vapor permeable coatings [34].

Collagen powder, polyurethane, carbon black, Desmophen® (trademark of Bayer for polyethers or polyesters), a catalyst and an isocyanate are ground together, then applied onto a surface [46]. Thermoplastic polymers, like polyurethanes or PVC, become hydrophilic by the addition of powdered silk [18].

Applications after Film Forming. A fabric raised on both surfaces is treated in the stretched stage with an adhesive coat. Then leather fibers are spread on it. The material is pressed and shrunk in hot water [24, 25]. Solutions or dispersions are applied on textile substrates. Hygroscopic fibres are flocked onto these substrates. The powdered fibers are swollen. After evaporation of the dispersing liquid and the swelling agent a porous coating is created [26].

A PVC layer with open pores is powdered with cork dust, laminated to a textile substrate and pressed to reduce its thickness to 1/2 or 1/10 of its original size [23].

A coating mixture for nonwoven or split leather consists of a polyurethane dispersion and 10 – 15 % of an organic filler like leather dust or dyed elastomeric PVC particles [14, 16].

Coatings or impregnations may be obtained as hydrophilic by a polyreaction of polymers of an opposite charge. Especially suited are polyelectrolytes like sulfonated polyvinyl alcohol, polyethyleneimine or polyacids [38].

Applications as Bonding Agents. Nonwovens consisting of polyamide fibers are impregnated with solutions of polyurethanes containing 1–500-μm fibers of cellulose, polyvinyl alcohol, starch, or particles of a polyurea or an inorganic filler. Afterwards these nonwovens may be coated to leather [8].

Nonwovens produced by a spinning process (see Chap. 21) consisting of 0.5 – 10-dtex polyester, polyamide-6, -6.6 or polyolefin fiber may be bonded by a latex consisting of a vinyl ester, polyurethane or butadiene–acrylonitrile copolymer containing carboxylmethyl cellulose or natural gum as a hydrophilizing agent [15].

A bonding agent for fabrics or nonwovens consists of a synthetic or natural rubber in a perchloroethylene solution containing powdered cork, glass, cement, etc. [21]. A nonwoven may be treated with polyethyleneimine prior to a latex [20]. A latex containing 20 – 100 % of polyvinyl alcohol, which may be acetylated afterwards becomes waterproof and water vapor permeable [30].

A nonwoven containing hydrophilic fibers [51] is treated with humidity, bonded by an elastomer solution and a solvent, then the humidity is evaporated. Then the nonwoven is coated with a paste containing hydrophilic fibers also [36].

Nonwovens containing 2 – 50 % of a bonding agent may become water vapor permeable by treatment with a water-based gelatin solution. Afterwards the gelatin is hardened with glutaric aldehyde [39].

Nonwovens are able to absorb water vapor in a reversible manner as long as the bonding agents contain a polymer with at least 15 mol% carboxylic groups [41].

Aziridine propionic ester or succinic ester whose alcohol component consists of a C_2-C_6 glycol or a diamine with 6-100 ethylene oxide units polymerized to it, may be used, together with the usual bonding agents, to bond nonwovens with a high reversible water vapor absorbency [45].

18.2
Hydrophilic Polyurethanes

Hydrophilic properties should ensure that a polymer can be applied as a homogenous coating. Water vapor could migrate through this homogenous film by means of adsorption and desorption. Such polymers could be applied to a textile substrate as an outer coating and be used to manufacture thin and extremely lightweight garments. Hydrophilic polymers take up water, which acts as a plasticizer, and weakens the resistance of the film to abrasion and rubbing. Worse yet, when drops of rain come into contact with the polymer, they cause localized swelling and produce unsightly blisters at the point of contact (see Fig. 18-1). These often remain once the rain has stopped, a situation which would lead to complaints from consumers.

The technical difficulties associated with the production of microporous coatings do not apply to hydrophilic products, which can be processed like conventional products, e.g. in the form of solutions, high solid brands etc. On the other hand, fastness properties with exposure to water play a decisive role when using hydrophilic products: swelling of the polymer up to total solubility, the formation of blisters on exposure to droplets of water and a reduction in rub fastness, scratch resistance and flexing fastness after exposure to water would reduce or even rule out the usefulness of such products. The water vapor permeability of simple hydrophilic polyurethanes is retained by incorporating a polyethylene oxide ether chain in the soft segment. The longer the ethylene oxide chain in the polyurethane, the higher the water vapor permeability.

With an increasing amount of polyethylene oxide soft segments in the polyurethane, the hydrophilic property of the polymer also increases [15, 18 (2), 20, 31, 32, 35, 38 - 40, 43, 50 - 54, 56, 57, 72, 79] (see Fig. 18-2). With increasing hydrophilic properties the swelling of the polymer also increases. One possibility to reduce the swelling of a polymer is to mix it with a hydrophobing agent [45].

The thickness of a hydrophilic film plays an important role in the transportation of water vapor through it: The thicker the film the slower is the transportation of water vapor which results in a lower water vapor permeability.

Polyurethanes with polyethylene oxide soft segments can be produced as high solids (18 [2]) or as two-component reactive systems [34, 46] or as soluble in organic solvents [69].

It has been reported that thermoplastic polyurethane coatings [77] containing polyethylene glycol soft segments can be transferred to films by extrusion or blown film techniques. The films are laminated to textile substrates by screen printing of a hot melt adhesive. The resulting product is especially suited in the production of protective garments [78].

Hayashi and Ishikawa examined ([58], 25.4 [22]) the influence of different hard segments and the molecular weight of the polyethylene oxide soft segments

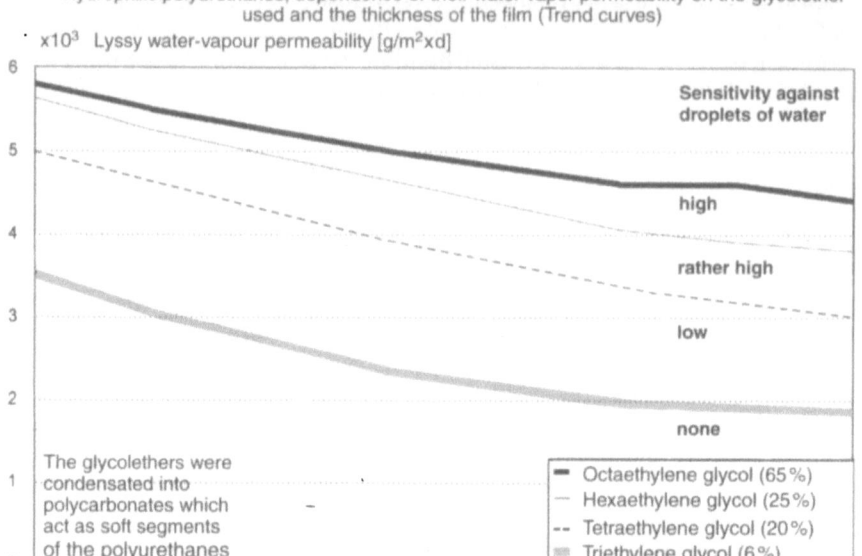

Fig. 18-2. Polyurethanes differing in their hydrophilicity. The water vapor permeability is demonstrated as a function of the glycol ether polycarbonate in one-component polyurethanes and the thickness of the film (18 [2]). It is demonstrated that the swelling-resistance against water droplets of films produced by one-component polyurethanes is bad as long as their water vapor permeability is high. Only films of a medium water vapor permeability have a sufficient resistance against swelling by water droplets

on the physical properties, such as tensile strength, etc. of the resulting polyurethane. They found that the degree of water vapor permeability depends directly on the quantity of polyethylene oxide units in the polymer. An increase in the molecular weight of the polyethylene oxide units also contributes positively to the water vapor permeability. Surprisingly, the degree of water vapor permeability, measured above the glass-transition temperature (T_g) of the polymer, was higher than a measurement at a temperature below the T_g. The authors explain this phenomenon with the Brownian movements of the soft segments being higher above the T_g. Due to the fact that in polyurethanes with polyethylene oxide soft segments only these soft segments are responsible for the water vapor transport. A higher oscillation of these segments results in higher water vapor permeability.

A model for a polyurethane hydrophilized by polyethylene oxide soft segments alone is shown in Fig. 18-3. To get a good water vapor permeability, a large quantity of these polyether soft segments is needed. On getting wet the hard segments consisting of urethane or urea groups are unable to keep the polymer in shape because the soft segments swell under the influence of water.

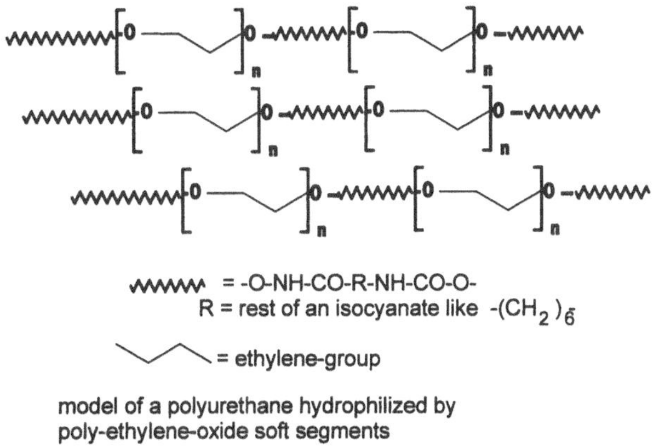

Fig. 18-3. Model of a polyurethane hydrophilized by polyethylene oxide soft segments

www = -O-NH-CO-R-NH-CO-O-
R = rest of an isocyanate like $-(CH_2)_6$

= ethylene-group

model of a polyurethane hydrophilized by
poly-ethylene-oxide soft segments

To reduce the swelling of an hydrophyilic polyurethane by liquid water the idea was born to build polyurethane(polyurea)s with two different hydrophilizing elements. Polyurethanes with polyethylene oxide soft segments received a second hydrophilizing element in the form of sulfonate groups in the hard segments. It was discovered that incorporating even comparatively weak hydrophilic components into *both* the soft *and* the hard segments resulted in a remarkable enhancement of water vapor permeability with low swelling and good drop repellence. *Dual hydrophilic* polyurethanes were created:

- Few polyethylene oxide groups in the soft segments create a slight hydrophily which is increased by
- ionic groups in the hard segments.

By this effect the soft segments are no longer as hydrophilic that under the influence of water they swell so much that the polymer loses its shape. The dual hydrophilizing in the soft and the hard segments has a synergistic effect. With – compared to the quantities needed – much fewer hydrophilic components the hydrophilicity is generated. A model of dual hydrophilic polyurethanes is shown in Fig. 18-4.

To get a better understanding of the principle of dual hydrophilicity three different products were investigated (Tab. 18-1, Tab. 18-2, Fig. 18-5). Three defined quantities of water, i.e. 10, 20, and 30 %, were added to films of the three polyurethanes in Fig. 18-5. Calorimetric measurements show how much water is absorbed by the different films and where the absorbed water is localized:

- One part of water is dissolved in the polymer in a molecular form; the amount increases according to the amount of polyethylene oxide groups in the polymer. The type PUR 1 may absorb higher amounts of water in the dissolved form than the other two types of polyurethanes.
- Another part is absorbed in the form of clusters. The amount of absorbed water depends on the amount of sulfonate groups present in the polymer.
- If the quantity of water offered exceeds the amount which can be absorbed, this water is present in the polymer in the form of free water, like fine droplets or a thin water film on the surface.

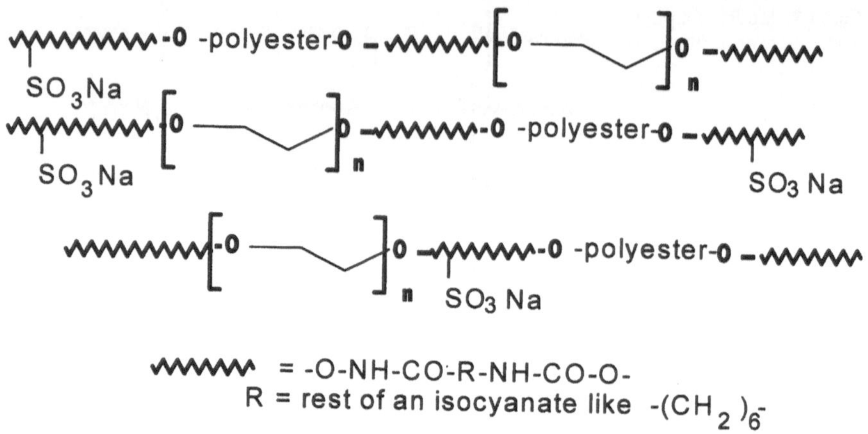

$$\text{WWWW} = \text{-O-NH-CO-R-NH-CO-O-}$$
$$R = \text{rest of an isocyanate like } -(CH_2)_6^-$$

Fig. 18-4. Model of a dual hydrophilic polyurethane

Table 18-1. Investigated polyurethanes

Polyurethane	Quantity of hydrophilic poly-ether (poly-ethylene-oxide)	Quantity of ionic groups	Water vapor permeability	Swelling by a droplet of water
PU 1: only hydrophilized in the soft segment	very high	none	high	very poor
PU 2: dual hydrophilized	high	high	high	good
PU 3: dual hydrophilized	lower	lower	sufficient	excellent

Depending on the type and manner of hydrophilicity a typical degree of water absorption takes place (Fig. 18-6). This is proof of the different water vapor permeability and the different sensitivity against liquid water (drop repellence; see Fig. 18-7).

Dual hydrophilic polyurethanes need less hydrophilic groups than polymers with only one type of hydrophilicity (see Fig. 18-8).

Water is adsorbed rather rapidly by a dual hydrophilic polyurethane (see Fig. 18-9). The same occurs in the desorption (see Fig. 18-10) [30]. Dry films absorbed approx. 10 % by weight of water, with saturation occurring after approximately 100 min. Water vapor was also desorbed very quickly, in less than 75 min. These effects are very important for the wear comfort of clothing.

Dual hydrophilic polyurethanes are not only of interest for shoes and apparel – they may also be applied to protect glass, e. g. goggles, against "misting" (see 27 [25]).

Table 18-2. Comparison of the water vapor permeability and the drop repellence of a normal hydrophilic polyurethane and a dual hydrophilized one

	Application in g/m^2 (type of product)	Solable PU strongly hydrophilic (Polyethylene-oxide type)	PU weakly hydrophilic (Polyethylene-oxide-type)	Dual modified PU2 and PU3	
Water vapor permeability of the film (g/m$^2 \cdot$ d) measured at Bayer AG by the silica gel rapid test = SST methods, which is similar to the method specified in publication DS 2109 TM 1 of the British Textile Technology Group, Manchester, UK		14,500	1,300	16,800	10,700
$R_{et} \cdot 10^3$ (m$^2 \cdot$ mbar/W)	40 (liner) 60–65 (transfer coating) 30 (direct) coating)	70	540	50 75 80	140
Drop repellence (1 = good; 8 = poor)		7–8	1	2	2

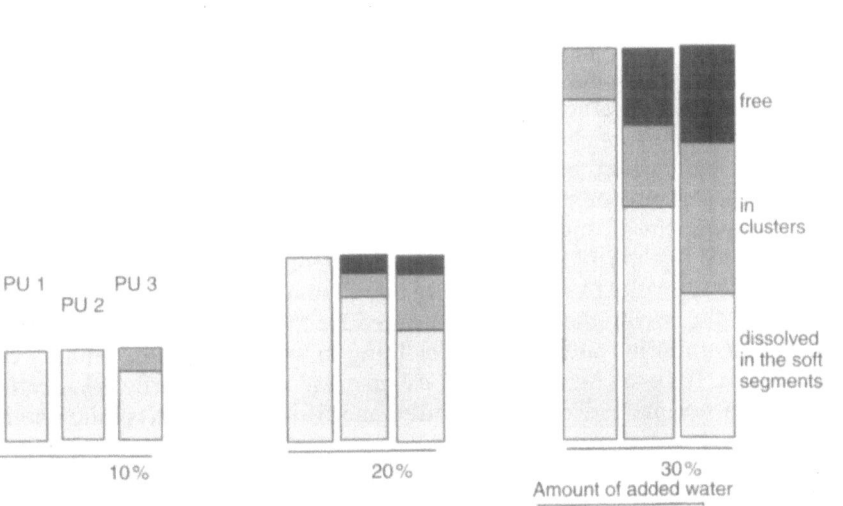

Fig. 18-6. The distribution of the water

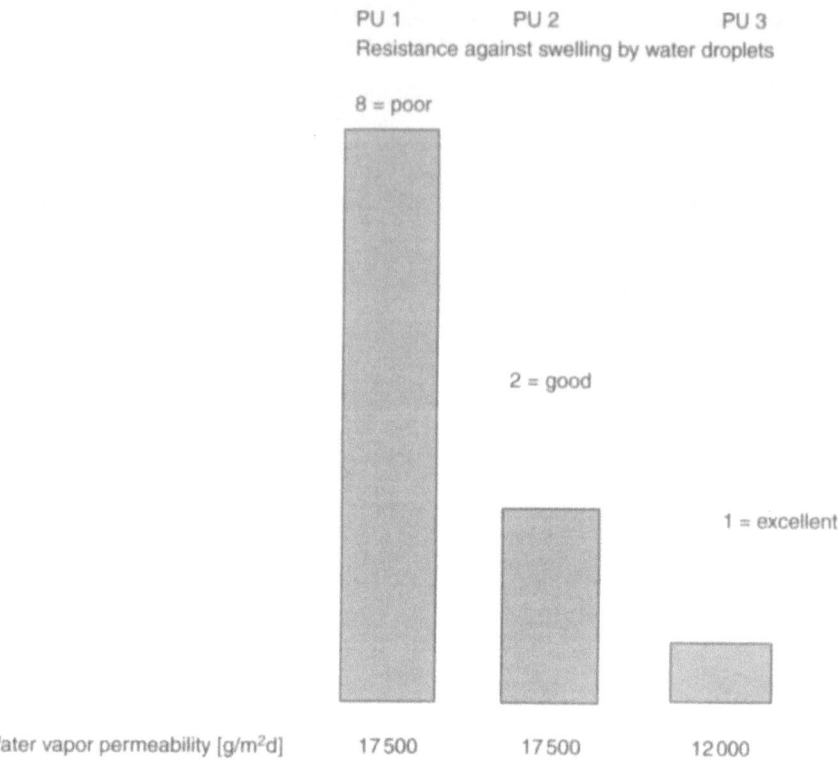

Fig. 18-7. Drop repellency and water vapor permeability

Petrik et al. [2] investigated polyurethanes based on 4,4′-diisocyanatodi-phenylmethane (MDI) and polyethylene oxide and/or polypropylene oxide. They found that the quantity of water which can be absorbed directly depends on the amount of polyethylene oxide present in the polyurethane. The water is absorbed in the form of clusters in the polymer.

Coatings with hydrophilic polyurethanes are advantageous due to their water vapor permeability combined with an impermeability against air and wind, "wind stopping effect" and their easy application. Hydrophilic polyurethanes can be applied like normal coating products. There is no need for coagulation baths and no difficulty in preparing and handling of inverse emulsions and so on. Hydrophilic polyurethanes are well suited for the production of functional clothing like uniforms and protective clothing in the construction, foundry or hospital area. They can be washed, dry cleaned and sterilized easily [5]. A comparison of a normal hydrophilic polyurethane with a dual hydrophilic one is given in Fig. 18-5.

To demonstrate the influence of water on the physical properties of poly-urethane films, films can be prepared from water-based polyurethane dispersions. Tensile strength and elongation at break in the dry and the wet were

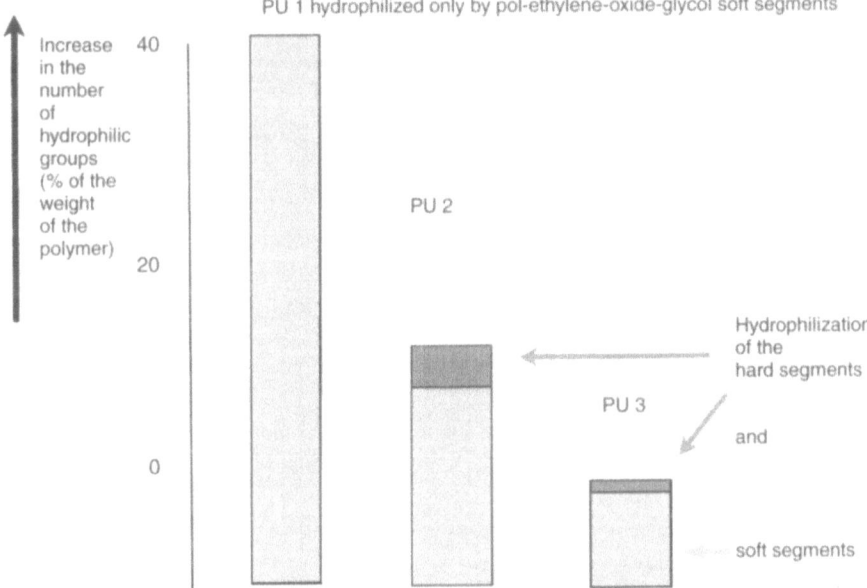

Fig. 18-8. Quantity of hydrophilicity. PUR 1 only hydrophilized by polyethylene glycol ether soft segments has the same water vapor permeability as the dual hydrophilic PUR 2. The quantity of hydrophilic groups of a dual hydrophilic polyurethane can be much lower to obtain the same quantity of water vapor permeability. This may be one of the reasons why a dual hydrophilic polyurethane has a better resistance against swelling by water droplets

measured (see Fig. 25-5). There should not be a large difference between dry and wet tensile strength. Large differences indicate a high water absorbency and – often – insufficient wet properties (for details, see Chap. 25.5). Polymers not linear in structure do not absorb water as much as linear ones. To reduce linearity in structure polymers may be crosslinked during application.

During the synthesis, the polyaddition reaction to produce the polyurethane already trifunctional compounds, like triols [42], may also be used to reduce a swelling of the polymer in water.

To control the hydrophilicity of a polyurethane a well-tempered mixture of hydrophilic polyethylene oxide and siloxane chains are used in the formation [62].

Thermoplastic polyurethanes with polyethylene oxide chains are mixed with silica and a wax with amine parts [73] to form a extruded water vapor permeable film [1].

A ricinole-adipate ester is prepolymerized with MDI then blocked with caprolactam and dissolved in xylene. After mixing with polyethylene glycol ether a textile is coated and the polyurethane is deblocked by heating and then polymerized [14].

A hydrophobic and a hydrophilic polyurethane are coated as a water-based dispersion on a textile [70]. The hydrophilic polyurethane consists of a diiso-

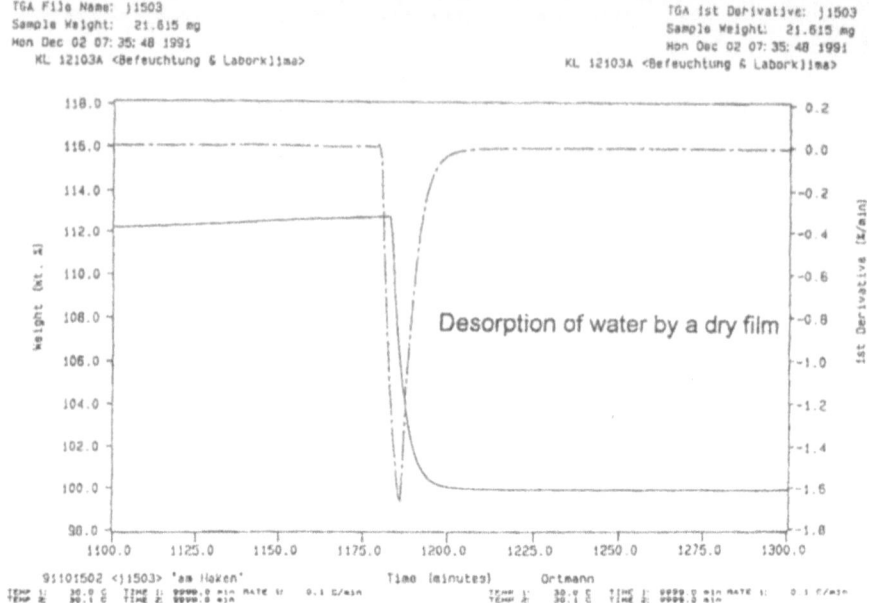

Fig. 18-9. Desorption of water

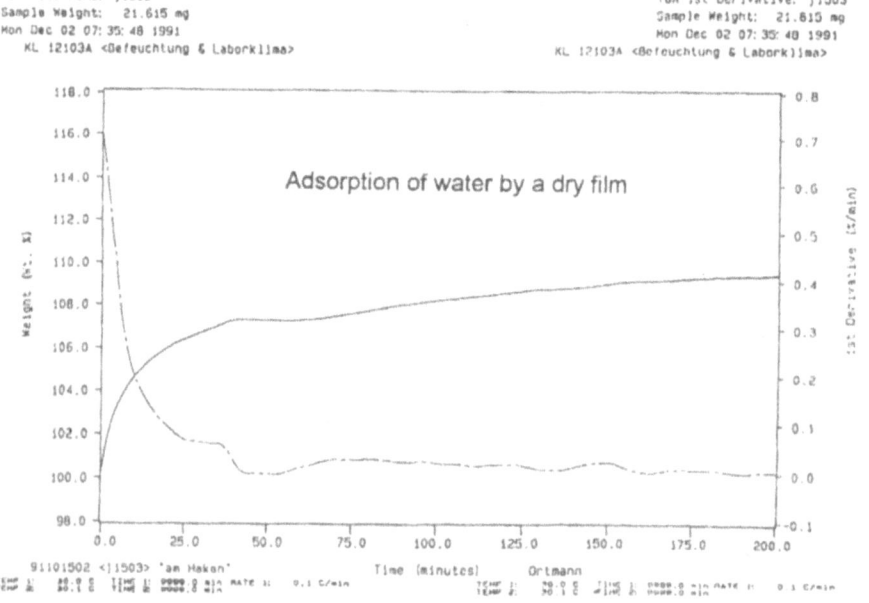

Fig. 18-10. Adsorption of water

cyanate and a hydrophobic polyol and a chain extender containing a double bond. A hydroxy-group-containing hydrophilic monomer is graft polymerized on the hydrophobic polyurethane [33]. The hydrophilic polyurethane may also contain a chain extender based on a condensation product of itaconic or fumaric acid with hydrazine or ethylenediamine [34]. Hydrophilic polyurethanes containing a carboxylic acid group as well as a hydrophilic polyether can be applied as a low viscous solution. The polyurethane has improved adhesion, hydrophilicity and a good washability [80].

Polymers, preferably polyurethanes, containing amino, hydroxy, keto or carboxy groups able to build intermolecular hydrogen bridges, are treated with the salt of an organic acid to form a gel [3]. Acidic solutions of polyamides, polyurethanes or polyacrylonitrile are neutralized, salt is added and precipitated to produce microporous membranes suited for use in the medical sector [4].

Polyurethane ureas containing polyethylene glycol ether in the soft segment and possibly mixed with polybutadiene or polyisobutylene may be applied as adhesive coats for artificial leathers [17].

Hydrophilic mixtures of polyurethane, ethylene propylene copolymers and fillers like stearic acid can be calendered to a film usable for garment or hygiene articles [27].

Polyurethanes containing polyethylene glycol ether in the soft segment [12] can be vulcanized on fabrics [8]. These products may also contain fibers or fillers. They may be crosslinked with formaldehyde [9, 10] or vulcanized with sulfur [16].

Microporous membranes with a gel-like structure are made from associated linear cationic and anionic polymers [7]. Dispersions of cationic and anionic polyurethanes are used to coat a polyamide fabric. After heating at 60 °C, this mixture coagulates to water vapor permeable adhesive or top coats [26].

Alkaline- and water-resistant, hard coatings can be produced from aminoplasts and hydrophilic polyurethanes [59].

Combinations of a hydrophilic polyurethane top layer and a foamed polyurethane intermediate coat on a textile fabric may be used in the manufacture of sport clothing [19].

Mixtures with excellent water vapor transmission can be obtained from 10–90 % polyurethane and 90–10 % of a vinyl copolymer. The vinyl copolymer consists partially of a carbonic acid and unsaturated ester groups. The acid groups are crosslinked via metal ions like Zn, Na, Ca or Cd. The polyurethane contains 30–100 % of polyethylene glycol ether in the soft segments [13].

Blends of PVC with a hydrophilic polyurethane may be moisture permeable [29].

Outside coatings of textile or other substrates with highly hydrophilic polyurethanes have such bad wet properties that they are scarcely used. Only medium permeable polyurethanes are normally used as outside coatings [76]. Hydrophilic polyurethanes are mainly used as adhesive coats ([71]; see Chap. 18). Hydrophilic polyurethanes can also be used as foams [47, 48]. These foams can be used as intermediate coats in man-made leathers [60].

Mixtures of Polyurethanes with Hydrophilic Products. To increase or modify the hydrophilicity of a polyurethane, mixtures with hydrophilic components have been discussed. The previous statements about *breathability providers* in connection with polyurethanes (see Chap. 18.1) is mostly overrated.

Polyurethane containing mixtures with hydrophilic products have been discussed as being suitable for the coating of leather, wood and textile substrates. To increase the water vapor permeability of polyurethanes prepared from dispersions therefore casein [41], powdered collagen [65], hairs or feathers [64], leather dust [42, 49] Irish moss [63] may be added. In microporous polyurethane films, leather, silk, chitin, cellulose or bamboo powder may also be mixed in.

For medical applications mixtures of proteins with polyurethanes find increased interest [23].

Chitin, which improves the water vapor permeability of coagulated polyurethane microporous layers, is also good for hydrophilic polyurethanes. Polyurethanes containing chitin and applied as water-in-oil emulsion are not influenced in their water vapor permeability [19].

DMF solutions of a polyurethane mixed with powdered hairs or leather are suited for the production of man-made suede [6].

The addition of conductive titanium dioxide results in a material with semiconductive properties with an electrical resistance on the surface of less than 10^{11} Ohm which may be reduced by the further addition of Cu, Ag or Al [68].

It is possible to produce films with a high abrasion and flexing resistance from a polyurethane mixture, soluble in DMF. The polyurethane with a polyethylene–polypropylene oxide (with 30 – 70 % of ethylene oxide) in the soft segment is mixed with coconut or linseed oil, and a surface active substance, like sorbitan tristearate [36].

A polyurethane in granular form, like Estane® (trademark of Goodrich) is mixed with poly-*N,N*-dimethylacrylamide then melted and extruded into water to produce a vapor permeable film [44].

Hydrophilic fibers of polyvinyl alcohol or regenerated cellulose and 10 – 20 % of a polymeric binding agent may be transformed to soft, elastic films [21]. Diacetone alcohol(meth)acrylate is copolymerized with vinyl monomers, (meth)acrylate, etc. or graft polymerized on polyurethanes or polyamide. The resulting product may be used to coat a nonwoven to water vapor permeable man-made leather [24].

Hydrophilic isocyanate prepolymers are blocked [11] with sodium bisulfite and mixed with methyl cellulose. A Jersey fabric is treated with this mixture and

$$\text{wwww N=C=O} + \text{NaHSO}_3 \longrightarrow \text{wwww—NH-CO-SO}_3^{\ominus} \ \text{Na}^{\oplus}$$

isocyanate + sodiumbisulfute forms a carbamoylsulfonate

$$+ \text{H-R}^{\ominus} \Bigg| \begin{array}{l} \text{which reacts with} \\ \text{a nucleophilic} \\ \text{compound} \end{array} \qquad\qquad (18\text{-}1)$$
$$- \text{HSO}_3$$

wwww NH-CO-R $\quad\downarrow$

if the nucleophilic compound is an amine
a urea is formed

Blocking and deblocking of an isocyanate with sodium-bisulfite by heating the polyurethane is deblocked. A highly water-absorbing textile is created [42] (see Eq. 18-1).

With glycerol as the starting component a trifunctional polyethylene glycol ether is produced which is prepolymerized and blocked. This product is – together with fibers – transferred to a water vapor material on a paper machine (see Fig. 21-1) [61].

Liners from hydrophilic polyurethanes are produced from an anionic, water-based polyurethane dispersion mixed with glycerol, polyethylene glycol and PVA [25].

DMF solutions of a polyurethane are mixed with a protein and used to impregnate a textile in the coagulation process [67].

A polyurethane containing a polyamino acid can be used for coating [18].

Powdered hairs from fur are mixed in a solution of a polyurethane in DMF. The mixture is used to coat a fabric and coagulated in water. After buffing a man-made suede is obtained [28].

18.3
Polyglutamic Acid Containing Compounds

Poly-γ-glutamic acid is a product which originated in the food industry. Almost exclusively, Japanese authors discuss the manufacturing processes based on poly-γ-glutamic acid and the chemistry done with this product as a starting material.

$$\begin{array}{c} COOCH_3 \\ | \\ CH_2 \\ | \\ CH_2 \\ | \\ \left(\!\!-NH\!-\!CHCO-\!\!\right)_n \end{array}$$

Poly-γ-methyl-glutamate

Poly-γ-glutamic acid in the form of C_1–C_4 alkyl esters can be used as a water vapor permeable coating for textile substrates [1, 5, 14]. These coatings may be colored [4, 15]. Poly-γ-glutamate dissolved in alcohol can be coagulated in alcohol, glycol or acetone [18].

Films consisting of poly-γ-methyl-glutamate containing 1,2-dichloroethane are described as looking like leather [11].

Introduction of Functional Groups to Poly-γ-glutamic Acid. Poly-γ-glutamic acid methyl ester (see above) of a molecular weight of 400,000 containing up to 60 % N-substituted amino acid esters like amino malonic acid or 2-pyrrolidone-5-carbonic acid may be used for coatings [2, 3]. N-Carboxy-poly-γ-glutamic acid methyl ester and epichlorohydrin in mixture with triethylaluminum as catalyst are possible coating products for leather substitutes [12].

An NCO prepolymer or a polyisocyanate is polyadded to poly-γ-glutamic acid [22, 27, 29] and used for impregnations or coatings [6–8, 13]. An NCO prepolymer is added to N-methyl-γ-L-glutamate and used as a water-in-oil emulsion to coat textile substrates with a microporous layer [30].

An isocyanate prepolymer is made by adding MDI to polytetramethylene glycol ether which is chain lengthened with N-methyl-γ-L-glutamate-N-carboxy anhydride.

HOOC-CH₂-CH-CH₂-C=O

Glutamic-γ-ethyl ester of N-carboxy anhydride

The resulting product is coated onto a fabric and coagulated [40]. Similarly, HDI prepolymers may be modified by γ-methyl-glutamate and used for coagulation [41]. Other amino acids, like D,L-alanine, L-aspartic acid, L-lysine, etc. may also be used [42]. If barium or strontium ferrite is added to the microporous layers, a magnetic conductivity is induced which helps the circulation of blood in sportswear [43]. A lacquer consisting of polyurethane and a compound which has ions is coagulated between two electrodes. Then a finish consisting of an isocyanate modified γ-methyl-glutamate is applied [44]. Copolymers of polyurethanes and amino acids are dissolved in DMF or dioxane. They are used for the impregnation of textile substrates. After coagulation they are washed and impregnated with a silicon [45].

A polyurethane with polyethylene glycol ether soft segments and terminal amino groups which is mixed with L-glutamic acid methyl ester N-carboxy anhydride may be used for water vapor permeable artificial leathers [24].

Copolymers resistant to UV rays consist of N-carboxylated amino acids, like aspartic acid, glutamic acid, glycine or alanine and an amino-terminated urethane prepolymer. These copolymers are used together with a crosslinker to coat a water vapor permeable textile [39].

Mixtures of Poly-γ-glutamic Acid with Other Polymers. Mixtures of poly-γ-glutamic acid and polyurethanes [9, 20, 23, 25, 26, 28] or a rubber elastomer [19] are discussed for the coating of textiles.

Poly-γ-methyl-glutamate or its methyl ester is dissolved in 1,1-dichloroethane. During rapid stirring the solution of a polyurethane together with a nonsolvent is added. The solvent/nonsolvent mixture is applied to form a microporous film by selective evaporation [16].

Polyethylene adipate, MDI and ethylene glycol are polyadded to form a polyurethane dissolved in DMF. Prior to the termination of the polyaddition an amide containing carboxylic groups is added. The resulting solution is used for the impregnation of cotton [17].

Nonwovens consisting of poly-γ-methyl-glutamate fibers and elastomeric binding agents are treated with 1,2-dichloroethane. The poly-γ-methyl-glutamate has an α- and β-form. The α-form is dissolved by this procedure [10].

Products based on poly-γ-methyl-glutamate give a leather substitute a dry, leather-like touch. Often these kind of products are described as a top finish for leather and its substitutes [21].

18.4
Other Hydrophilic Products of Natural Origin

In this chapter various amino acids other than glutamic acid derivatives are described. Due to the fact that often one acid is mentioned, others, however, are not excluded, the subjects of this chapter and the forgoing one overlap a little. Also processes and products discussed in Chap. 20.1, about regenerated collagen, may overlap with the subjects described here.

Polyamino acids are hydrophilic. They cannot be used for flexible products alone, because films or coatings consisting of polyamino acids alone usually are not elastic. Therefore, in most cases, they are modified by other polymers either by physically mixing or by a chemical reaction. The polymer mostly used is polyurethane or an isocyanate derivative. Some amino acids, like hydrolyzed proteins, chitin or chitosan, are byproducts from the food, leather or other industries. Most of these products are the basis of products of a low commercial value or waste. Therefore a lot of experiments have been carried out to find better uses for these byproducts. In this chapter some results of these experiments are cited: For a nonpolluting material technology, chitin or chitosan based products were investigated [11]. Amino acids are potential chain extenders for polyaddition reactions of isocyanate-containing products. Hydrophilic, biodegradable polyurethanes were produced using as a chain extender a diester of L-phenylalanine and 1,4-cyclohexane dimethanol. These products may be used in biomaterial applications [12].

An isocyanate prepolymer consisting of polypropylene glycol ether and TDI is polyadded to N-carboxybenzoxy-L-lysine-N-carboxyanhydride in tetra-chloroethane. The resulting product is dissolved in dimethylacetamide. The solution is used for a water vapor permeable coat on a textile substrate [1].

A water-based polyamino acid solution is mixed with a polyurethane and stearyl titanate: The mixture may be used in the production of man-made leather [8].

Powdered, liquid or fibroid protein is transferred to a film which is treated with trans-glutaminase and/or lysiloxidase or fungi producing these enzymes able to crosslink the film consisting of protein. The resulting product is resistant to heat and able to permeate or adsorb humidity [2].

Proteins like gelatin, casein or collagen are esterified with polyfunctional alcohols. The resulting products are mixed with polyethylene, polypropylene, polyester, polyamide or polyisocyanate and used to coat a textile with a leather-like, permeable coating [3].

A coating mixture for man-made leather consists of collagen, glutaric dialdehyde, (meth)acrylic acid and a polyurethane dispersion [4]. To modify an artificial leather tanned, powdered collagen may be used [5]. Protein of vegetable origin, for instance from soybeans, is treated under pressure with water and heat, then cooled and extruded and tanned to an artificial leather [6].

β-Chitin, dispersed in water, may be used to coat a textile by the transfer process [7].

Stockings, etc., consisting of fibers of polyamide and polyurethane become hydrophilic by a treatment with chitosan [9]. A suede substitute is produced with polyester fibers of 0.01 – 2 den and polyurethane as bonding agent. The material is treated with 0.01 – 3 % of chitosan or collagen and with waterproof and hydrophilic binders. The resulting suede substitute is waterproof and antimicrobial [10].

CHAPTER 19

Production of Synthetic Suede

Genuine suede leather is produced by finely buffing tanned and dyed leather from sheep or goats. Another source for suede is split leather from cattle hides which is produced by splitting the thick hides horizontally (see Fig. 20-2). Suede leather is mainly used in the production of garments. Genuine suede has excellent wearing behavior due to its high water vapor permeability. Its water repellence is not very good and its sensitivity to dirt is high. Genuine suede cannot be washed or dry cleaned without a loss of softness.

Bonded nonwovens very often look like suede leather. However, their softness, water vapor permeability and water repellency are usually not good enough for use in the production of garments. In the following chapter production methods are described by which synthetic, suede-like products can be obtained.

Nonwovens bonded with polyurethanes treated with coagulated products consisting of NBR, SBR, PVC or polyamide are regarded as synthetic suedes [42]. A specially treated nonwoven consisting of fibers of PVA behaves like a suede [10]. A typical bonded nonwoven consists of a polyethylene terephthalate fiber fleece bonded by a latex of a copolymer of 25–50% styrene, 75–50% butadiene and 1–5% (meth)acrylic or itaconic acid. A surface active agent, like a silicone oil [1, 3], an aliphatic or naphthenic hydrocarbon [2, 3] or a C_4–C_{30} alkyl oleate or a C_8–C_{30} alkyl sebacinate, may be added. After drying, the material is buffed to produce the suede-like surface. By needling, shrinking and pressing under heat a fiber fleece is increased in its density by 15–20%. Then it is impregnated with a latex whose polymer only has a weak adhesion to the fibers. The latex is coagulated in the nonwoven [4].

A raised knitted or woven textile is coated with a crosslinked polyurethane like Desmolin® N and Desmodur® L (both trademarks of Bayer) [20] or PVC [7]. The raised side is impregnated with a polyurethane dispersion and then buffed [5].

Nonwovens are filled by a polyurethane dispersion, then buffed [15] and split [6, 8]. Nonwovens with ultrafine fibers on both sides are treated with a polyurethane, PVC or latex and then buffed [21, 23].

A textile fabric consisting of a polyamide is coated on one side with a polyurethane dispersion and on the other with a butadiene–acrylonitrile latex. Uncovered fibers are dissolved in 6 molar hydrochloric acid to produce a man-made suede [11].

A 100 g/m² nonwoven consisting of polyamide and rayon (80:20) is treated with an emulsion of a polymer of 23 parts styrene, 75 parts butadiene and 2 parts itaconic acid. The polymer contains calcium carbonate. The latex is coagulated in a calcium chloride solution. The nonwoven is washed, dried and buffed [19].

A raised fabric consisting of ultrafine fibers is treated with an emulsion of an amino resin, then dried and hardened with a melamine resin [24]. Woven or knitted fabrics or nonwovens are impregnated with a two-component polyurethane then covered on both sides with 0.2 den fibers. The article may be used as artificial suede [25].

Fibers of cotton, pigment, nitrocellulose, a plasticizer and a filling agent are mixed to obtain a paste. This paste is treated on rollers at different speeds to give an artificial leather without any textile fabric as a support [13].

Bonding by a Coagulation of DMF Solutions. Fabrics consisting of fibers with different solubilities are treated with a solution of a polyurethane. Then the product is coagulated and the fiber component which is soluble is dissolved [44]. A polyester–polyurethane, soluble in DMF, should contain a polyester with a molecular weight of 800 – 4000 [32].

Fabrics consisting of cotton and fine rayon fibers are impregnated with a polyurethane, coagulated and then buffed [35].

A polyester nonwoven with a coagulated dyed polyurethane is coated twice with a glossy printing ink, embossed and then buffed [26].

A nonwoven consisting of ultrafine fibers with a sea component (see Chap. 4.1), able to be removed by dissolving, is impregnated with an elastomer, coagulated, buffed and dyed [27]. A nonwoven is manufactured using island sea fibers consisting of polyamide or polyolefin and PVA. The nonwoven is impregnated with a polyurethane which is coagulated. Then the substrate is impregnated with an amino silicone and raised to form an artificial suede [33]. Woven fabrics with fibrillated polyamide-6 and polyester fibers are raised and treated with an emulsion of benzyl alcohol then with a solution of a polyurethane in DMF, like Crisvon® (trademark of Dai Nippon Ink), coagulated, dyed, dried and raised [36]. Nonwovens consisting of polystyrene–polyethylene terephthalate fibers are impregnated with a solution of polyvinyl alcohol, then the polystyrene is removed. These steps are followed by an impregnation with a polyurethane/DMF solution which is then coagulated [37].

Nonwovens consisting of polyester and polyolefin or polyamide fibers are bonded by a polysiloxane-containing polyurethane in such a way that the fibers are covered by the polymer. The polyurethane is coagulated [38, 41].

A similar process consists of the steps in the manufacture of a needled nonwoven consisting of composite fibers. After impregnation and coagulation, the nonwoven is shrunk and split in the planar direction. Then the article is treated with a solvent or a swelling agent for one of the components of the fibers and then buffed [28].

A nonwoven of cotton fibers is treated with a reactive polyurethane (Elastron® W-11, trademark of Dai Ichi Kogyo Seiyaku) and a melamine resin (Sumitex® M-3, trademark of Sumitomo) for 3 – 5 min, crosslinked at 110 °C and tumbled. Then it is impregnated with a polyurethane (Crisvon®, trademark of Dai Nippon Ink), coagulated, dried and buffed [34].

Torn Off Layers. It is apparent that the removal of a polymer coat from a textile article by buffing results in a velvet-like surface of this fabric. Normally, the polymer coat also penetrates a little into the surface of the textile substrate. By buffing, parts of the fibers of the substrate are raised which are surrounded by the residues of the polymer. Another method of creating a suede-like surface is to cover a textile substrate with a foamed and a compact layer. After the removal of the compact layer and possibly slight buffing, the remaining foamed surface of this substrate looks like a velvet or a suede.

Laminates consisting of a textile support, a foamed and a compact layer become suede-like if the compact layer is torn off or removed by buffing [16].

A foamed PVC, vinyl chloride–vinylidene chloride copolymerisate, etc., is laminated with a knitted or woven fabric. Then the surface is treated with a heated roller, so that the surface is removed and the foamed layer can be seen [12].

Another way to form a suede-like article is to mix a 30 % polyurethane solution in butan-2-one (100 parts) with sodium bicarbonate powder (150 parts). The mixture is coated on a release paper, adhered to a textile and dried. Then the textile is torn off the paper in such a way that a part of the polymer remains on the paper. Finally the salt is leached [18].

A composite material consisting of two layers is torn off. The resulting porous layer is then laminated with a textile substrate to form a suede-like article. The composite material consists of a porous and a compact PVC layer [14].

Mixtures of polyurethanes and acrylates also containing a wax and powdered cellulose can be used to impregnate a textile substrate. After drying the product is buffed [31].

Tearing of Coagulated Layers. On a textile substrate a coat of a polyurethane is coagulated. A second film is laminated on the coagulated one. Then the compact layer is torn to create a velvet surface [9, 22]. Two textile substrates adhered together by a coagulated polyurethane film are torn off each other to get two artificial suedes [45, 46].

Two textiles are laminated with each other by a partially coagulated polyurethane. Then they are split, the coagulation is terminated and dried [39, 40]. The same can be done with two nonwovens: they are laminated with each other by a polyurethane which is coagulated. Then the laminate is split and the textile is buffed [43].

A substrate containing polyester and polyamide fibers is impregnated with a polyvinyl alcohol solution, then with a polyurethane solution and coagulated. The polyester consists partially of sulfophthalic acid and polyalkylene oxide; these components can be removed by extraction with an alkaline liquid and the PVA by the coagulating bath. By the use of PVA the adhesion of the polyurethane to the fibers is reduced and, therefore, the article will be softer [29]. A nonwoven may also be treated prior to a coagulation with sodium dodecylbenzyl sulfonate [30].

CHAPTER 20

Leather Board

Waste is produced during the manufacture of leather: Waste such as cuttings, leather dust and shavings. The amount of this waste may be around 3% of the weight of the leather produced. To reduce or correct the thickness of leather, the material is usually shaved with knife rolls, normally after the chrome tannage (see Fig. 1-8, step 11). These shavings can be treated with a latex and possibly further tanning agents on a paper-making machine to produce leather board. Leather board is a material which can be used as a lining material in shoes or trunks or for bookbinding. Leather board is called Lederfasermaterial or Lefa in German and cuir reconstitué in French.

Shavings can be used as they are; but other leather waste must be chopped and possibly mixed with other fibers [13, 14]. The shavings or chopped cuttings are suspended in water and deposited on a Fourdrinier felt, in the same way that they are used in paper technology [23]. Tanning agents and latex bind the shavings together; by pressing, the water content of the fibers is reduced and the density of the material increased [2, 7].

A continuous process consists of mixing a fiber suspension with a latex under pressure, treatment with a solution of an aluminum salt and forming onto a paper felt [9]. A schematic view of a paper machine is shown in Fig. 20-1.

Leather board still has some properties similar to genuine leather: Leather board absorbs and desorbs water vapor. The manufacturing process for this kind of material has been known for more than 50 years. Published surveys exist for obtaining information about the specific items of this material [1, 3, 5].

Specific newer modifications of the manufacturing process are: A suspension of the leather fibers is mixed with a latex, glue or a PVC dispersion in a kneading equipment like a pigment grinder. On a porous felt a panel is formed by pressing and extracting the water. Then the material is dried and pressed again in a hot metal form [6].

Mixtures of chrome- and aluminum-tanned leather fibers can be used [20]; mixtures of shavings from vegetable-tanned leather [21] or shavings from tanning with synthetic organic tanning agents are also suitable [22].

Mixtures of 50–80 parts of leather fibers with 50–20 parts of polyamide fibers may be bonded with dispersions of acrylic polymers [11].

Chrome shavings (80 parts) and other leather waste is mixed with a urea-formaldehyde condensate (20 parts) and a hardening product. After forming to a sheet the mixture is pressed at 10 bar and 70 °C to obtain a regenerated leather

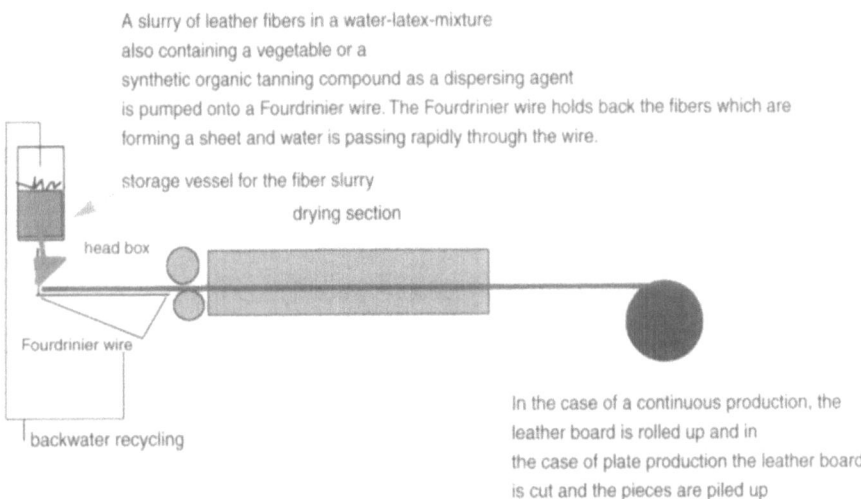

Fig. 20-1. Schematic view of a paper producing machine

[12]. A mixture of 40 – 80 % flax and 60 – 20 % leather fibers is cooked in a buffer solution, mixed with a latex, pressed and dried [8].

An ethylene–vinyl acetate copolymer (100 parts) and chrome leather shavings (10 – 60 parts) are mixed in water with 7.5 parts of azodicarbonamide. The mixture is transformed into regenerated leather with improved water vapor permeability [15].

A textile substrate is coated on one side with a microporous layer of a polyurethane or a polyamide and the other side consists of leather fibers [19].

Ground leather is mixed with a siloxane able to polymerize further under the influence of high energy rays (> 100,000 eV and < 2 megarad). The mixture is used for coating [16]. Leather waste is treated in a grinding mill and then deposited in an electrostatic field on nonwovens [17].

20.1
Regenerated Collagen

The hide and skin of mammals consist of collagen. Leather is tanned collagen. All the various types of waste which come from the manufacturing process of leather are a possible source of collagen. Preferably untanned waste of beamhouse operations is used in the production of glue, gelatin or collagen. Glue and gelatin are unstructured, strongly degraded proteins. Processed collagen still has the primary fibroid structure of the living organism. The only things missing are the linked protein fiber bundles while the secondary and tertiary structures still exist [20]. Processed collagen can be used to manufacture sausage casings, wound covers or hygiene articles.

There is a great amount of literature covering collagen as a subject. In this chapter only uses where water vapor permeability is essential are discussed and only literature covering this is cited.

Collagen may be used to obtain a mat surface with anti-slip properties [39]. A collagen mass is precipitated on a gauzy textile. Then it is pressed and tanned by zirconium salts or aldehydes [26]. Nonwovens may be impregnated with a solution of collagen and tanned with trivalent chromium salts [38].

Using microwaves gelatin becomes insoluble and is usable in the manufacture of man-made leathers [35].

Tanning waste is partially hydrolyzed under heat and pressure. Under kneading the products are treated with a polyisocyanate, like hexamethylene diisocyanate, and a polyester. Then the product is hardened after drying by the action of formaldehyde [13].

Water-soluble collagen produced by a thermal, a hydrolytic, an alkaline or an enzymatic action is treated with a monoisocyanate; it may be used as an additive in the production of leather substitutes [32].

A water-based solution of a protein is treated with a crosslinking, an alkylation agent, an acid or a Schiff reagent. The material may be used to produce textiles, microcapsules, coating agents for agricultural products and fertilizers with better hydrophilic properties [36].

Mixtures with Other Polymers. Collagen is mixed with a PVA solution. Water is removed with sodium chloride. The material is transferred into panels and tanned [3].

Solutions of tanning wastes are mixed with more than 200 % (based on solid content of the solution) of an anionic polyurethane dispersion. The mixtures may be used in the manufacture of artificial leather [14].

Modifying Collagen and other Proteins (Besides Tanning). Natural proteins from soybeans, casein or bacterias are graft polymerized with mono-vinylic compounds, acrylates, or glycidyl esters thereof. The result is an emulsion able to be used for water vapor permeable coatings [40].

Casein is treated with acetic acid anhydride and benzoyl chloride to become soluble in methanol. The resulting solution, together with formaldehyde, can be used to finish leather and leather substitutes [17].

Protein becomes soluble in dichloroethane, methanol and DMF (1:1:1). Acrylonitrile and glycidyl methacrylate is graft polymerized onto the protein. The resulting product may be used to increase the structure strength of artificial poyurethane leather [20].

Swollen animal hypodermic connective tissue is treated with sound waves (80 Watts/l) at a pH of 9. Using this treatment collagen may be separated from the other byproducts. Then it is treated at a pH of 3 with ultrasonic waves of 100Watts/l. In this way collagen may be produced without the use of an enzyme [31].

Nonwovens and Other Textile Substrates with Collagen Fibers [37]. A lot of publications [19] and patent applications discuss the dissolution of animal protein and its transformation into collagen fibers [1, 7, 18, 22]. These collagen fibers can be used in the production of nonwovens for man-made leather. The fibers can be tanned, e.g. with formaldehyde [18] or chrome tanning salts [5, 6, 15, 16]. Mixtures of these fibers with other polymers like polystyrene [27] is also possible [2, 23 – 25].

Solutions of collagen can be transferred to fibers. Threads are made out of these fibers to give leather-like textiles [4]. Proteins may also be isolated with the help of hypercritical carbon dioxide [33].

Nonwovens consisting of collagen [11] and polyamide fibers are bonded with an acrylate-based dispersion. After finishing, these nonwovens may be used directly as man-made leathers [8 – 10].

Collagen fibers (60 parts) with a length of 2.5 – 4.5 cm containing 1 – 1.5 % chromium are mixed with 20 – 30 parts of natural or synthetic fibers having a longer staple length than the collagen fibers. Three parts of raw collagen are also mixed in. The fibers are transferred into a nonwoven, which is needle punched and treated with hot water. After a further treatment with formaldehyde the production of the nonwoven not containing any other binding agent is terminated [28]. The tanning can also be done with a vegetable tanning agent [29]. These kind of nonwovens have a high water absorption.

By the addition of collagen fibers, e.g. from chrome-tanned split leathers, the water vapor permeability of a nonwoven consisting only of synthetic fibers is increased remarkably [34].

A water-based suspension of collagen fibers containing 1 – 5 % collagen fibers is deposited on a nonwoven. Then the fibers are freed of water. Thereby the collagen fibers aggregate. Then the product is treated with a suspension of synthetic polymers and a tanning agent. By this action a coated leather-like product is manufactured [30].

A layer of collagen fibers is treated with a water-based solution of a protein, e.g. gelatin. By heating the protein up to 55 °C it gels. By cooling the temperature down below 0 °C the water freezes. The ice crystals tear the gel. The product is warmed up again, water removed and the protein is precipitated with chromium salts [21].

A nonwoven is produced with collagen and synthetic fibers, like polypropylene. The nonwoven is bonded by a solution of an isocyanate prepolymer in DMF [12].

20.2
Coating of Leather

Prima facie coating of leather does not seem to be covered by the subject of this book. The fact that leather has an excellent water vapor ad- and desorption is one reason why split leathers, instead of a textile fabric or nonwoven, are well suited as substrates for leather substitutes even if coated by a homogeneous polymer which may have only a poor water vapor permeability. The water vapor ad- and desorption of the split leather also guarantees in a composed article like coated leather a good wearing property.

In most cases an animal skin is much thicker than it should be. Therefore the hides – and in many cases skins – are split in the horizontal direction (see

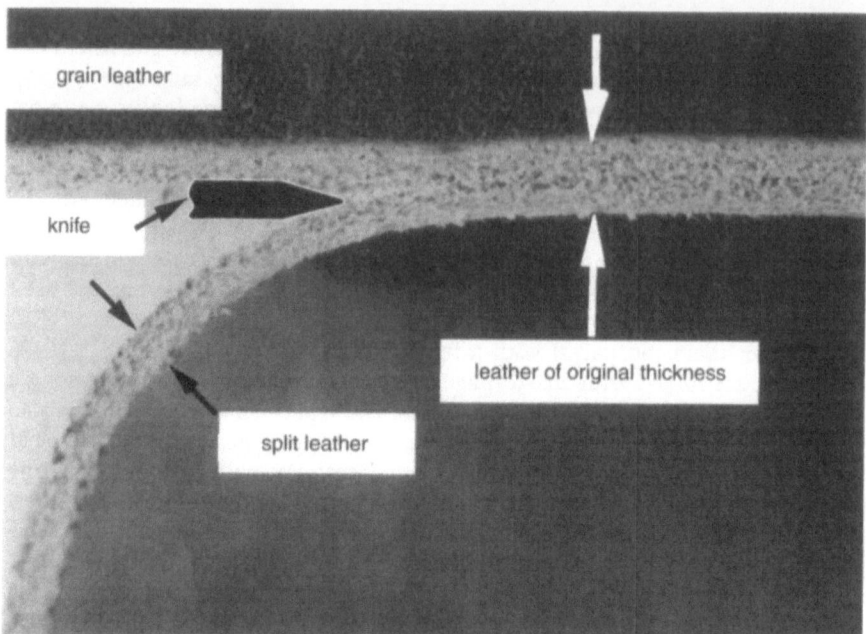

Fig. 20-2. Hides are split in the horizontal direction

Fig. 20-2): The result is a grain split and split leather without a grain. The grain split (see Fig. 20-3) is normally only called leather. It is the material well known for shoe uppers, upholstery and garments. The split leather (see Fig. 20-4) is normally used as "suede" for garments, gloves, etc. Split leather is a byproduct of the leather production. The broad use of suede depends on fashion. Due to quality and fashion, the quantity of splits produced often exceeds the demands of the market. If for fashion reasons suede leather cannot be sold in large amounts, tanners cannot get rid of it easily. Split leather has the same water ab- and desorption and an even higher water vapor permeability than genuine leather.

There has been a lot of experimental work done to transfer a split to a product which looks and behaves like grain leather.

It is possible to create a synthetic grain on split leather so that it can be used like a grain split, as long as the properties allow it. When it is possible to create a synthetic grain which looks natural, then tanners have no problems in selling their split leather.

The simplest way to do so is to get a finish by using acrylic binders on top of the split. This finish is usually carried out by plushing, spraying or curtain coating. After one or two base coats a grain pattern is usually applied by embossing. Then a top finish with a nitrocellulose lacquer is applied. Due to the thermoplasticity of the acrylic binder, the embossed grain does not remain in exact shape on the surface of the "new grain" of the split.

Fig. 20-3. Grain leather

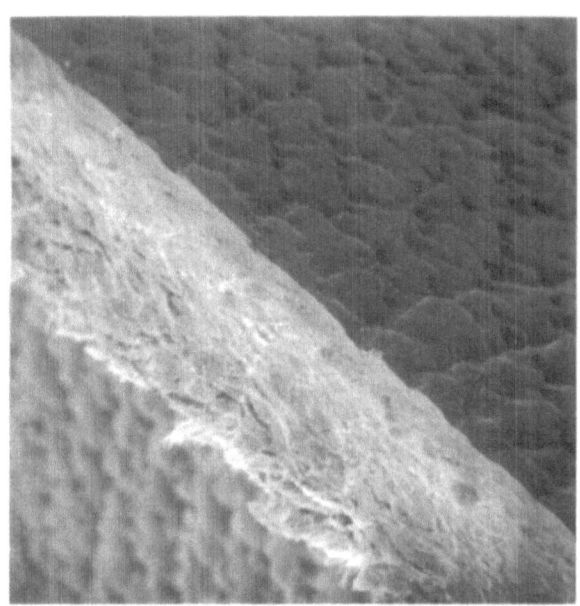

Fig. 20-4. Split leather

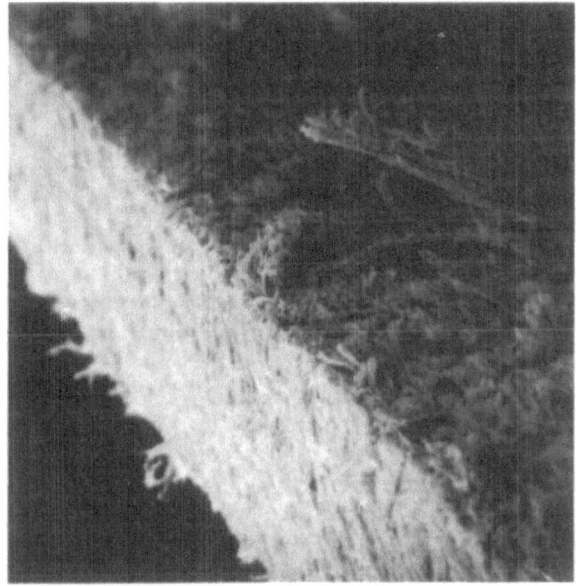

An advantage of this method is that it is a cheap finishing method which can be carried out with the equipment that an average tannery has. The disadvantage of this method is that the look of these splits is average, the physical properties are poor, and the split loses thickness in the embossing step.

The first improvement made over this "classical" finishing method was the application of oligomeric butadiene copolymer based products, (EUDERM® resin, trademark of Bayer) which were able to terminate their polymerization during the embossing step. The advantage of this is that the pattern of the grain embossed rests fixed in the finish because the polymerized binder keeps its shape. Butadiene copolymers have another advantage: The cold flexing is very good and butadiene copolymers can be applied as using acrylics with the average tannery equipment.

The properties of butadiene copolymer finished splits are good enough to be used in the manufacture of shoes. They have the same loss in thickness as acrylic polymer based finishes (see Fig. 20-10).

These finishing operations are a direct process, i.e. the products are directly applied onto the split leather. The finishes need a pressing operation by which the grain is embossed into the surface. A problem with these classical finishing methods is that the artificial grain created by this finish is "empty" and has no "body". Even a non-expert can realize just by looking at these finished splits that they look different from genuine grain leather.

The next step that was taken to improve the method to obtain a more grain leather-like finish was the application of foam systems [14]. Here three systems exist:

- chemically induced foams by the addition of blowing agents, well known from the artificial PVC leather,
- foams by evaporation of low boiling liquids or,
- mechanically produced foams.

Blowing agents need a rather high temperature to be decomposed so they are able to split off nitrogen which induces the foam. These temperatures are normally too high in leather production.

Optionally, fluorochlorocarbons were used as low boiling liquids. Nowadays the use of these fluorochlorocarbons should be avoided because they are believed to damage the ozone layer in our upper atmosphere. Furthermore, the storage time of a finished foam mixture is limited due to the steady evaporation of the low boiling liquid.

The best system today seems to be a mechanically produced foam (for the equipment needed, see [21]). Foam finishes may be applied directly onto the split. In order to create a nicer grain a foam finish should be applied indirectly by using a silicone rubber mold or a release paper.

There are several advantages with foam finish applications: The foam finish increases – especially when the indirect method is used – the thickness of the split. Another advantage is that the grain pattern is better formed, the finish has fullness and the artificial grain looks like genuine leather. The level of physical properties depends on the polymers applied. Mixtures of acrylics with butadiene copolymers and polyurethanes result in a sufficiently high level of properties. These splits can be used in the manufacture of shoes, upholstery, etc.

There has been a lot of research using foils to cover the surface of splits [1]. Foamed PVC or polyurethane foils have been used to laminate the splits. The process of applying foils consists of covering the spilt with an adhesive, normally by roller coating. The split is transferred to the foil and pressed, Thermoplastic polyurethane foils and PVC foils result in thick coated splits which do not look leather-like. The properties are comparable with grain leathers. The softness and touch, however, allows the use of these products in cheap articles only.

In the 1970s two processes were introduced in the leather industry; (1) the coating of splits with a microporous polyurethane foil (DESMODERM® foil, trademark of Bayer) [4], and (2) the coating of splits by the LEVACAST® process (trademark of Bayer)

Foil Lamination. Lamination with microporous foils results in articles being extremely leather-like in regards to water vapor absorbency and permeability. The delivery of foils of different colors, mat and glossy effects and grain structures by the chemical supplier may create logistic problems [7, 8, 18]. Due to the rather high price of the process, foil lamination took much longer to become successful in the market. Due to the uneven structure of the splits, foil lamination and indirect processes have a product loss of at least 25 % (see Fig. 20-5).

In the 1970 s a material consisting of a split coated with foamed PVC able to be transferred into a finished shoe upper by high frequency molding was introduced by Freudenberg (Germany) under the trade name Frequenta® [2] (see Fig. 20-6).

Due to the thermoplasticity of the PVC, waste from the lamination of PVC foils onto textile substrates or leather is easier to recover than polyurethane waste.

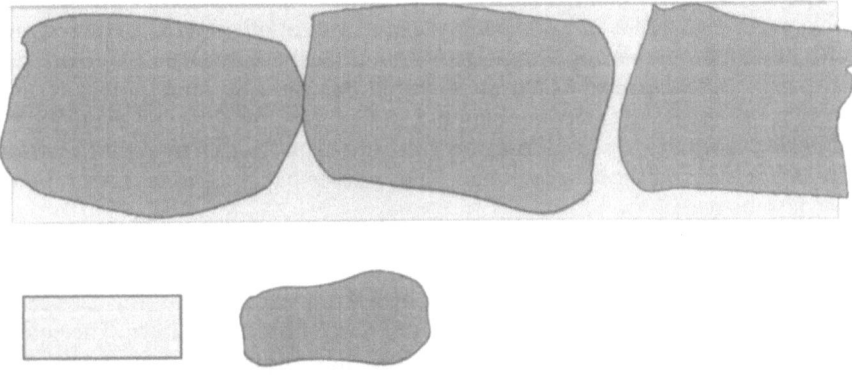

indicates a foil   demonstrates the shape of a split leather

Fig. 20-5. Loss of products by foil lamination or indirect processes. Similar to an overspray in foil lamination or an indirect coating of split leather there is a loss of polymer due to the irregular size of split leathers. Due to the fact that the price of split leathers is higher than that of the polymers used the loss of foil or coating material is economical and accepted

Fig. 20-6. Split leather
coated with a PVC coating;
0.8 mm foamed and a
0.3 mm compact PVC layer

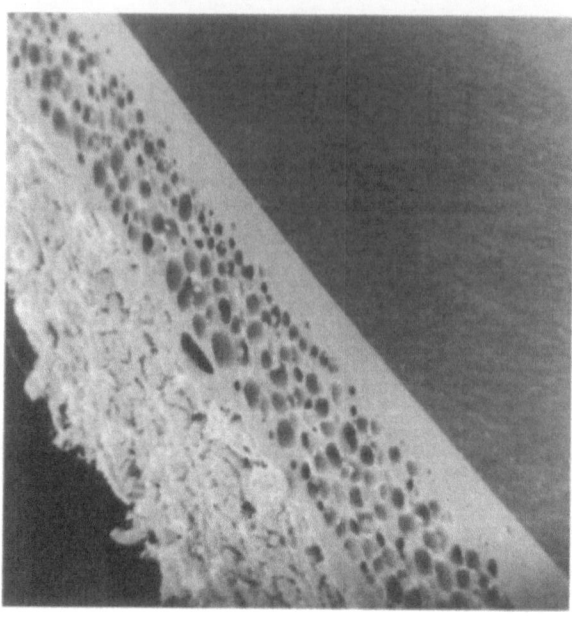

As mentioned before splits have an uneven surface. Therefore, a treatment of the split surface prior to lamination seems to be necessary. To improve the adhesion, a treatment with polyglycol fatty acids [4] or polyurethane dispersions [5] prior to lamination is recommended.

Indirect Coating of Leather. Leather can be coated by the indirect coating process in the same manner as a textile substrate with polyurethanes [12] (see Chap. 2.2). Polyurethanes can be used as two-component systems in organic solvents, as high solids or as dispersions. Mixtures of oligomeric urethanes and bifunctional monomers able to polymerize under the influence of UV-rays [16] have also been discussed as a coating system for splits. In most cases a release paper is used as an intermediate support. To facilitate the stripping of the finished leather from the release paper polyurethane dispersions may contain salts of fatty acids like zinc stearate [17]. Instead of release paper silicone rubber molds may also be used [13].

The Levacast® Process. Reactive polyurethane systems can be used easily for leather coating [3, 22]. Worldwide the best known process is the Levacast® process. In Fig. 20-7, a Levacast coating unit is schematically shown. The equipment consists of two dosing pumps. With these pumps an isocyanate prepolymer and a chain-lengthening agent, i.e. a special diamine, are pumped into a mixing chamber moving back and forward over a release paper (Fig. 20-8). The products are mixed and sprayed out instantaneously onto the release paper. They start to react on the release paper. The initial liquid mixture solidifies. In the solidifying polyurea a split is laminated. After passing through a drying

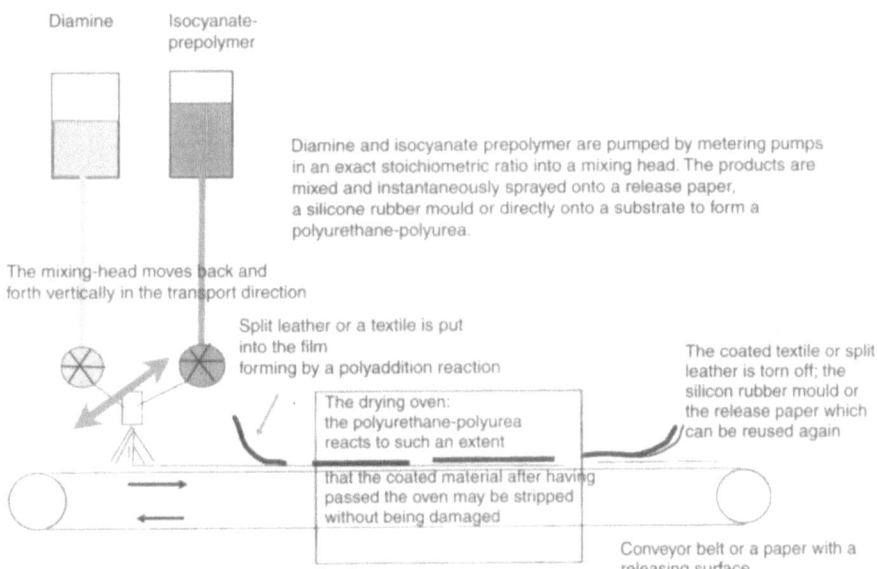

Fig. 20-7. Schematic view of a Levacast coating unit

tunnel the coated split can be stripped off. In Fig. 20-9 a coated split manufactured by the Levacast process is shown [10]. The reactive layers may possibly be applied in two steps [19].

Indirect coating systems are advantageous as there is no reduction in the thickness of the split as found in normal finishing applications. Figure 20-10 shows a comparison of a split finished in two different ways: Classical application with curtain coater and embossing afterwards and also the Levacast® process. The split used was the same for both applications. The reduction in thickness by the normal finishing method can be seen easily.

There are some advantages with the Levacast coating process:

- a gain in thickness,
- irregularities of the split surface are equalized (see Fig. 20-11); for comparison a foil laminated split still shows small holes between the foil and the substrate where an irregularity of the surface has been (Fig. 20-12), and
- abrasion resistance and adhesion of the coating onto the split are outstandingly good.

In addition, complicated surface patterns may also be reproduced if silicone rubber molds are used. Even the scale of a snake's skin may be reproduced (Figs. 20-13 – 20-15). Silicone rubber molds are the only method to reproduce overlapping structures of a model.

There are also disadvantages with the Levacast coating process:

- The chemical process of polyurethane formation is no longer carried out at the chemical supplier: A tannery needs to apply the polyaddition itself.

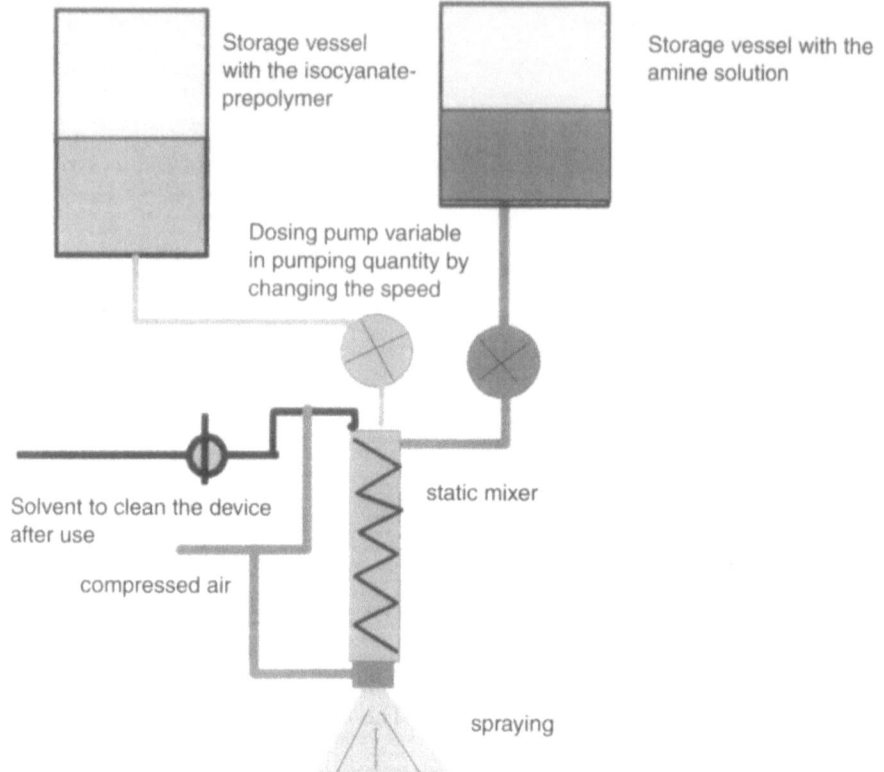

Storage vessel with the isocyanate-prepolymer

Storage vessel with the amine solution

Dosing pump variable in pumping quantity by changing the speed

static mixer

Solvent to clean the device after use

compressed air

spraying

Fig. 20-8. Detail of a Levacast® mixing head and spraying gun

- The final properties are achieved at the earliest the next day.
- The polyaddition process needs to be carried out exactly.
- The inevitable overspray (see Fig. 9.2) results in a loss of product of about 25 %.

Table 20-1 summarizes the different split finishing/coating systems in relation to their characteristics.

Direct Coating of Leather. An easy way to coat leather with a polymer seems to be to use a PVC plastisol, easy because the plastisol does not contain any solvents. A plastisol is a mixture of PVC with a plasticizer. A PVC plastisol, however, needs a gel temperature of above 170 °C to obtain a homogeneous soft PVC layer, which is a temperature leather does not resist without damage. It has been suggested that a first coat of an isolating layer with aluminum powder in a PVC plastisol should be applied and then another coat only with PVC plastisol. It has been proven that leather will then withstand the gel temperature [6].

Fig. 20-9. A split with a Leva-cast® coating [10]

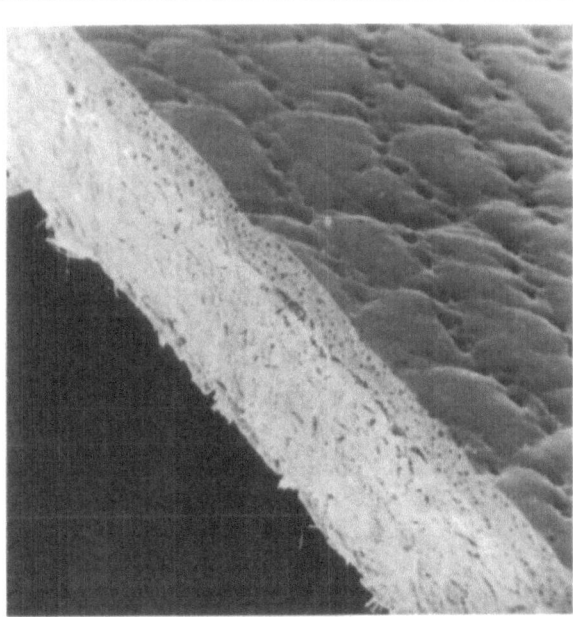

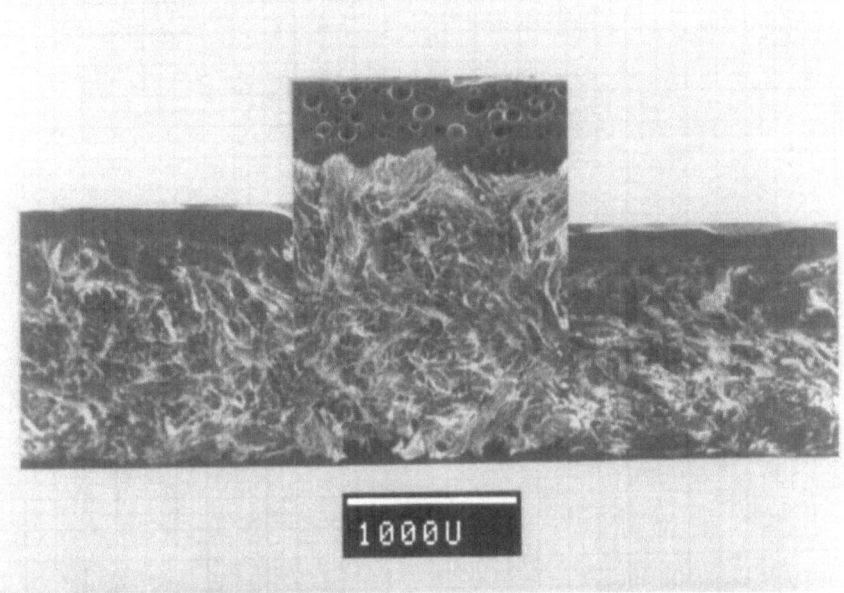

Fig. 20-10. Comparison of the finish of a split in two different ways: Classical application with a curtain coater and embossing afterwards (*right and left picture*) and by the Levacast® process (*middle*). The split used was the same for both applications. The splits finished in the usual way lose more than half of their thickness

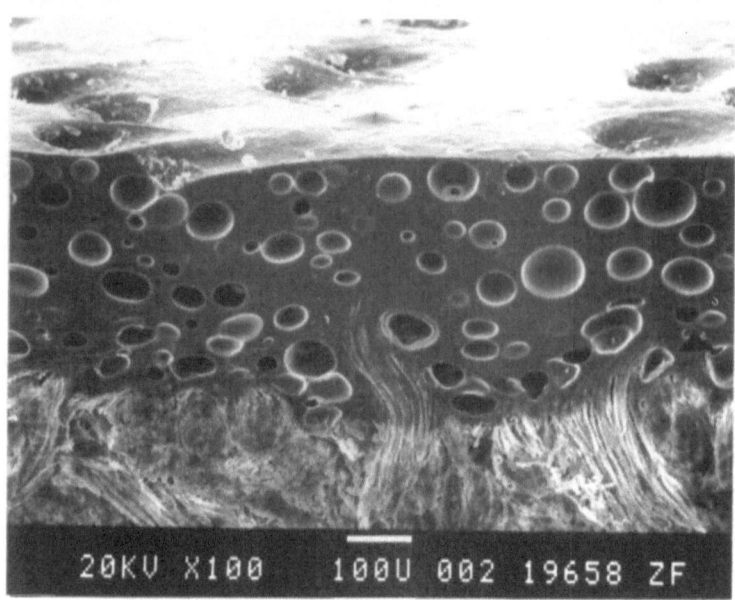

Fig. 20-11. Coating by the Levacast® process: the uneven surface of the split is equalized by the coating product

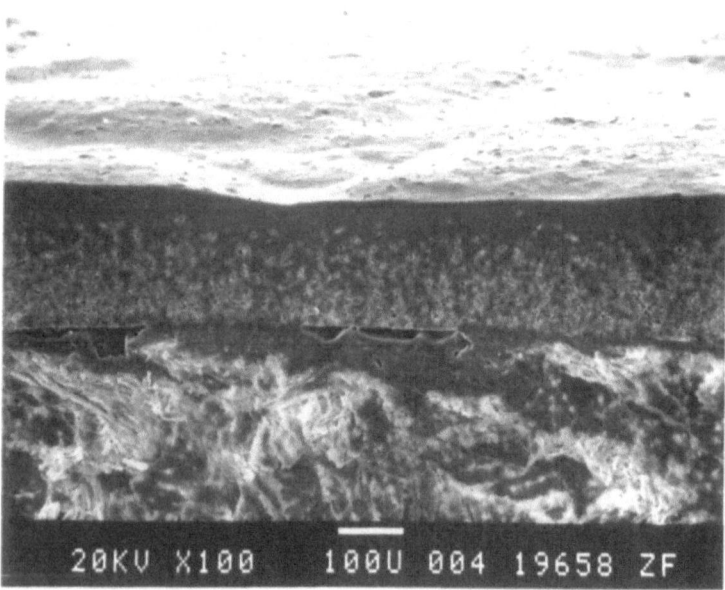

Fig. 20-12. A foil laminated split still shows small holes between the foil and the substrate where an irregularity of the surface has been

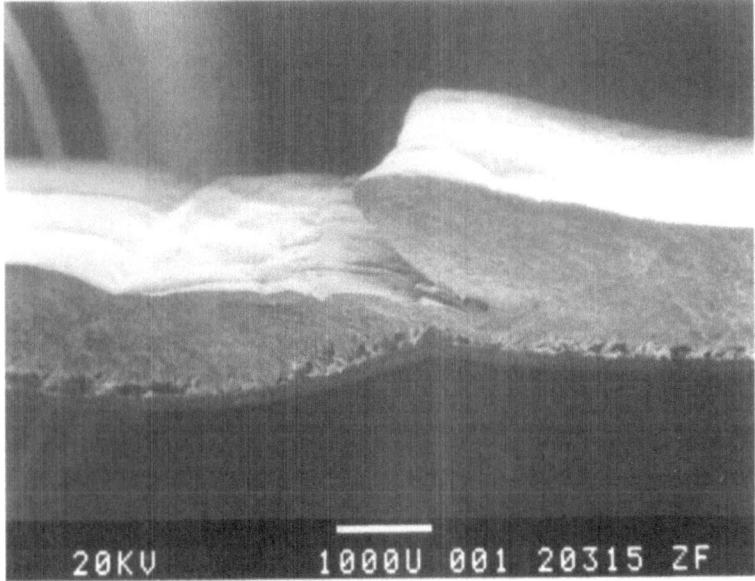

Fig. 20-13. A picture of a snake's skin

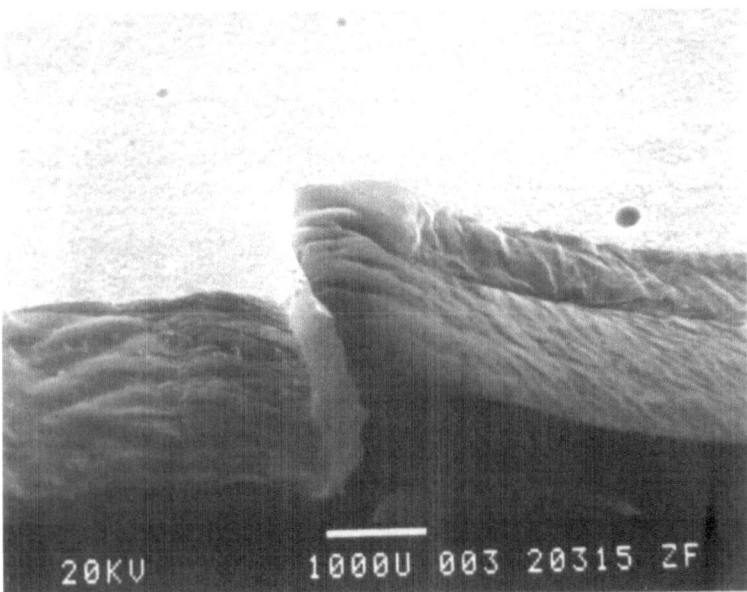

Fig. 20-14. A negative reproduction of a snake's skin in silicone rubber

Fig. 20-15. The final positive reproduction of the snake's skin in polyurethane; there is the haptic impression that it is possible to touch behind the scale – an impression which in the case of an embossing is not possible

Reactive systems able to react at ambient temperatures are solvent free or contain low amounts of solvent [15]. However, reactive application systems usually need rather a complicated equipment [3]. As long as polyurethanes are used the systems are similar to the polyaddition in a reactive selective evaporation (see Chap. 10.2).

Foamed dispersions of polymers like alkylacrylate, halogenated vinylic compounds or styrene copolymerisates may be used for coating of leathers which can be embossed easily [9].

As a summary it can be stated that coated splits may have the positive properties of leather, i. e. high water adsorption and desorption, hygrodynamic property (see Chap. 1), like the change in area due to the influence of wearing and recovery overnight and the high abrasion and rub resistance of a polyurethane-polyurea. For a long time Levacast coated splits have been the only materials used where it has been possible to adhere PVC and polyurethane soles onto the shoe shafts without additional working steps.

The different methods of split finishing or coating are summarized in Table 20-1.

Table 20-1. A comparison of the different split finishing/coating processes

Method	Polymer	Type of polymer	Equipment	Resistance to flexing	Rub and abrasion resistance	Thickness	Grain structure	Look	Price level
curtain coating	acrylate	water-based	curtain coater embossing press	<20000	medium	half of the split thickness	typical embossed grain; not distinct and stable	medium	low
curtain coating	butadiene	water based	curtain coater embossing press	>20000	medium	half of the split thickness	grain structure better and stable	medium	low
curtain coating	mixture of acrylate, butadiene and PUR	water based	curtain coater embossing press	>30000	medium	half of the split thickness	grain structure better and stable	medium	medium
transfer process	acrylate, butadiene and PUR	water based	transfer coating equipment	>50000	good	increase of split thickness	very good, stable grain structure	good	high due to the investment for the equipment
foil	polyurethane	solid adhesive water based or solution	roller coater	>100000	good	increase of split thickness	very good, stable grain structure	good	medium
LEVACAST	polyurethane	low solvent content	Levacast machine	>100000	excellent	increase of split thickness	very good, stable grain structure	excellent	high due to the investment for the equipment
high solid	polyurethane	low solvent content	transfer coating equipment	>100000	good	increase of split thickness	very good, stable grain structure	very good	high due to the investment for the equipment

CHAPTER 21

Nonwovens

According to a Japanese publication, the production of nonwovens has an annual growth rate of more than 10 % in Japan. The growth rate for nonwovens based on polyester and polypropylene are responsible for this high growth rate, that for polyamide is stagnant. Polypropylene is cheap, resistant to chemicals and may be used in the medical and hygiene fields [97].

Nonwovens are also important substrates for water vapor permeable products. Very often their manufacturing process is linked closely to the production process of vapor permeable top coats or impregnations. In their manufacturing process typical reactions are often used as for the production of water vapor permeable foils or coatings.

A general overview of the most important production processes for nonwovens is given. For more detailed information it is necessary to look into surveys [1–7, 13].

Nonwovens can be produced principally in five different ways:

- mechanically by carding of staple fibers,
- like paper from a fiber slurry treated in paper making,
- pneumatically by means of an air stream,
- by spinning fibers, or
- by a melt blown process [97].

The production of knitted or woven fabrics needs thread. Nearly all nonwovens are produced with staple fibers. Besides the spinning process all the other processes are carried out with more or less short staple fibers.

Carding Process. Cards are machines which have been used for a long time in the processing of wool fibers. For nonwovens staple fibers are unified on a card, possibly mixed with other fibers and orientated in the running direction of the card (Fig. 21-1).

The orientated fibers are then laid by a cross lapper (Fig. 21-2) onto a running belt and situated at 90 °C at the running direction of the card. The cross lapper moves back and forth over the belt and produces different layers of fibers whose orientations are no longer in the running direction but in angles. The aim of this is to finally get a textile substrate which is anisotropic (see Fig. 21-5) in its physical behavior.

The resulting material is needled, shrunk, bonded, split and buffed. These operations will be discussed later.

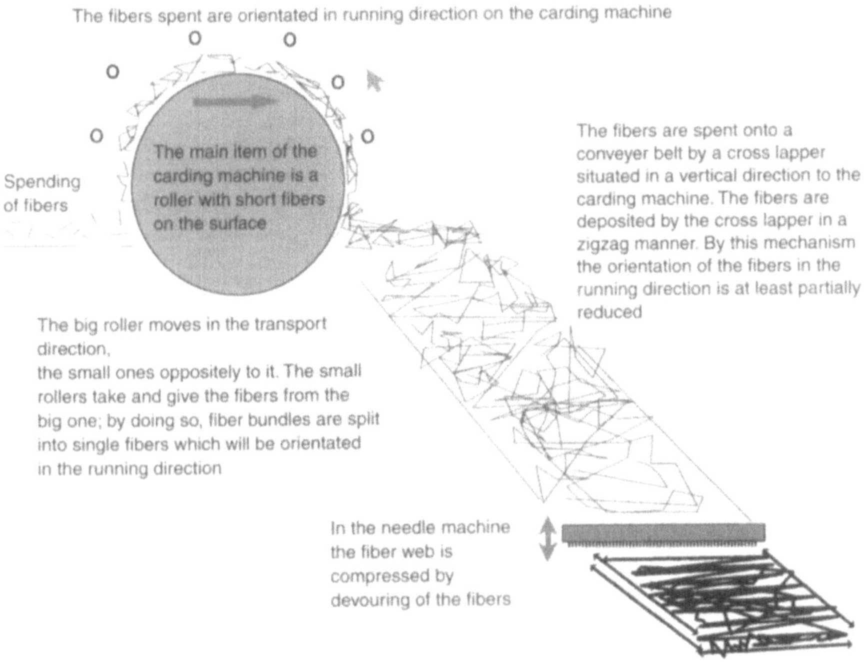

The fibers spent are orientated in running direction on the carding machine

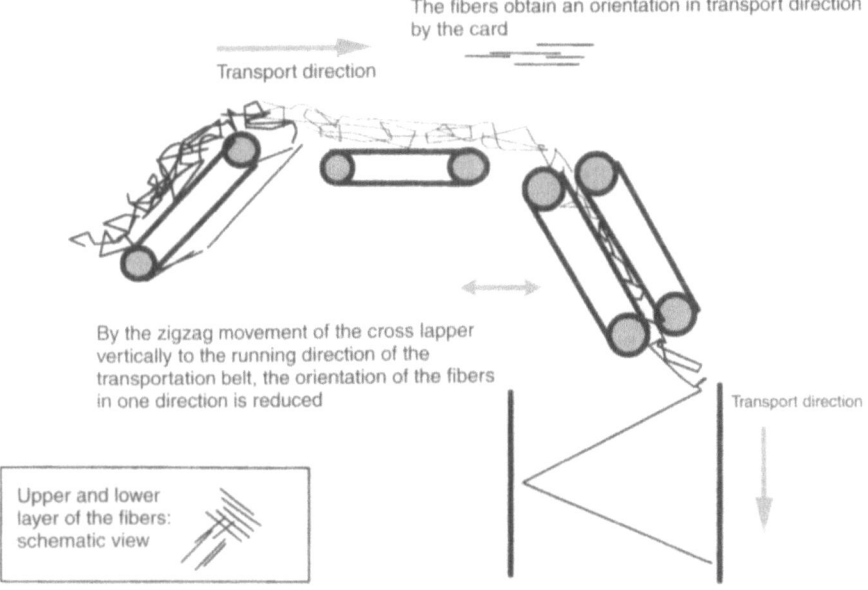

Spending
of fibers

The main item of the
carding machine is a
roller with short fibers
on the surface

The fibers are spent onto a
conveyer belt by a cross lapper
situated in a vertical direction to the
carding machine. The fibers are
deposited by the cross lapper in a
zigzag manner. By this mechanism
the orientation of the fibers in the
running direction is at least partially
reduced

The big roller moves in the transport
direction,
the small ones oppositely to it. The small
rollers take and give the fibers from the
big one; by doing so, fiber bundles are split
into single fibers which will be orientated
in the running direction

In the needle machine
the fiber web is
compressed by
devouring of the fibers

Fig. 21-1. Schematic view of a card

The fibers obtain an orientation in transport direction
by the card

Transport direction

By the zigzag movement of the cross lapper
vertically to the running direction of the
transportation belt, the orientation of the fibers
in one direction is reduced

Transport direction

Upper and lower
layer of the fibers:
schematic view

Fig. 21-2. Cross lapper

Wet Process. Similar to the production of paper textile, fibers are suspended in water at a concentration of 1–5%. Then the fibers in a paper-making machine (e.g. Hydroformer®, trademark of Voith, Germany) are deposited on a paper felt, a Fourdrinier paper felt (see Fig. 20-1) [20, 29, 59, 70].

The deposition of the fibers may occur in the presence of a polymer dispersion [62], e.g. a polyurethane dispersion [66], or in a mixture with thermoplastic fibers [42, 43, 87]. Wet produced nonwovens are nearly impossible to needle due to their bad internal consistence. Therefore the compression wanted can be done with bonding fibers. These are fibers which are stretched during manufacture and cooled under their glass transition temperature (T_g). If these fibers are warmed up to their T_g, they shrink and increase the density of the nonwoven.

Another possibility which increases the strength of the final product is to include in the nonwoven a finely powdered polymer which is able to swell in the solvent of the later applied bonding agent, like polyurethane [63].

Thermoplastic fibers or fibrids of polyurethane which are precipitated with the fibers of the polymer able to sinter or melt under heat create punctual bonding points in the nonwoven [71].

To guarantee a nonwoven without any non-homogeneous parts, fibers for card or wet processing should not exceed a certain maximum length. Fibers for a card process should not exceed a length of 6 cm and for a wet process fibers longer than 4 cm can only be processed in an extremely low concentration.

In contrast to the paper-making procedure, fibers may be deposited from a suspension onto a porous drum by suction (Rotoformer®, trademark of Sandy Hill Corp Hudson Falls NY, USA). Wet formed nonwovens have a lack of firmness, so it is difficult to apply a bonding agent after forming. It is impossible to do so by impregnation. Wet processed nonwovens are normally bonded by spraying and fixed under heat [74].

A special wet process for nonwovens is to prepare a suspension of flocked polyurethane particles, leather and synthetic polymer fibers, a dispersing agent, like naphthalene sulfonic acid condensate and a flocking agent, such as sodium aluminum sulfate in water, which is processed in a paper-making machine to produce a wet formed nonwoven [2] (see also chapter 20). Prepolymers of polyethylene glycol ether and toluylene diisocyanate and stark isocyanate are processed with polyamide fibers in a paper-making machine to form a uniform nonwoven [90].

Pneumatically Produced Nonwovens. A vacuum cleaner filter evidently shows that the textile fibers precipitated on that filter are randomly orientated and do not have any preferable orientation. Evidently in a technical application, fibers being deposited by air should have the desired anisotropic behavior. The Rando-Webber® (trademark of Curlator Corp., East Rochester NY, USA) is a piece of equipment which produces nonwovens by an air stream (for principle, see Fig. 21-3).

Spun Bonded Nonwovens. The combination of the spinning of fibers with the manufacture of nonwovens are spun bond nonwovens. The polymer which produces the fiber is melted and extruded by small nozzles to spin an endless

Fig. 21-3. The principle of a Rando-Webber®

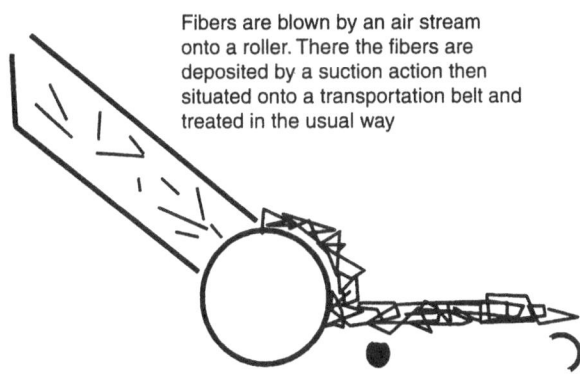

Fibers are blown by an air stream onto a roller. There the fibers are deposited by a suction action then situated onto a transportation belt and treated in the usual way

fiber. The nozzles may be round, oval or Y-shaped [54]. The fibers are drawn out of the nozzle by an air stream, or in an electrostatic field [80]; the speed of drawing may vary [65]. The deposited fibers build a nonwoven. The fibers of this nonwoven are extremely long. Therefore it is difficult or impossible to needle to compress it. Very often an additional compression is not necessary because the freshly prepared fibers are sticky – and therefore bond with each other [67]. Another possibility is to use two-component fibers which may bond with each other when heated to above 200 °C in a calender. Core-shell fibers consisting of a core of polyhexamethylene adipate and a shell of poly-γ-caprolactam may act as a suitable two-component fiber [75]. Another possibility is to use two polymers where one of them melts at a lower temperature than the other [32]. Spraying with a solvent for one of the fibers after the manufacture of the nonwoven and a pressing operation under heat or a treatment with latex also results in a bonded nonwoven.

Polypropylene and polyamide-6 are mutually spun and laid into a nonwoven. With hot air the fibers are stretched during the spinning process [41].

Fibers of a high titre are laid on a spun bonded nonwoven. Then the material is needled and impregnated. By this method an even surface is created [77].

Melt Blown Process. This process is rather new: The fibers produced are very thin within 0.001 – 0.03 den and, therefore, they range in the ultrafine class. The polymer is melted in an extruder by heat and pressure. Then it is forced through a melt blown die and holes at the tip of the die. The holes at the tip are much smaller than the holes of a spinneret in the production of synthetic fibers. Then the fibers are exposed to high-pressure streams of air. Since the polymer is attenuated under a high temperature and pressure as well, it maintains its melted form until it is naturally cut into fine fibers. The fibers are finally blown by cold air so that they may be delivered onto a collecting conveyor belt with a screen stucture to produce a uniform melt blown web [97].

The Compression of Nonwovens. Nonwovens which look like wadding directly after production need to be compressed to reach the necessary firmness and physical properties. Compression is achieved by using fibers able to shrink under the influence of heat [50, 60], IR rays or high frequency [78], by needling, or by a combination of both processes.

A vibrating action [64] or needling with water beams [71, 95] or beams of other liquids [14, 96], or with strong air streams [19], are variations of the above mentioned processes.

The Fibers. The fibers are the dominant part of the character of the nonwoven. Cellulose based fibers, for example, improve the water vapor adsorption as well as the runnability of the fibers on the card. Due to their adsorption of water there is a decrease in the tendency of the fibers to build up electrostatic charges on their surface. Cellulose based fibers decrease physical properties, like flexing and tear strength. In many cases a mixture of polyamide, compressible polyester and cellulose fibers are a good base for nonwovens, such as leather substitutes for shoes.

The thickness of the fibers used and the titre also determines the character of the nonwoven. A general rule is that the finer the titre the softer the nonwoven. It is difficult to spin extremely fine fibers because the nozzles of the spinning unit need to be extremely fine as well. The danger of plugging the nozzles during the production process therefore increases with decreasing titre. Fibers of a titre < 1 den are therefore difficult to produce by the classical processes.

If the fiber should shrink, e.g. to an degree of about 60%, the titre of the fiber prior to shrinking should be even finer than 1 den in order not to become too thick after shrinking. Mostly in Japan new ways of fiber technology have been developed since the 1970 s: these are composite fibers – i.e. mixed spun fibers with a component which can be eliminated later (see Chapt. 21.1).

It is also possible to bind nonwovens by adequate fiber mixtures as we have seen with the spun bonded fibers. Mixtures of polymers like polystyrene and polyamide, polyamide and polyolefin, PVC and polyacrylonitrile or polyester and polyamide are transferred to fibers. The fibers are cut to the desired length. Then a nonwoven is made in a wet or dry process. The nonwoven is treated with a solvent for one of the fibers and slightly heated up. After evaporation or washing the solvent out a leather substitute is obtained [52].

Nonwovens containing more than 50% cellulose fibers and compressible PVA fibers are impregnated with a latex able to coagulate at heat. These nonwovens are hydrophilic [73].

Textile substrates containing mixtures of fibers whereby one component cannot adhere to polyurethane, such as polyolefin, are treated with a polyurethane. By this action the substrates become microporous [90].

The Bonding of Nonwovens by a Latex. As previously explained nonwovens still have a certain isotropy in their physical behavior. Tensile strength and elongation at break differ if they are measured in the running direction or perpendicular to it. With an increasing amount of bonding agent the difference of the tear strength in the two directions is minimized (Fig. 21-4). Therefore bonding is a necessary step in the manufacturing process of nonwovens to obtain isotropic products. Bonding additionally has the effect that the bonded

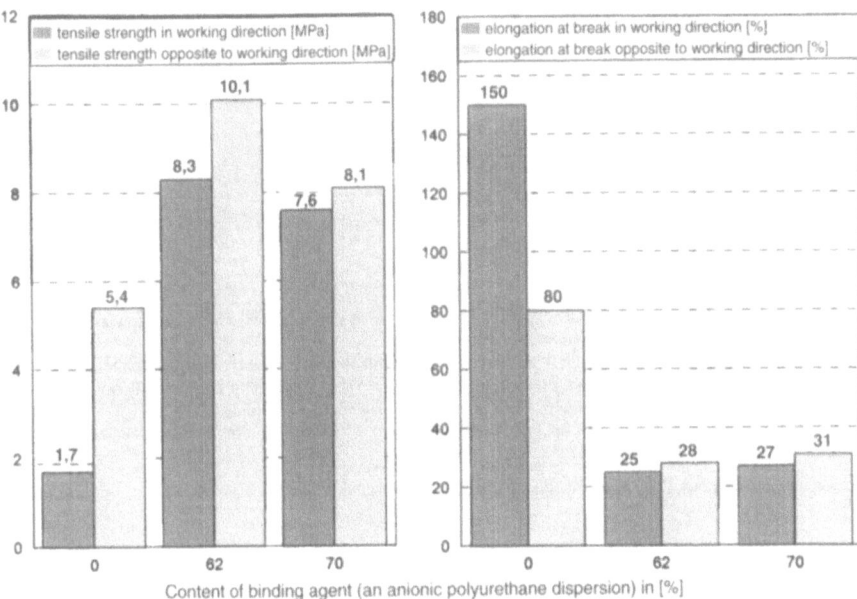

Fig. 21-4. The amount of the bonding agent in the nonwoven decreases the anisotropy of the physical properties. An anisotropic behavior in a nonwoven is negative due to the strong differences of elongation at break, tear propagation strength, etc. The bonding of the nonwoven decreases the anisotropy. The Graph demonstrats that the difference of the tear strength of the nonwoven with 0% bonding agent is more than three times higher if it is measured in a direction opposite to the original working direction. With 62% bonding agent this difference decreases to 20% and with 70% bonding agent to approx 10%

product looks more like a textile and less like a sheet of paper because the nonwoven gains fullness and textile characteristics.

Nonwovens compressed by compression fibers and/or needled are normally bonded with 50–100% of a latex with a natural or synthetic polymer [12, 39, 47, 68, 76]. Spraying or impregnating devices, well known in the textile industry, are used to impregnate the nonwovens [8, 9, 11] (see Fig. 21-5). Auxiliaries like emulsifiers such as arylsulfonic acids or diarylsulfonyl sulfonic acids may be added to the latex [92].

The bonding agent should not fix the fibers to the extent that their internal movement is impossible or difficult [88]. If this bonding agent adheres too much to the fibers, the resulting nonwoven is stiff in these cases. Its tear strength is usually also poor. To reduce the adhesion of the fibers to the bonding agent prior to impregnation the unbound nonwoven is treated with a releasing agent containing a polysiloxane [72]. Mixtures of fibers having a different adhesion towards the bonding agent have a similar effect [91].

Todays state-of-the-art is a *thermocoagulation* of a latex in the nonwoven [93]. In the thermocoagulation a product which becomes insoluble in water at high temperatures is added to the latex. A crosslinker is also added to this mixture and possibly a pigment. If the mixture is heated, this product precipitates. The poly-

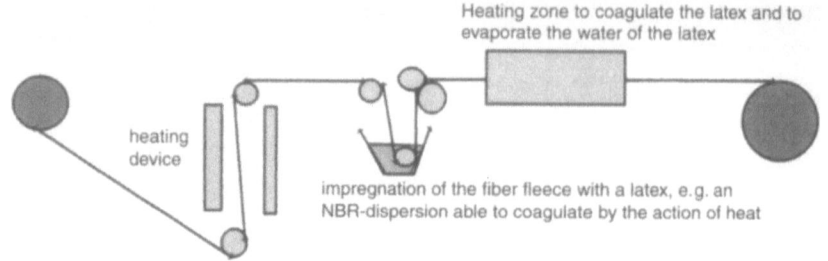

Heating zone to coagulate the latex and to
evaporate the water of the latex

heating
device

impregnation of the fiber fleece with a latex, e.g. an
NBR-dispersion able to coagulate by the action of heat

The unbonded fleece contains mostly cellulose fibres to improve the hydrophilic property, polyamide fibres
to increase the tear strength and polyester.
The polyester fibers are produced and stretched at a temperature below their glass transition temperature
(Tg). By passing into a zone of a higher temperture (than their Tg) the polyester fibers shrink and compress
the nonwoven: The shrinking of the fibers increases the density of the nonwoven further.
The resulting bonded nonwoven can be split in different layers. The surface usually is grinded to become
smooth.

Fig. 21-5. The bonding of a nonwoven

mer in the latex also becomes insoluble and precipitates when heated. A thermo-coagulated latex has a microfoam-like structure similar to a wet coagulated polymer (see Chap. 8). The nonwoven is washed and dried, possibly in a field of high frequency [79].

As bonding agents, water-based dispersions such as natural rubber or co-polymers of the following monomers are suitable: 4–6% methacrylic acid-amide, 5–15% vinyl alcohol, 1–45% vinyl ester like vinyl acetate, and 46–90% C_4–C_{18} alkylacrylate [21] or 10–70% of an ethylene–propylene copolymer [37]. Copolymers of 1,3-diolefins, esters of acrylic or methyacrylic acid and N-sub-stituted, basic compounds of methacrylic amide are also well suited products, due to the fact that they may be crosslinked over a wide pH range [38]. Anionic polyurethanes as dispersions can also be used. The polyurethane coagulates by immersing it in a salt solution [94].

A mixture of a latex able to be thermocoagulated with a PVA solution results in a nonwoven with an improved water vapor uptake [10, 69].

The degree of softness is influenced by the type and make-up of the fibers and the bonding agent. Thermocoagulation is easy and does not need a solvent; therefore, most of the nonwovens nowadays are bonded by a thermocoagulation process.

The use of soft polymers should also result in a soft bonded nonwoven. There are, however, difficulties. Soft polymers – even when they are able to be thermo-coagulated – do not remain in the original foam structure obtained by the pro-cess during drying due to a collapse of the pore by the action of heat. Soft polymers do not normally have the necessary physical properties needed for the final bonded nonwoven.

Polyacrylonitrile and its copolymers with (meth)acrylic acid or esters in a water-based dispersion in a mixture with a solvent such as ethylene carbonate, DMF, butyrolactone, etc., are used to impregnate a nonwoven. The product is dried at 20–200 °C and further treated by heat and pressure. The result is a soft nonwoven [27].

A nonwoven consisting of polyamide fibers is sprayed with 2–20 % of a plasticizer such as benzenesulfonic acid methylamide. After heating to 120–160 °C the nonwoven is solidified [31, 40].

Nonwovens containing 10–50 % cellulose fibers bonded by thermoplastic non-water-based dispersions of vinyl (co)polymers are treated with a plasticizer [33].

The Bonding of Nonwovens by Fibers. Besides bonding with latex other ways have been developed to produce soft nonwovens. One way is to use thermoplastic fibers [44–46, 48, 49, 55] which by leading the nonwoven between two heated steel rollers bond it punctually [23]. Besides steel rollers one roller may consist of steel and the other of silicone [25]. A row of rollers with an increasing temperature [26] hot air or hot steam can also be used [30]. If the nonwovens are under stress they may be bonded thermally with high temperature [36].

Another way is to add to the fibers a fiber able to be dissolved in a solvent or water. After the nonwoven is bonded it is treated with water or solvent. By this process big pores are created in the nonwoven resulting in an excellent increase in softness [16].

The nonwoven may be bonded by applying heat if during the manufacturing of the nonwoven in addition to the original fibers other fibers are used which have a lower melting point, e.g. two polyamides like polyamide-6 or -11 with polyamide-6.6, then the nonwoven may be bonded by applying heat [18].

Partially spun bonded polyamide nonwovens are treated with a 1–2 % solution of zinc chloride. The solution is able to dissolve partially the polyamide and bond the fibers. The remaining salt is washed out after drying [83].

A nonwoven is produced with non- or partially soluble fibers of natural or synthetic origin. This nonwoven is mixed with 1–3 % of PVA. The PVA is insoluble in cold water but is soluble in hot water. The nonwoven is sprayed or impregnated with hot water and is dried at 150 °C. By this method the nonwoven is bonded voluminously [22].

A fabric containing PVA, polyamide and polyolefin fibers spun together is raised and then compressed in hot water. Polyamide is extracted with a solvent to give a voluminous product [51].

A textile substrate on one side containing many fibers which can be dissolved or swollen in a solvent and on the other side only a few of these fibers is needled, then treated with a plasticizer and pressed [35].

Bicomponent fibers consist of mixed spun polymers. They may consist of polyamide-6.6 and polyamide-11, polyethylene terephthalate and polyamide-11 or polyethylene and polypropylene. A nonwoven is prepared from these fibers. Under heat the fibers curl and the components begin melting at lower temperatures and adhere the nonwoven together [53].

Cellulose ether fibers able to swell in water and soluble in alkaline media are mixed with other fibers. A nonwoven is made of this mixture. The nonwoven is then moistened with soda and substances dissolved in water which are able to crosslink, such as formaldehyde. The nonwoven is bonded by this method. The resulting product is voluminous [28].

Nonwovens consisting of polyamide fibers are treated at 50–225°C with polyhydroxylic benzene derivatives like resorcinol, hydroquinone or pyrocatechol. These hydroxylic benzene derivatives are able to bond the fibers punctually [34].

In a nonwoven thermoplastic powders of a particle size of 20–200 μm are shot in by an electrostatic field and fixated thermally [82].

Processes with Nonwovens. The coating of nonwovens is not discussed here due to the fact that all common coating processes have been explained in the previous chapters. Coatings do not correlate with the purpose of this book if they are not water vapor permeable [24]. A coating which may be water vapor permeable consists, for example, of a silicone oil containing, crosslinked polyurethane on a nonwoven impregnated with polyurethane [89].

A layer of 4 mm, 1 den PVA fibers is needled onto a spun bonded polyethylene terephthalate fiber nonwoven with water beams [85]. The same may be done on both sides of the nonwoven with a fabric consisting of fine fibers. The bonding of the composite material is done by liquid injections [86].

A nonwoven is laid onto both sides of a foamed film. The material is needled in such a way that part of the fibers are exposed on the surface [61]. After treatment with heat the nonwoven is compressed [56].

21.1
Special Fibers

Leather consists of extremely fine fibers of collagen which become coarser from the surface (grain) to the reverse (flesh side). To adapt the appearance of a genuine suede leather it seems to be necessary to also have extremely fine fibers on the surface of a synthetic material. The most important invention regarding substrates for man-made leather is the technique using *multicomponent fibers* consisting of fibers of different solubilities. With this technique it is possible to obtain extremely fine fibers resulting in soft, regular and even anisotropic nonwovens or fabrics [13]. Fine fibers have a titre of 1–2. dtex (2 [4]) and ultrafine fibers have a titre of less than 1 dtex. With decreasing titre the surface of a nonwoven becomes increasingly denser and softer. The physical properties increase with decreasing titers. Nonwovens with ultrafine fibers can be needled in the same manner as conventional nonwovens. If the needling is done with water beams in a spun laced technique an increase in firmness of 40 % at a nearly perfect isotropy can be achieved [11].

There are two types of fibers which result in *ultrafine fibers*:

- *Core-shell* fibers: two polymers which are extruded together in a way that the fiber to be removed later is on the outside.
- *Island-sea* fibers: two polymers which are spun together with nozzles in a star-, Y-, V- or cross-shaped structure.

The fibers are transferred to a nonwoven by the previously discussed processes: Normally they pass a card and, if necessary, a thermal compression step is

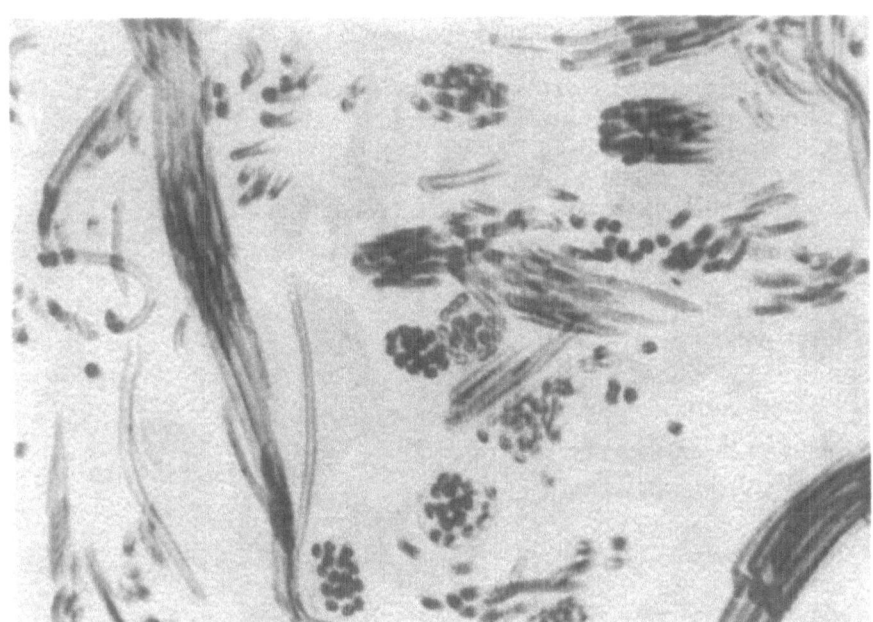

Fig. 21-6. The original fiber bundles can still be seen

followed and then the nonwoven is impregnated. For high quality products the coagulation process with DMF solutions of polyurethanes is still the state-of-the-art. After coagulation the polymer component of the fiber which is no longer necessary is extracted with a solvent for that polymer. Very often the remaining fibers still have the original bundled structure (Fig. 21-6).

If the extraction is only partially carried out the most soluble polymer can also act as a bonding agent and bond the fiber net punctually [2].

Examples of this process are: 70 % polyamide shell and 30 % polystyrene in the core [1] or a shell consisting of polyamide-6, -6.6, -6.10, -11, etc., and a core of polyethylene or polypropylene [3]. A nonwoven is produced with fibers of polyamide-6 and polyethylene (1:1) bonded with a polyurethane solution in DMF. After coagulation, polyethylene is extracted with hot toluene. The product is dried and buffed [7].

Hollow fibers consist of polymer with different solubilities where the most soluble component is dissolved and extracted [8].

Polyamide-6 and polyethylene terephthalate are co-spun in such a way that one polymer is V-shaped and the other radial. The primary fiber has a titre of 30 den and a knitted fabric is produced. By chemical or physical methods the fibers are finally spliced to get a man-made suede [9].

Fibers consisting of the sea component polystyrene (10 parts) and an island component of polyester or polyamide (90 parts) fibers are transferred to a nonwoven which is impregnated with a polyurethane dissolved in DMF. As the polyester component, polyethylene terephthalate has been especially mentioned

[4]. The polyurethane is coagulated and polystyrene is extracted with chloroform [6].

Fibers are produced by co-spinning two high molecular substances with different solubilities. The fibers are dispersed in water and then transferred in a paper machine to a nonwoven. By immersing the nonwoven in a solvent one of the components swells and adheres the fibers together or is dissolved.

Combinations of polymers with different solubilities are:

- polyamide and polystyrene,
- polyamide and polyolefin [31, 32],
- PVC and polyacrylonitrile,
- polyester and polyamide,
- PVA and PVC, and
- PVA and polyacrylonitrile.

Solvents for the fiber combinations are [19]:

- calcium chloride in methanol,
- titanium tetrachloride,
- lithium chloride,
- cyclic esters,
- aliphatic hydrocarbons, and
- ketones

Nonwovens consisting of fibers with two high molecular weight polymers are bonded with polyurethane or a latex of synthetic rubber. Polymer A of the two fiber-building polymers consists of ultrafine fibers of 0.005 – 0.5 den in the core and in the shell of polymer B with 1 – 20 den. Polymer B, e. g. polystyrene, can be dissolved by toluene [20]. As core-shell fibers, polyethylene terephthalate or polyamide in the core and polystyrene as the shell [21, 22] or polyester amide in the core and polyolefin, polyacrylonitrile or polyurethane in the shell can be used [23]. Other fibers consist of a core of polyester, polyamide or polypropylene and a shell of 95 – 15 % polystyrene or PVA [24].

Figure 21-7 shows an artificial leather manufactured from fibers where one component is dissolved. For comparison, split leather (Fig. 21-8) is also shown.

In Fig. 21-9 a nonwoven consisting of fine fibers is shown which is compared to the surface of a genuine split leather (Fig. 21-8). This comparison shows that beside the fibre stucture there is almost no difference in the look of both materials. Figure 21-10 shows the principle of manufacturing with these kinds of processes (7.1 [14]).

A nonwoven consisting of multi-component and thermoplastic fibers is heated, melted and compressed. Then it is impregnated with an solution of an elastomer able to swell the thermoplastic fibers. After solidifying the elastomer, one of the components of the fiber is leached out. The multi-component fiber may consist of polyamide-6 and polystyrene. Polystyrene may be easily dissolved by toluene [17].

Nonwovens consisting of polyamide and polyethylene fibers are impregnated with polyurethane. Polyethylene is dissolved with toluene with the assistance of the action of a brush roller [18].

Fig. 21-7. Astrino® (trade-mark of Kuraray)

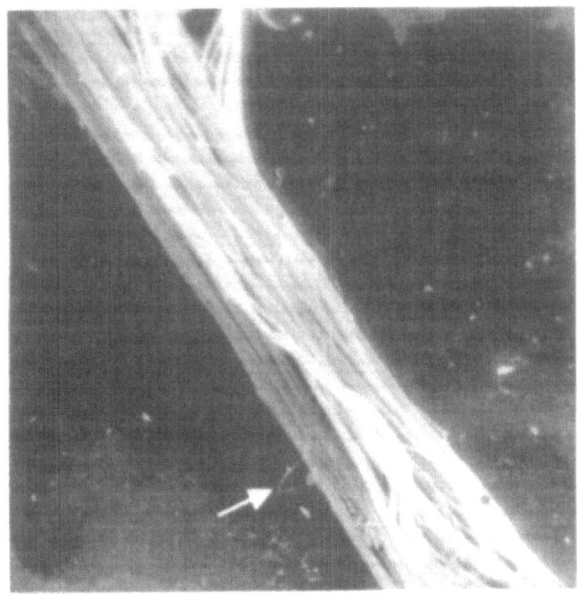

20 µm

Fig. 21-8. Split leather as a comparison

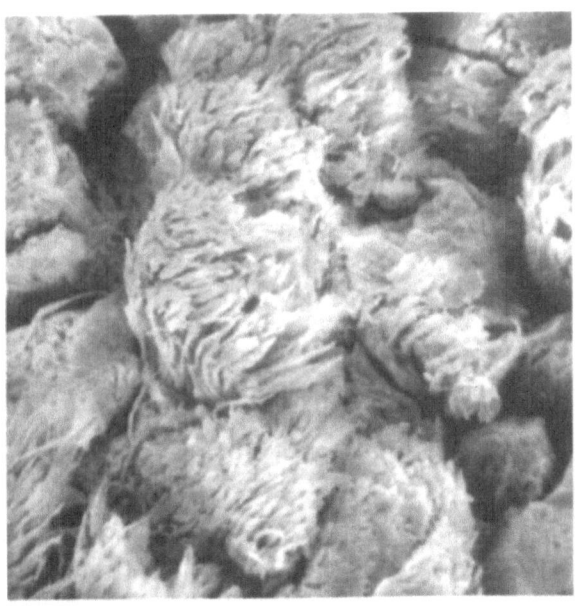

50 µm

Fig. 21-9. Clarino® nonwoven
(Clarino® is a trademark of
Kuraray)

50 µm

A nonwoven consisting of polyamide-6 and low-density polyethylene co-spun fibers is impregnated with polyurethane, coagulated and then the polyethylene is extracted with toluene [10, 14].

Extremely soft substrates can be produced by using core-shell fibers for the nonwoven. The nonwoven is impregnated with a water-soluble polymer, then the shell component of the fibers is extracted and an impregnation with a DMF solution of polyurethane follows. By coagulating the polyurethane the water-soluble polymer is also extracted [15].

Fibers still able to react further are partially polymerized. A fabric or a non-woven is produced with these fibers. With a solvent the unreacted parts are leached out and the product is stretched in various directions [5].

Lyocell® {trademark of Courtaulds, UK and Lenzing, Austria (29.1 [17])} fibers are able to be spliced if they are treated in DMF, DMAc or NMP under high shear stirring. Nonwovens may be produced with these fibers or the fibers can be added to other fibers as additives [12].

Another form of fine fibers – sometimes called *fibrids* – can be produced by precipitating dissolved polymers in the presence of a non-solvent by stirring under high shear forces. According to the literature this kind of fiber production is expensive due to the energy needed [29]. Therefore it is only carried out when the special effect the fibrids have cannot be achieved by other means. For functional textiles, for instance, the use of polyurethane fibrids offers an advantage in water vapor permeability, high impermeability against water and a high crease recovery angle.

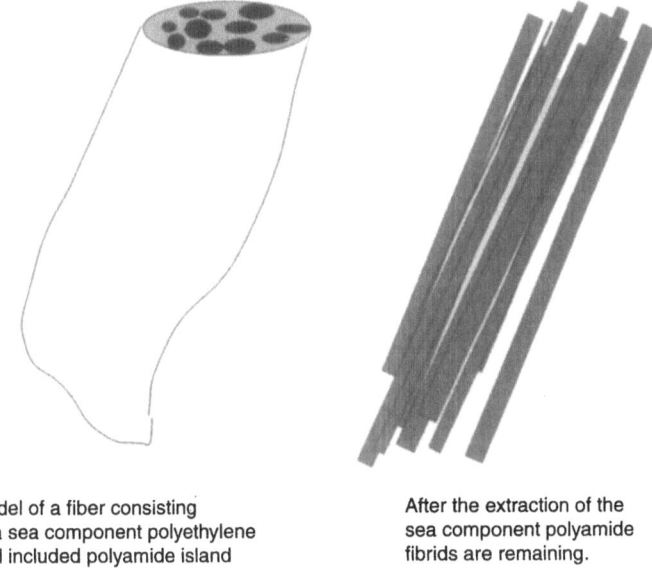

Model of a fiber consisting
of a sea component polyethylene
and included polyamide island
fibrids.

After the extraction of the
sea component polyamide
fibrids are remaining.

Fig. 21-10. Principle of the production technique of a nonwoven using ultrafine fibers [25]

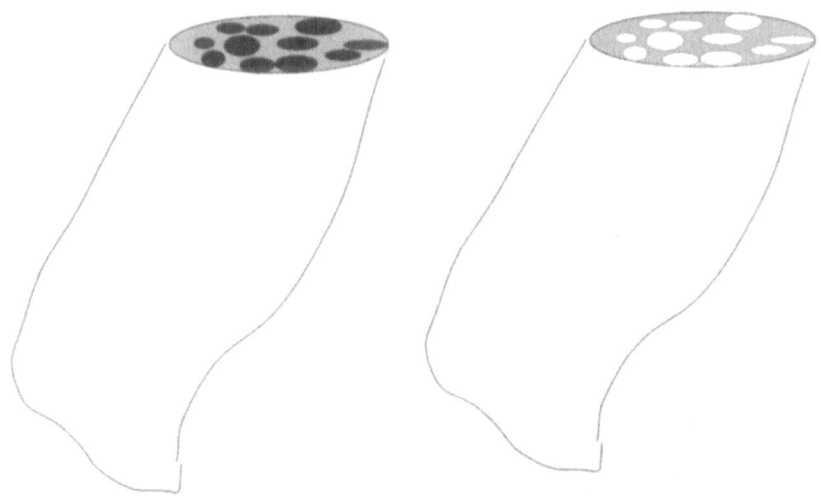

A fiber consisting of a polyamide sea
and a polyethylene island component

After the extraction of the island components a fiber
is remaining with hollow chanels inside. These
hollow chanels effect extreme softness of the fibers
containing them.

Fig. 21-11. Model of an island-sea fiber; extraction of the island component

1.) A nonwoven is produced with cospun fibers consisting e.g. of the sea component polyamide and the island component polystyrene.

4.) The polyurethane is coagulated in water and – simultaneously – the polyvinylalcohol is washed out.

5.) Finally the sea component of polystyrene is extracted with toluene. By this action only the extremely fine polyamide fibers remain in the nonwoven. If fibers should be produced where the island component should be extractable, then the mixture of the fibers is opposite to this example. The island part in this case should then consist of polystyrene.

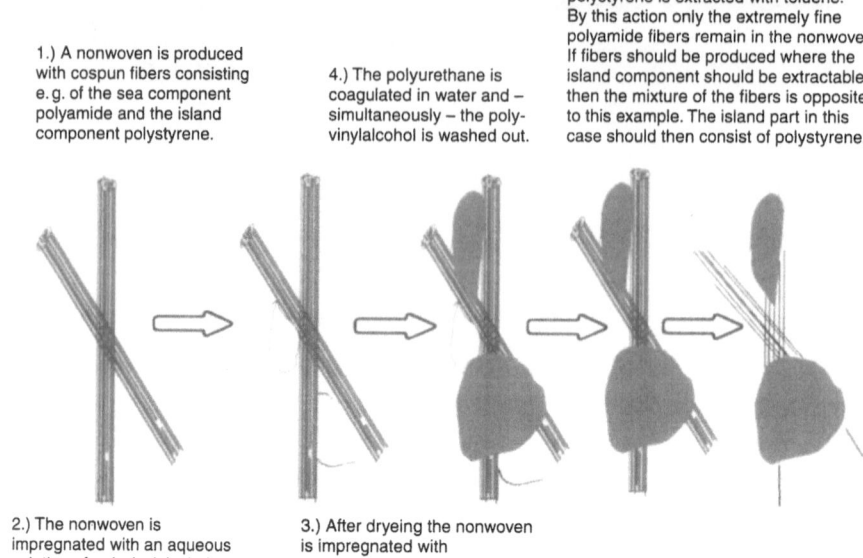

2.) The nonwoven is impregnated with an aqueous solution of polyvinylalcohol.

3.) After dryeing the nonwoven is impregnated with a solution of a polyurethane in DMF.

Fig. 21-12. Reduction of the adhesion of the binding agent to the fibers by an intermediate binding agent which is dissolved during the coagulation process

A film produced by a polyurethane solution is unraveled under shear forces into fibrids which possibly mixed with other staple fibers can be transformed into man-made leathers [26].

A nonwoven containing up to 50 % fibrids is produced by a wet process. The fibrids melt at least 20 °C below the other fibers. After forming a nonwoven and drying a treatment by heat and pressure follows and the nonwoven is bonded punctually. The fibrids are produced from solutions of polyethylene glycol terephthalate under high turbulence in a bath of non-solvent [27].

Fibers consisting of an easily soluble sea component and a thermoplastic island component able to fibrillate are used in the production of a synthetic suede [28].

The stonewash effect on jeans – usually done by mechanical action and/or enzymes – can also be achieved by using conjugate polyester and polyamide fibers. A fabric produced by using these fibers which can be fibrillated on the surface and afterwards impregnated with polyurethane, embossed and dry-milled, looks like it has been stonewashed [30].

References to Part 3

11
Blowing and Foaming

See also: 8.1 [3, 12], 15 [20]

1. FR 1 113 905 (Degussa; D Prior. 31.10.53; 27.10.54/5.4.56)=BE 532 919
2. Neumaier HH (1974) Aqueous dispersions of polyurethane ionomers for coating and laminating. J Coated Fabr 3:181; Zimmermann B (1975) Mechanically frothed urethane foams in simulated leather composites. J Coated Fabr 5:124
3. OS 2 137 048 (Et. Hutchinson Co. Nat. du Caoutchouc; C Canat; FR Prior. 24.7.70 and 10.2.71; 23.7.71/30.3.72)
4. OS 2 210 721 (Tenneco Chemicals; J Winkler; US Prior. 5.3.71 and 27.12.71; 6.3.72/2.11.72)
5. OE 229 047 (O Zwoboda; 14.10.60/15.1.63)
6. RA 183 373 (Stern et al.; 18.5.61/Jan.67)
7. GB 1046 355 (Borg-Warner Corp.; JA Johnston; 13.2.64/26.10.66)
8. FR 1 513 193 (Dynamit Nobel AG; P Spielau; D Prior. 8.3.66; 7.3.67/9.2.68)=NE 6 703 579
9. JA 55 137 277 (Matsumoto Yushi Seiyaku; 16.4.79/25.10.80); see also JA 55 137 276
10. EP 123 966 (W. L. Gore & Assoc. Ind.; SJ Brinton, JL Manniso; 6.4.84/7.11.84; US Prior. 3.5.83)
11. Varhanik J (1977) Lösemittelfreies PUR-System und seine Anwendung für Poromere. Kzarstvi 27(8): 229
12. DE 2 009 070 (Uniroyal Inc., EC. van Buskirk; 26.2.78/17.9.70; US Prior. 26.2.69) addition to DE 1 469 578
13. JA 79–28 444 (Kuraray; 10.5.72/17.9.79)=JA 72–46 683

11.1
Production of Foams

1. OS 1 619 240 (C Freudenberg; HD Krug et al.; CH Prior.1.11.65; 8.10.66/2.10.69)=OE 259 509 =SZ 467 378
2. OS 1 947 124 (ICI Ltd.; CW Phelps; GB Prior. 17.9.68; 17.9.69/16.4.70)
3. OS 2 121 227 (Dunlop; JA Jenkins, JO Wood; GB Prior. 1.5.70; 29.4.71/11.11.71)
4. NE 6 815 671 (Dunlop; GB Prior. 4.11.67 and 2.4.68; 4.11.68/6.5.69)
5. OS 1 806 121 (Dunlop; MW Higgs, DI Clarke; GB Prior. 4.11.67 and 2.4.68; 30.10.68/19.9.69) addition to OS 1 619 222
6. BE 713 947 (Dunlop; GB Prior. 21.4.67; 19.4.68/16.9.68)
7. OS 1 919 804 (Dunlop; RSt Wharton et al.; GB Prior. 20.4.68; 18.4.69/17.9.70) addition to OS 1 619 222
8. US 3 355 535 (Union Carbide; EWR Hain et al.; 13.12.63/28.11.67)
9. FR 1 382 197 (Dunlop; GB Prior. 17.10.62 and 9.9.63; 17.10.63/9.11.64)
10. SZ 418 284 (C Freudenberg; 1.11.63/15.2.67)=DOS 1 444 152=OE 257 534
11. JA 23 857/64 (R Kawaguchi; 2.8.63/24.10.64)
12. JA 12 073/67 (Kokoku Chem. Ind. Co. Ltd.; 3.9.63/10.7.67)

13. JA 9 025/67 (Y Mizuno; 31.8.67/1.5.67)
14. JA 27 275/67 (Kokoku Chem. Ind. Co. Ltd.; 15.8.64/23.12.67)
15. DDR 50 120 (Vogtländ Kunstlederfabrik; W Schubert, W Schumann; 16.9.65/25.10.66)
16. OS 1 900 975 (Dunlop; RJ Speight, AG Marriott; GB Prior. 10.1.68 and 28.3.68; 9.1.69/ 4.9.69)
17. BE 712 582 (Dunlop; GB Prior. 21.3.67; 21.3.68/31.7.68)
18. DOS 2 425 553 (Rohm and Haas Co.; J. F. Levy et al. US Prior. 9.7.73; 27.5.74 / 27.3.75)
19. Levy JF et al. (1975) Die Anwendung von zusammengepresstem Acrylschaum auf Leder. Rev Techn Ind Cuir 67:6, 194
20. OS 1 957 889 (W Glander; 18.11.69/3.6.71)
21. JA 48 039 602 (Kawaguchi Rubber Ind. Co.; 30.9.71/11.6.73)
22. GB 885 084 (ICI; A Lowe, JF Wood; 5.7.57/16.6.58/20.12.61)
23. US 3 085 896 (Interchemical Corp.; NG. Britt et al.; 10.2.58/16.4.63)
24. FR 1 491 627 (ICI; GB Prior. 10.8.65; 10.8.66/3.7.67)
25. JA 4 672/68 (Kokoku Chem. Ind. Co. Ltd.; 16.7.65/20.2.68)
26. FR 1 320 997 (Marles-Kuhlmann-Wyandotte; KC Frisch, SL Axelrood; US Prior. 9.4.65 and 19.8.65; 8.8.66/5.1.68)
27. SZ 476 153 (C Freudenberg; 27.10.66/15.9.69)
28. FR 2 016 224 (Bayer; D Prior. 22.8.68; 22.8.69/8.5.70)=BE 737 724
29. US 3 642 563 (Burlington Ind. Inc.; DA Davis, SR Avvette, GD Voss; 23.9.69/15.2.72)
30. OS 1 629 275 (Bukflex Processes Ltd.; E Bukalder; GB Prior. 28.1.65; 27.1.66/23.4.70)
31. BE 682 956 (Genset Corp.; JL Zuckermann; US Prior. 2.2.65; 21.6.66/22.12.66)
32. GB 1 223 624 (DuPont; A Mitchell; US Prior. 6.11.67; 6.11.68/3.3.71)=CA 844 665
33. JA 11 754/65 (Fujikura Rubber Works Ltd.; 29.7.63/11.6.65)
34. OS 1 951 977 (Kalle; KD Hammer, H Porrmann; 15.10.69/22.4.71)=BE 757 409
35. JA 48 48 605 (Dainichiseika Colour and Chem. Mfg. Co. Ltd.; 15.10.71/10.7.73)
36. JA 48 039 602 (Kawaguchi Rubber Ind. Co.; 30.9.71/11.6.73)
37. FR 2 126 932 (Soc. Naphthachimie; C Bovis, A Gibier-Rambaud; 28.1.71/13.10.72)
38. OS 1 815 527 (Hooker Chem. Corp.; AC Tieniber; US Prior. 18.12.67 and 20.11.68; 18.12.68/ 14.8.69)
39. JA 5762/64 (Shibata Gomu Kogyo KK; 24.5.61/28.4.64)
40. FR 1 377 774 (BF Goodrich; WHV Larner, CF Schollenberger; US Prior. 19.12.62; 14.12.63/ 28.9.64)=GB 1 061 655=US 3 214 290
41. JA 25 473/69 (Rubber Ind. Co. Ltd.; 24.12.63/27.10.69)
42. JA 23 858/64 (Sumitomo Chem. Ind. Co. Ltd.; 22.8.63/24.10.64)
43. JA 73/09 585 (Asahi Chem. Ind. 2.12.69/26.3.73)
44. NE 6 800 305 (Gebr. Holzapfel; 9.1.68/11.7.69)
45. US 3 661 671 (Nippon Kakoh Seishi KK; K Katagiri et al.; JA Prior. 21.2.69 (44/12 513); 20.2.70/ 9.5.72)
46. BE 697 159 (Celanese; SE Jarrison, CH Anthony; US Prior. 18.4.66; 18.4.67/18.10.67)
47. GB 1 130 773 (Dunlop; PI Wilson, JF Yardley; 16.7.66/16.10.68)=BE 701 412
48. US 3 438 829 (Rohm and Haas Co.; B Coe; 28.6.66/15.4.69)
49. US 3 483 069 (AD Little Inc.; WJ Cairns, RS Lindstrom; 23.8.65/9.12.69)
50. RA 225 805 (SA Peltex; L Collet, H Chentrier; 25.8.66/Juni 69; FR Prior. 10.9.65), FR 1 464 822 (Soc. An. Peltex; 24.11.65/28.11.66)
51. JA 73/41 523 (Toyo Rubber Chem.; 26.5.70/7.12.73)
52. JA 48 025 066 (Dainippon Ink. and Chem. Co.; 3.8.71/2.4.73)
53. FR 2 131 796 (Et. Pennel et Flipo; 2.3.71/17.11.72)
54. US 3 713 868 (Gen. Latex and Chem. Corp.; PL Gordon; 6.1.71/30.1.73)
55. JA 48 008 902 (M Nakamura; 17.6.71/3.2.73)
56. Durante AJ, Zimmermann B (1975) Mechanisch geschlagener Urethanschaum für Lederersatz-Verbundmaterialien. Ein Überblick über die Technologie der Textilbeschichtung und die Erfordernisse für Schaum-Verbundstoffe. J Coated Fabr 5:124
57. JA 50/018 602 (Toyo Cloth KK; 1.6.73/27.2.75)
58. JA 75 033 121 (Sekisu Chem. KK; 30.5.66/28.10.75)
59. JA 50 025 870 (Unitika KK; 3.7.73/18.3.75)

60. JA 50 013 506 (Mitsubishi Rayon KK; 11.6.73/13.2.75)
61. JA 50 018 603 (Toyo Cloth KK; 22.6.73/27.2.75)
62. OS 2 528 947 (Rohm and Haas Co.; CT Arkens; US Prior. 3.7.74; 28.6.75/22.1.76)
63. JA 50 053 502 (Tsuyaei Kogyo Co. KK; 22.9.73/12.5.75)
64. JA 50 039 364 (Mitsubishi Petrochem KK; 13.8.73/11.4.75)
65. FR 2 012 904 (ICI; GB Prior. 12.7.68; 11.7.69/27.3.70)
66. Ullmann's Encyclopedia of Industrial Chemistry vol.A23; W. v. Langenthal Rubber, 5.Technology p 427
67. WO 9321263 (AH Hides & Skins Australia Pty.; IB Portaya, ES Vainerman; 22.11.93/28.10.93) =NZ 93–0250249=EP 612791
68. EP 295 677 (VEB Vogtländische Kunstlederfabrik Tannenbergsthal; M Olschewski, R Becker et al.; 17.6.87/21.12.88)=DDR 87–303 882
69. US 3 939 021 (Toyo Cloth; Y Nishibashi; 17.2.76)
70. Yen M-S, Yeh T-C (1996) Effect of foaming methods on moisture transfer and waterproof properties of foamed finished Nylon fabrics. Sen i Gakkaish 52:639

11.2
Treatment of Foams

1. US 3 709 752 (Pandel-Bradford Inc.; R Wistozky, RE Petersen; 20.1.71/9.1.73; cip 17.8.67)
2. JA 38 918/70 (Honey Otafokuwata; 25.12.65/8.12.70)
3. JA 20 789/70 (Kurashiki Rayon; 18.5.66/15.7.70)
4. US 3 694 530 (Goodyear Tyre; JD Wolfe; 17.11.69/26.9.72)
5. JA 13 556/67 (Tokyo Toyo Rubber; 5.9.63/1.8.67)
6. JA 2 589/69 (Toyo Gomu Kagaku Kogyo KK; 30.12.65/3.2.69)
7. FR 1 587 855 (Dunlop; GB Prior. 5.4.67 and 4.11.67; 4.4.68/3.4.70)
8. US 3 622 435 (AF Cacella; 3.11.67/23.11.74)
9. A survey with many references is, e.g., Bennett B (1955) Progress in foamed latex sponge. Rubber Age 78:67 (103 references)
10. Schmidt P Beschichten mit Kunststoffen. München, 1967; Schmidt P Die Beschichtung von Textilien mit PVC Schaumstoffen (1963) Melliand 76:186, 391, 1251, 1373; Schmidt P, Polte A (1967) Beschichten von Textilien mit offenporigen PVC-Schaumstoffen. Kunststoffe 57:25
11. JA 7 471/68 (Kyowa Leather Co. Ltd.; 9.4.64/21.3.68)
12. FR 2 041 450 (C El-Baz Nouchy; 24.4.69/29.1.71)
13. GB 923 207 (US Rubber Co.; US Prior. 8.3.60; 28.2.61/10.4.63)
14. GB 930 438 (SW Alderfer; US Prior. 30.1.61; 18.4.61/3.7.63)
15. JA 7 711/67 (Asahi Electrochem.Co. Ltd.; 22.1.63/29.3.67)
16. JA 9 035/67 (Asahi Denka Ind.Co. Ltd., 18.6.63/1.5.67)
17. SZ 426 250 (Union Carbide; DE Peterson, RR Cosner; US Prior. 26.7.63; 24.7.64/15.6.67)=GB 1 071 824=FR 1 418 071=BE 650 907
18. US 3 360 415 (General Foam Corp.; C Hellmann, JB Lodi; 21.10.63/26.12.67)
19. JA 11 755/65 (Kokoku Kagaku Kogyo Co. Ltd.; 27.8.63/11.6.65)
20. JA 468/69 (Kokoku Chem. Ind. Co. Ltd.; 9.11.63/10.1.69), JA 469/69 (Kokoku Chem. Ind. Co. Ltd.; 11.11.63/10.1.69)
21. JA 9 038/67 (Toyo Rubber Co. Ltd.; 17.10.63/1.5.67), JA 9 037/67 (Toyo Rubber Co. Ltd.; 8.8.63/1.5.67)
22. OS 1 901 613 (Kay-Metzeler Ltd.; WJ Geddes, GD Walmsley; 14.1.69/6.8.70)
23. JA 46 891/72 (Nippon Art Paper Mfg. Co. Ltd.; 13.8.69/27.11.72)
24. US 3 793 414 (Tenneco Chem.; F Buff et al.; 9.12.71/19.2.74; cip 22.1.69)
25. GB 1 240 664 (Dunlop; RSt Wharton et al.; 20.4.68/28.7.71) addition to GB 1 177 483
26. OS 1 930 585 (Dunlop; JM Lowe, RF Morris; GB Prior. 29.6.68; 16.6.69/8.1.70)
27. OS 1 917 023 (Dunlop; DI Clarke, TG Evans; GB Prior. 2.4.68; 2.4.69/23.10.69)
28. GB 1 237 397 (Dunlop; DI Clarke, MW Higgs; 16.4.68/30.6.71); addition to GB 1 177 483
29. NE 6 905 404 (Genset Corp.; US Prior. 5.4.68/8.4.69/7.10.69)

30. OS 1 925 567 (Scott Paper Co.; Ch Shoustal et al.; US Prior. 20.5.68; 20.5.69/27.11.69)
31. JA 39 322/72 (Toyo Rubber Chem. Ind. Co. Ltd.; 6.5.68/4.10.72)
32. FR 1 550 790 (Dunlop; GB Prior. 11.1.67; 10.1.68/20.12.68)
33. FR 93 925 (Dunlop; GB Prior. 11.1.67 and 21.3.67; 11.1.68/6.6.69)=NE 6 800 346
34. OS 1 794 033 (Toray Ind. Inc.; K Matsumoto, K Mihira; JA Prior. 29.8.67 (=JA 54 999–67) and 27.2.68 (=JA 11 995–68); 29.8.68/10.2.72)=FR 1 602 701
35. BE 706 897 (BASF; D Prior. 22.11.66; 22.11.67/22.5.68)
36. FR 1 522 613 (Dunlop Co. Ltd.; GB Prior. 6.4.66; 6.4.67/26.4.68)=NE 6 704 881; FR 1 518 133 (Dunlop Co. Ltd.; GB Prior. 6.4.66; 6.4.67/22.3.68)=NE 6 704 882; FR 1 518 134 (Dunlop Co. Ltd.; GB Prior. 6.4.66; 6.4.67/22.3.68)=NE 6 704 883
37. NE 6 709 262 (Genset Corp.; US Prior. 5.7.66;4.7.67/8.1.68)=BE 700 939=FR 1 530 302
38. JA 20 800/70 (Kurashiki Rayon; 24.5.66/15.7.70)
39. BE 708 657 (Kalle AG; D Prior. 30.12.66; 27.12.67/27.6.68)
40. OE 277 591 (Semperit; J Schwab, K Kromp; 15.9.66/29.12.69)
41. FR 1 428 287 (Agricola Reg.Trust; 5.3.65/3.1.66)
42. JA 12 077/67 (Bridgestone Tire Co. Ltd.; 11.3.65/10.7.67)
43. FR 1 493 185 (Genset Corp.; F Buff et al.; US Prior. 14.9.65; 12.9.66/17.7.67)=NE 6 612 957
44. GB 1 128 611 (ICI; FW Lord; 3.2.65/25.9.68)
45. BE 682 972 (TV Peters; US Prior. 6.7.65; 22.6.66/1.12.66)
46. FR 1 451 374 (R. Koepp & Co.; D Prior. 7.7.64; 6.7.65/25.7.66)
47. JA 4 672/69 (Sekisui Chem. Ind. Co. Ltd.; 3.12.64/26.2.69)
48. US 3 776 790 (Inmont Corp.; GN Harrington, FD Civardi; 4.12.70/4.12.73)
49. US 3 788 882 (Richardson Co.; ThM Noone; 23.6.72/29.1.74; div. 18.2.70)
50. Brodbeck M (1972) Flamm-Bondieren und -Laminieren mit PUR Schaum. J Coated Fabr 2(10):57
51. OS 2 230 308 (Gebr. Holzapfel & Co. KG; N Holubova; 21.6.72/10.1.74)
52. OS 1 900 975 (Dunlop; RJ Speight, AG Marriott; GB Prior. 10.1.68 + 28.3.68; 9.1.69/4.9.69)
53. OS 1 919 804 (Dunlop Co. Ltd.; RSt Wharton et al.; GB Prior. 20.4.68; 18.4.69/17.9.70) addition to OS 1 619 222
54. NE 6 815 671 (Dunlop; GB Prior. 4.11.67 + 2.4.68; 4.11.68/6.5.69)
55. OS 1 806 121 (Dunlop; MW Higgs, DI Clarke; GB Prior. 4.11.67+2.4.68; 30.10.68/19.9.69) addition to OS 1 619 222
56. JA 12 073/67 (Kokoku Chem. Ind. Co. Ltd.; 3.9.65/10.7.67)
57. BE 713 947 (Dunlop; GB Prior. 21.4.67; 19.4.68/16.9.68)
58. US 3 355 535 (Union Carbide; EWR Hain et al.; 13.12.63/28.11.67)
59. Oertel G (1993) Polyurethane. In: Becker, Braun (eds) Kunststoffhandbuch, vol 7. München, p 209
60. US 4 454 191 (H v Blücher; 17.8.82/12.6.84; D Prior. 17.8.81)=DE 3 132 324
61. US 4 515 844 (Nylco Corp.; 19.3.84/7.5.85)
62. US 4 122 223 (Inmont; FP Civardi, FC Loew; 30.1.75/24.1.78)=US 75–544548
63. GB 1 600 839 (Inmont; 19.9.77/21.10.81)=US 77–83457

12
Microporosity by Controlled Melting Processes

1. BE 664 168 (Soc. Rhodiaceta; 20.5.64); BE 657 849 (Balamundi; 7.1.64)
2. DOS 2 314 512 (Bayer; K Noll; 23.3.73/17.10.74)
3. BE 663 102 (Wyandotte; 27.4.64/27.4.66)
4. DAS 1 097 678 (DuPont; 30.6.53/19.1.61); US 2 968 575 (DuPont; 21.5.54/17.1.61)
5. NE 6 805 896 (Bayer; D Prior. 28.4.67; 25.4.68/29.10.68)
6. BE 714 305 (Bayer; D Prior. 28.4.67; 26.4.68/19.9.68)=NE 6 805 988=US 3 622 527=DOS 1 694 148
7. BE 716 587 (Bayer; D Prior. 16.6.67; 14.6.68/4.11.68)=NE 6 808 422
8. BE 714 304 (Bayer; D Prior. 28.4.67; 26.4.68/16.9.68)
9. SZ 377 532 (SC Johnson and Son Inc.; HR Leeds; US Prior. 14.1.57; 13.1.58/15.5.64)

10. US 3 879 515 (M Yoshihige; 18.10.73/22.4.75)
11. DOS 2 350 765 (Bayer; KA Weber et al.; 10.10.73/24.4.75)
12. OS 1 964 059 (Kalle; E Hutschenreuther et al.; 22.12.66/24.6.71)
13. OS 2 226 526 (Bayer; H Reiff et al.; 31.5.72/13.12.73)
14. OS 2 259 614 (Aktiebolaget Bofors; ChL Haakanson, HP Schmid; SE Prior. 8.12.71; 6.12.72/14.6.73)
15. FR 1 276 161 (Motte-Bossut SA; 7.10.60/9.10.61)
16. FR 1 354 611 (Rogers Corp.; JJ Abell; US Prior. 5.9.62; 19.4.63/27.1.64)
17. US 3 348 991 (Rogers Corp.; JJ Abell, RC Berry; 12.1.66/24.10.67, div. ex 20.12.62)
18. JA 26 896/63 (Takasago Toryo Co. Ltd.; 11.6.62/27.12.63)
19. OS 1 917 223 (Kalle; K Andrä et al.; 3.4.69/15.10.70)
20. OS 1 964 061 (Kalle; H Porrmann et al.; 22.12.69/24.6.71); addition to OS 1 917 223
21. NE 6 817 872 (Licencia Talalmanyokat Ertekesitö Vallalat; HU Prior. 16.12.67; 12.12.68/18.6.69)
22. SZ 470 525 (C Freudenberg; L Hartmann; D Prior. 26.10.66; 7.8.67/14.5.69)
23. GB 1 157 457 (Rhone Poulenc; P Rouault; FR Prior. 1.7.66; 30.6.67/9.7.69)=NE 6 708 804 =FR 1 491 556=BE 700 814
24. BE 719 431 (Thorne; CA Prior. 10.5.65; 13.8.68/16.1.69)
25. BE 669 712 (Ludlow Corp.; RC Wilkie et al; US Prior. 21.9.64 and 25.6.65; 16.9.65/13.3.66) =NE 6 512 242
26. FR 2 079 922 (Aquitaine-Organico; C Kaspar, P Lescaut; 17.2.70/ï2.11.71)
27. OS 2 134 779 (Philip Morris Inc.; NB Rainer, PA Wilson; US Prior. 10.8.70; 12.7.71/17.2.72)
28. JA oo 535/73 (Toray Ind. Inc.; 25.6.70/9.1.73)
29. DAS 1 273 802 (RC Buchmann; 17.5.67/25.7.68)
30. DAS 1 127 072 (C Freudenberg; Th v Schachowskoy; 18.6.60/5.4.62)
31. US 4 248 552 (Inmont Corp., FP Civardi et al.; 23.5.79/3.2.81); US 4 341 581 (Inmont Corp.; FP Civardi, MJ Getting; 23.5.79/27.7.82)=US 79–41780
32. WO 84/03848 (Schaetti & Co.; J Schaetti; 14.3.81/11.10.84; CH Prior. 28.3.83)
33. DE 3 425 553(RM Ind. Products Inc.; JC Zuckeret et al.; 11.7.84/7.2.85; US Prior. 11.7.83)
34. Müller J-P (1981) Luftdurchlässige atmungsaktive Trockenkaschierung. Chemiefasern Textilind 31/83 12, 946
35. EP 678, 618 and EP 678, 619 (Sarna Patent und Lizenz AG; J Vogt, H Unold; 19.4.94/25.10.95)
36. Wehlmann J (1997) Alternative Umkehrbeschichtung durch Pulversintern auf einer Stahlbandanlage. Coating Feb, 30, 32, 34
37. DOS 2 657 943 (Vsecjusnyj nautschno-issledovatelskij institut sintetischeskich smol VI-25 A.; I. Larionov et al.; 21.12.76/22.6.78)
38. Miles DC (1997) Reactabond – New Adhesive Technology for Textile and Automotive Applications. J Coated Fabr 26:221
39. JA 92 01 909 (Kao Corp.; S Sato, T Mansuki, F Kikuchi, A Gunji; 26.1.96/5.8.97)=JA 96–11 594
40. PCT WO 9745259 (DuPont; Procter & Gamble; NL Carroll et al.; 29.5.96+6.11.96/4.12.97)=US 96–655 046 and 96–744487
41. JA 09 286 085 (Kao Corp.; S Sato, T Masuki, F Kikuchi, Y Sakai; 24.4.96/4.11.97)=JA 96–102 318

13
Perforation

See also: 11.1 [24, 61], 18 [82]; Survey: Grosse W (1993) Mikroperforationstechnik für Kunststofffolien. Coating June, pp 201–205
1. OS 2 201 192 (Celanese Corp.; SR Schulze; US Prior. 14.1.71; 12.1.72/3.8.72)
2. DBP 950 090 (Sauterer; 20.8.52); DBP 952 937 (Sauterer; 1.3.53); DOS 1 719 236 (Tiefenbacher & Co.; 2.3.68)
3. DGM 1 683 786; 26.3.53 (Pfälz.Gummiwerke); DBGM 1 680 835; 27.11.52 (Dornbusch) (1955) Mod. Plastics 32:102, (1953) Mod. Plastics 30:30; DBP 958 518 (Peiler; 3.8.55); DBP 1 008 700 (Peiler; 10.12.55)
4. Trademark of Balamundi (NL)
5. FR 1 502 324 (ICI; 26.11.65)

6. DOS 1 629 403 (C Freudenberg; 7.4.66)
7. FR 1 325 137 (American Viskose Corp.; GG Whytlaw; 3.4.62/18.3.63)
8. JA 20 113/66 (Electro-Chem. Ltd.; 27.12.63/22.11.66)
9. GB 1 042 089 (Toyo Gomu Kogyo KK; JA Prior. 27.12.63)=JA 70 432/63; 23.10.64/7.9.66=FR 1 414 421
10. OS 1 469 541 (Griffine; L Vinovsky; FR Prior. 21.6.63; 19.6.64/30.10.69)=BE 649 493
11. NE 6 502 039 (Siemens-Schuckert-Werke AG; D Prior. 3.3.64; 18.2.65/6.9.65)
12. OS 1 719 236 (P.Tiefenbacher & Co.; R Steiger, G Adank; 2.3.68/12.8.71)
13. OS 2 108 785 (Kuraray Co. Ltd.; K Harada; JA Prior. 27.2.70+12.6.70; 24.2.71/9.9.71)
14. JA 58 036 270 (Asahi Chem. Ind. KK; 24.8.81/3.3.83)
15. JA 61 152 876 [Toppan Printing KK; 25.12.84/11.7.86 (277 246)]
16. Päßler H, Körber H, Grundke K (1996) Bildung von Mikroporen in Folien durch Bestrahlen mit Feststoffpartikeln für Wetterschutzbekleidung. Melliand 3:131
17. DOS 3 111 340 (Benecke; P Schaefer, G Hillebrandt; 25.3.80/28.1.82)
18. JA 80–45 670 (Lonseal; 15.4.78/19.11.80)=JA 78–44 438
19. JA 9 188 975 (Toray Ind. Inc.; H Takahashi, U Yasutaka, W Koji; 28.12.95/22.7.97)=JA 95-352 699

14
Discontinuous Coatings – e.g. by Printing

1. FR 1 135 860 (US Rubber; HT Nelson; US Prior. 10.9.54; 17.8.55/6.5.57)
2. FR 1129 332 (J. Votteler's Nachf. GmbH; D Prior. 10.8.54; 9.8.55/18.1.57)
3. DAS 1 070 586 (A. Schoeler; 22.6.56/10.12.59)
4. DDR 17 714/8h (HJ Schmidt, W Berthold; 1.12.56/23.11.59)
5. GB 898 072 (KH Barnard; US Prior. 18.4.59; 17.4.59/6.6.62)
6. JA 5 679/68 (Nippon Leather Industry; 17.9.65/1.3.68)
7. FR 1 211 884 (Goodrich; A Kelly; US Prior. 18.12.57; 18.12.58/12.10.59)
8. SZ 350 632 (Gummi-Werke Richterswil; D Henzi; 13.12.57/15.12.60)
9. GB 982 105 (Degussa; D Prior. 21.11.62; 18.4.63/3.2.65)=BE 638 200
10. BE 662 065 (Jung & Simmons; D Prior. 22.4.64; 5.4.65/2.8.65)=NE 6 504 847
11. FR 1 400 910 (Kurashiki Rayon Co. Ltd.; O Fukushima; 10.7.64/20.4.65; JA Prior. 15.7.63) =JA 38 395/63
12. JA 74 024 644 (Yamanashi Kasei Kogyo Co.; 1.8.70/25.6.74)
13. FR 1 175 764 (Textilausrüstungsgesellschaft Schroers u.Co.; D Prior. 10.3.72; 20.2.73/26.10.73) =OS 2 211 614
14. JA 59 094 685 (Toray; 19.11.82/31.5.84)

15
Flocking

See also 8.1 [27], 16.2 [26, 27], 19 [25]
1. General surveys: Pomeraniec J et al. (1963) Elektrostatisches Beschichten von Geweben. Mod Plastics 41:133; Müller J (1995) Flocktechnologie – eine oft verkannte leistungsfähige Technik. ITB Veredlung March, pp 26–30
2. FR 1 550 024 (Bayer; D Prior. 5.1.67; 5.1.68/13.12.68)=BE 708 987=NE 6 800 022
3. DDR 77 687 (W Neumann, K P Knecht; OE Prior. 5.8.68; 1.8.69/20.11.70)
4. NE 6 800 022 (Bayer; D Prior. 5.1.67; 2.1.68/8.7.68)=BE 708 987
5. OS 2 221 106 (Deering Milliken Research Corp.; EE Habib; US Prior. 30.4.71; 28.4.72/16.11.72)
6. GB 930 604 (USM; US Prior. 7.8.58; 4.8.59/3.7.63); FR 1 320 258 (GB Prior. 1.3.61; 28.2.62/ 28.1.63)
7. RA 126 861 (IV Plotnikov; 7.12.58/10.3.60)
8. DAS 1 000 009 (Palladium; P Hirschberger; F Prior. 30.1.52 and 21.4.52; 11.6.52/3.1.57) =US 2 715 074
9. OS 2 113 790 (Kufner Textilwerke KG; J Hefele; 22.3.71/28.9.72)

10. JA 11 076/65 (Daisei Kako Co.Ltd.; 27.6.63/3.6.65)
11. BE 663 617 (Dunlop; GB Prior. 9.5.64; 7.5.65/1.9.65)
12. JA 17 429/68 (Y Mizuno; 9.10.65/23.7.68)
13. JA 8 437/65 (K Ishizuka; 7.5.63/30.4.65)
14. JA 43 048/72 (Toray Ind. Inc.; 25.2.69/31.10.72)
15. NE 6 717 368 (Elie Adjiman; FR Prior. 22.12.66; 20.12.67/24.6.68)=FR 1 511 650=GB 1 190 830
16. FR 2 014 882 (Bayer; D Prior. 1.8.68; 1.8.69/24.4.70)
17. OS 1 635 707 (C Freudenberg; Th v Schachowskoy, J Fehlhaber; 28.2.68/19.8.71)
18. OS 1 937 863 (Roser Neudorf GmbH; W Neumann, KP Knecht; OE Prior. 5.8.68; 25.7.69/ 12.2.70)
19. FR 2 071 281 (Et. LaChaignaud; 23.12.69/17.9.71)
20. JA 46 893/72 (Kokoku Chem. Ind. Co. Ltd.; 15.9.69/27.11.72)
21. JA 46 888/72 (Kokoku Chem. Ind. Co. Ltd.; 16.5.69/27.11.72)
22. J A 7 324 429 (Daiichi Lace Co. Ltd.; 19.8.70/20.7.73)
23. OS 2 357 664 (Besnier-Flotex SA; P Kaufmann; SZ Prior. 20.11.72; 19.11.73/22.5.74)
24. OS 2 315 741 (P Kaufmann; SZ Prior. 15.4.72; 29.3.73/18.10.73)
25. DOS 2 944 063 [Kanebo Ltd., K Asano; 31.10.79/14.5.80; JA Prior. 31.10.78 (135 075)]
26. Kaimer M (1992) Wasserdichte, wasserdampfdurchlässige beflockte Membranen und Gewebe. Dissertation, Inst. für Textil- und Faserchemie der Universität Stuttgart
27. US 3 573 121 [(K Fukuoka et al. 8.4.68/30.3.71; JA Prior. 6.4.67 (42/22 045)]
28. DOS 3 820 296 (Lorica S.p.A.; G Poletto, 19.6.87/29.12.88)=IT 87-67534
29. JA 57 154 472 (Okabe Kinzoku Kogyo; 19.3.81/24.9.82)
30. CN 1 144 252 (Guizhou Prov. Inst. of Chemical Ind; H Peng, Y Xie, S Xiong; 9.5.94/5.5.97) =CN 94-105 401

16
Crystallization of Homopolymers

See also: 4.2 [11], 9.2 [11], 10 [3, 47], 11 [10], 18 [6–8, 13, 24, 44, 62, 86], 20.1 [16], 27 [62]
1. US 3 315 020 (WL Gore, US Prior. 21.3.62/18.4.67)
2. US 3 664 915 (WL Gore, US Prior. 3.10.69/23.5.72)
3. DOS 3 020 335 (Celanese Corp., JW Soehngen, K Ostrander; 29.5.80/11.12.80; US Prior. 1.6.79); DOS 3 020 372 (Celanese Corp; JW Soehngen; 29.5.80/11.12.80; US Prior. 1.6.79)
4. EP 0 227 384 [Japan Gore Tex. Inc.; 10.12.86/1.7.87; JA Prior. 11.12.85 (189 657) and 7.2.86 (23 973)]
5. Duncan J (1984) Wasserfeste wasserdampfdurchlässige Schuhobermaterialien. J Coated Fabr 13(1):161
6. US 4 443 511 (WL Gore & Ass.; 19.11.82/17.4.84); EP 0 110 627 (WL Gore & Ass.; DA Tibbets; 17.11.83/11.5.88; US Prior. 19.11.82)
7. US 4 194 041 (WL Gore & Ass.; WL.Gore, B Allen; 29.6.78/18.3.80)
8. US 3 953 566 (WL Gore & Ass.; WL Gore; 3.7.73/27.4.76)
9. US 4 942 214 (WL Gore & Ass.; DJ Sakhpara; 29.6.88/17.7.90)
10. EP 0 427 769 (WL Gore & Ass.; KR Driskill, RL Henn; 25.7.89/22.5.91; US Prior. 27.7.88)=WO 9 000 969-A
11. JA 5 91 58 252-A (Japan Gore Tex KK; 1.3.83/7.9.84)
12. EP 184 392 (3M; 4.12.84/11.6.86)
13. EP 615 779 (WL Gore & Ass.; W Buerger; 16.3.93/21.9.94)=DE 4 308 369
14. EP 637 610 (Himont Inc.; M Hallam; It Prior. 3.8.93/8.2.95)
15. WO 9300837 (Gore & Assoc. Inc.; KR Driskill, RL Henn, J Norvell; 12.7.91/21.1.93)=US 5 289 644; EP 595 941=JA 7500026
16. DOS 2 737 756 (Gore; 22.8.77/1.3.79)
17. Nago S, Mizutani Y (1996) Microporous polypropylene fibers containing $CaCO_3$ fillers. J Appl Polym Sci 62:81
18. DOS 3 833 705 (Minnesota Mining Mfg. Co.; WL Kausch; US Prior. 8.10.87/20.4.89)=US 87-106719

19. JA 10 34 726 (Tokuyama Soda KK; 30.7.87/6.2.89)=JA 87-188901; JA 88 136 153 (Tokuyama Soda KK; S Nagou, S Nakamura; 12.6.86)=JA 86-135 018
20. DOS 2 737 745 (Akzo; AJ Castro; 9.3.78)
21. OS 2 055 193 (Celanese Corp.; ML Druin et al.; US Prior. 13.11.69 and 28.10.70; 10.11.70/ 19.5.71)
22. US 2 757 100 (DuPont; VL Simril; 4.11.52/31.7.56)=FR 1 090 575=DAS 1 010 945=GB 763 604
23. DOS 19 625 389 (WL Gore & Ass. GmbH; W Zehnder; 25.6.97/2.1.98)

16.1
Mixtures of Incompatible Polymers

1. JA 43 150/72 (Toyo Spinning Co. Ltd.; 17.7.69/31.10.72)
2. JA 22 311/67 (Kyodo Kasei Kogyo Co. Ltd.; 29.12.62/1.11.67)
3. JA 5 677/68 (Shoshichi; 3.8.62/1.3.68)
4. GB 999 868 (US Rubber; US Prior. 11.6.63; 17.4.64/28.7.65)
5. US 3 536 638 (Uniroyal; LP Dosman; 8.11.68/27.10.70; cip 3.9.65 and 21.6.63)
6. JA 39 509/71 (Toyo Spinning; 8.10.68/20.11.71)
7. FR 2 022 928 (Uniroyal Inc.; LPh Dosman; US Prior. 8.11.68; 7.11.69/7.8.70)
8. JA 73-43 946 (Sumitomo Electric Ind.; 16.12.64/21.12.73)
9. JA 73-28 789 (Sumitomo Electric Ind.; 4.11.64/4.9.73)
10. JA 74-001 841 (Kuraray Co. Ltd.; 6.8.70/17.1.74
11. JA 73-5 882 (Mitsubishi Plastics Ind. Ltd.; 30.6.70/21.2.73)
12. US 4 483 900 (Oakwood Ind. Inc., R Goldfarb; 15.7.82/20.11.84)
13. JA 07,174,373 (95, 174, 373) (Nitto Denko Corp.; H Ootani, K Okada; 16.12.93/14.7.95)
14. JA 07 070 944 (Toyo Cloth; 2.9.93/14.3.95)
15. BE 656 583 (Courtaulds Ltd.; GB Prior. 3.12.63 and18.12.63; 3.12.64/1.4.65)=NE 6 413 997

16.2
Incorporation of Solid Compounds

See also: 27 [58]

1. OS 2 236 789 (Kureha Kagaku Kogyo KK; M Onozukau et al.; JA Prior. 23.7.71 and 18.11.71; 24.7.72/1.2.73)
2. OS 2 218 051 (Potters Ind. Inc.; JH Ettlinger, GM Knafo; US Prior. 15.4.71; 14.4.72/26.10.72)
3. OS 1 629 326 (Continental; K Brünger, F Koch; 17.5.66/28.1.71)
4. FR 1 251 677 (3M Co.; DL Johnson; US Prior. 19.3.59; 18.3.60/12.12.60)
5. JA 9 158 051 (Seiren Co. Ltd.; M Masahiro, M Yoshiki, O Kenji; 13.12.95/17.6.97)=JA 95-346 285
6. JA 61160480 (Toray Ind. Inc.; 7.1.85/21.7.86)=JA 85-164
7. OS 1 908 677 (Bayer; KA Weber et al.; 21.2.69/10.9.70)
8. JA 9 158 051 (Seiren Co. Ltd.; M Masahiro, M Yoshiki, O Kenji; 13.12.95/17.6.97)=JA 95-346 285
9. OS 1 957 759 (Et. A. Chromarat et Cie.; C Mao; FR Prior. 8.5.69; 17.11.69/8.4.71)=BE 745 881
10. JA 7 331 884 (Kawaguchi Rubber Ind. Co. Ltd.; 11.11.69/2.10.73)
11. JA 39 510/71 (Toyo Spinning; 8.10.68/20.11.71)
12. FR 1 591 286 (Montecatini-Edison; IT Prior. 7.11.67; 6.11.68/5.6.70)=BE 723 437=IT 815 167=NE 6 815 687=OS 1 806 990
13. FR 90 450 (A Baudou, G Joseph; FR Prior. 4.7.66; 4.7.66/8.12.67); addition to 89 410 and 1 453 978
14. BE 704 514 (Celanese; HS Bierenbaum et al.; US Prior. 15.8.66; 29.9.67/29.3.68)
15. JA 6 360/71 (Toyo Spinning; 28.10.66/17.2.71)
16. FR 1 487 779 (Celanese Corp.; HD Noether, H Brody; US Prior. 26.7.65 and 1.10.65; 26.7.66/ 7.7.67)
17. DDR 47 116 (W Schumann, G Kirscheis; 29.4.65/5.4.66) (WP 81/110 604)
18. GB 1 060 779 (Skin-Yamato Gomugaku Ma. Co. Ltd.; T Iseki; 15.4.64/8.3.67)
19. OS 1 635 631 (T. Takase Co.; T Iseki; 13.5.64/20.8.70)

20. JA 73-23 884 (Nippon Cloth Kogyo Co. Ltd.; 26.8.70/17.7.73)
21. JA 73-20 425 (Nippon Cloth Ind. Co. Ltd.; 31.7.70/21.6.73)
22. OS 2 249 199 (Kureha Kagaku Kogyo KK; M Onozuka et al.; JA Prior. 5.10.71; 5.10.72/ 12.4.73)
23. OS 2 426 193 (Celanese Corp.; HS Bierenbaum et al.; US Prior. 30.5.73; 29.5.74/19.12.74)
24. JA 50 061 459 (Kogoku Chem. Ind. KK; 1.10.73/27.5.75)
25. RA 170 459 (AS Chakoyan et al.; 22.10.62/Sept.65)
26. JA 4 477/65 (Y Kimoto; 19.12.62/10.3.65)
27. JA 11 753/65 (Teijin Co.Ltd.; 4.7.63/11.6.65)
28. GB 1 278 893 (Bayer; E Meisert et al.; D Prior. 21.5.69; 29.4.70/21.6.72)=OS 1 925 997
29. EP 49242 (Amoco Corp.; CW Bauer, SR Clingsman, P Jacoby, WT Tapp; 21.12.90 [US 90-633087 + 23.8.91 (US 91-749213)/1.7.92]=US 5 175 953=US 5 317 035
30. JA 61 152 876 (Toppan Printing KK; 25.12.84/11.7.86)=JA 84-277246
31. JA 61 160 480 (Toray Ind. Inc.; 7.1.85/21.7.86)=JA 85-164
32. JA 06 093 571 (Daiichi Lace KK; Hosokawa Micron KK, 26.9.91/5.4.94)=JA 91-0318 744
33. JA 060 93 571 (Daiichi Lace KK; Hosokawa Micron KK, 26.9.91/5.4.94)
34. JA 061 92 433 (Kyoeisha Kagaku KK; Osaka Prefecture; Tomen KK; 24.12.92/12.7.94)

17
Precipitation of Polymers in Water-Based Dispersions

1. (1961) Houben-Weyl, Methoden der organischen Chemie, Makromolekulare Stoffe Teil I, Thieme, Stuttgart, p 133ff
2. JA 50 040 869 (Kurashiki Spinning; 13.8.73/14.4.75)
3. DOS 2 348 662 (Bayer; H Träubel, E Komarek; 27.9.73/24.4.75)
4. FR 1 384 422 (Degussa; D Prior. 6.5.63; 26.2.64/23.11.64)=SZ 441 210
5. US 4 224 375 (Compo Ind.Inc.; MJ Veiga, HE Petersen; 22.5.78/23.9.80)
6. DE 3 127 228 (C Freudenberg; W Föttinger et al.; 10.7.81/3.2.83), EP 0 069 788 (C Freudenberg, W Föttinger et al.; 10.7.81/19.1.83)
7. DE 3 633 874 (Chem. Fabrik Stockhausen GmbH; K Dahmen, D Stockhausen et al.; 4.10.86/ 14.4.88)
8. Hemmrich J, Fikkert J, van den Berg M (Stahl) (1993) Porous structural forms resulting from aggregate modification in polyurethane dispersions by means of isothermic foam coagulation. J Coated Fabr 22:268
9. DOS 3 325 163 (Norwood Ind. Inc.; JR McCartney; 12.7.83/19.1.84; US Prior. 14.7.82)=US 82-398 260

17.1
Bonding of Nonwovens

See also: 8.2 [76]

1. DOS 2 264 935 (BASF; D Distler et al.; 30.5.72/17.7.75)
2. US 3 185 582 (AA Allegre; SP Prior. 17.12.53; 28.2.61/25.5.65)
3. DAS 1 044 021 (British U.S.M. Co. Ltd.; PHV Dawson, JA Hawkes; GB Prior. 10.12.55; 22.11.56/20.11.58)=FR 1 165 149
4. OS 2 034 536 (Bayer; O Koch et al.; 11.7.70/3.2.72)
5. BE 634 892 (DuPont; EB Fitzgerald, FE Jenkins; US Prior. 13.7.62; 12.7.63/13.1.64)
6. Sandomiirskii DM, Pilmenshtein ID (1972) Änderungen in den strukturellen und mechanischen Eigenschaften von Kautschuklatices während des Gelierens mit Natriumsilikofluorid. Soviet Rubber Technol 21:7
7. JA 24 798/68 (Toyo Cloth Co. Ltd.; 2.11.62/25.10.68)
8. FR 1 381 822 (Bayer; G Sinn, H Hornig; D Prior. 5.2.63; 4.2.64/2.11.64)
9. DDR 73 748 (Forsch. Inst. für Textiltechn.; 30.9.68/12.6.70) (WP 135 086)
10. OS 1 921 458 (Goodrich; G Leroy; US Prior. 29.4.68; 26.4.69/6.11.69)

11. OE 277 140 (C Freudenberg; A Petersik, A Gräber; D Prior. 26.9.67; 13.9.68/10.12.69)=DAS 1 619 054=GB 1 171 267
12. FR 1 547 257 (Shell; NE Prior. 20.12.66; 18.12.67/26.11.68)=NE 6 617 922=BE 708 157
13. OS 1 560 658 (BASF; G Wetzel, K Gans; 13.5.64/2.10.69)
14. FR 1 441 116 (Dunlop; GB Prior. 22.7.64; 21.7.65/25.4.66)
15. FR 1 508 072 (WR Grace; AM Sacerdote; US Prior. 9.10.64; 8.10.65/5.1.68)=DOS 1 544 987 =GB 1 118 444
16. OS 2 147 229 (C Freudenberg; E Fahrbach, HA Weber; 22.9.71/29.3.73); see also US 2 719 302 and OE 225 148
17. OS 2 417 705 (Bayer; A. Matner et al.; 11.4.74/23.10.75)
18. DAS 1 288 061 (AA Alegre; US Prior. 28.2.61;27.2.62/30.1.69)=DAS 1 204 618
19. US 3 265 527 (DuPont; RL Adelman; 27.5.63/9.8.66)
20. GB 1 029 651 (British United Shoe Mach. Corp.; PHV Dawson et al.; 1.3.61/18.5.66)=EP 1 029 651
21. US 3 663 266 (DuPont; JL Dye; 21.5.70/16.5.72)
22. JA 517 573 (Teijin Cordley Ltd.; 24.12.91/9.7.93)=JA 91–35 556
23. EP 167 189 (Stahl Chem. Ind. B. V.; DP Spek, LA van der Heyden; 21.5.85/8.1.86; NL Prior. 4.6.84)=NE 84–1 784
24. JA 09 95 869 (Kuraray Co. Ltd.; N Moriyasu, K Akamata, M Makimura; 29.9.95/8.4.97) =JA 95–253 580

18
Hydrophilic Polymers

See also: 9.3 [3, 11], 11.2 [60], 23.1 [5]
1. Träubel H, Tork L, Zorn B, Wenzel W (1987) Vernetzbare Bindersysteme. Das Leder 38:177
2. Schröer W, Schütze D-I, Thoma W (1992) Wasserdampfdurchlässige kompakte Textilbeschichtungen mit Polyurethanen. Coating September, pp 290–296
3. WO 95/30 793 (Gore; A Dutta, R Henn; 5.5.94/16.11.95)
4. JA 06 297 648 (94, 297, 648) (Fujimori Kogyo Co.; R Ichikawa, T Kobayashi, K Hashimoto; 19.4.93/25.10.94); JA 07 24,978 (95 24,978) (Fujimori Kogyo Co; R Ichikawa, T Kobayashi, H Terui, S Kishi; 7.7.93/27.1.95)
5. Saito N et al. (1996) Synthese und Hydrophilie multifunktionell hydroxylierter Polyacrylamide. Macromolecules 29:313
6. Haldankar GS, Spencer HG (1989) Properties of bound water in poly(acrylic acid) and its sodium and potassium salts determined by differential scanning calorimetry. J Appl Polm Sci 37:3137
7. Shibukawa M, Ohta N, Onda N (1990) Investigation of the states of water in water-swollen hydrogels by liquid chromatography and differential scanning calorimetry. Bull Chem Soc Jpn 63:3450
8. JA 91 04 828 (Toyo Ink Mfg. Co., K. Shiozawa, T. Sato, 3.10.94/23.4.96)
9. Kerres B (1996) Superabsorber für wäßrige Flüssigkeiten. Textilveredlung 31:238
10. DDR-Pat 235 470 (Akad. der Wiss. der DDR; 15.3.85/7.5.86); DDR-Pat.235 468 (Akad. der Wiss. der DDR; 15.3.85/7.5.86)
11. FR 1 161 575 (Ions Exchange and Chemical Corp.; M Mendelsohn, C Horowitz; US Prior. 25.10.55; 25.10.56/2.9.58)=GB 812 460
12. DOS 2 355 537 (Bayer; W v Bonin, H Striegler; 7.11.73/22.5.75)
13. DOS 1 444 056 (Eastman; JR Caldwell, CC Dannelby; US Prior. 20.11.62; 29.10.63/21.11.68) =BE 640 081=FR 1 405 830=GB 1 072 874=US 3 265 529
14. FR 1 427 852 (Westo GmbH; D Prior. 4.1.64 and 21.10.64; 23.10.64/3.1.66)=NE 6 415 169 =BE 657 661
15. RA 161 683 (AD Zaionckkovskij et al.; 25.4.61/July 64)
16. OS 2 257 393 (BASF; R Gäth, R Linke; 23.11.72/20.6.74)

17. Anon. (1977) Dichte und Wasserdampfdurchlässigkeit von formalisierten und Säure behandelten PVA. Membranen. Konbushu Ronbushu 34(7):475
18. OS 1 965 587 (BASF; R Gäth et al.; 30.12.69/15.7.71)
19. OS 2 114 371 (BASF; R Linke, R Gäth; 25.3.71/5.10.72)
20. OS 1 938284 (Kanegafushi Boseki KK; T Tensho; JA Prior. 27.7.68; JA 68-52 824; 28.7.69/26.3.70)=GB 1 257 095=US 3 627 567 and US 3 575 753 (Kanegafushi Boseki KK; M Maruya et al.; 15.11.68/20.4.71)
21. OS 1 812 746 (Kanegafushi Boseki KK; M Maruya et al.; JA Prior. 4.12.67; JA 77 782/67; 4.12.68/3.7.69)
22. DOS 19 653 301 (Akzo Nobel N. V.; MG Haderlein, JCW. Spijkers, A Zoepfl; 25.4.96/30.10.97)

18.1
Incorporation of Hydrophilic Materials

See also: 8.1 [29–32, 45, 64, 68], 8.6 [29, 30, 32, 34, 37–45, 60], 9.2 [20], 10 [11], 11.1 [13, 14, 64], 15 [2, 3, 6–9, 11, 12, 18]

1. DP Appl. 13 417/8h (Degussa; 25.10.52/12.4.56)=FR 1 091 060
2. DAS 1 008 706 (Bennecke; W Kühn; 20.4.54)
3. DAS 1 254 115 (Göppinger Kaliko; W Gräbner; 20.2.60/16.11.67)=GB 960 445=FR 1 281 070 =OE 213 367
4. US 3 198 690 (US Rubber; JC Starke; 10.7.59); cip 7.11.60/3.8.65
5. RA 127 232 (K M Skirodova; 18.5.59/25.3.60)
6. DOS 1 635 557 (Nino; O Honegger; 8.5.64/2.1.70)
7. BE 597 860 (Lantor Ltd.; GB Prior. 9.12.59; 6.12.60/30.12.60)
8. GB 1 160 237 (Kurashiki Rayon; Kubashiki Kaisha; JA Prior. 15.10.65; 13.10.66/6.8.69) =JA 63 436/65
9. US 3 393 188 (National Polychemicals; R Strauss et al.; 20.10.65/16.7.68)
10. JA 14 559 (Nisshin Vinyl Kogyo KK; 16.9.66/19.4.71)
11. JA 37 412/71 (Showa Rubber Co. Ltd.; 31.5.67/4.11.71)
12. RU 2 010 862 (Zantsev V. K., VI Gusev, VM Malov, VK Zantsev; 15.1.92/15.4.94)
13. DOS 2 502 468 (Holzstoff S.A.; JL Baron et al.; SZ Prior. 23.1.74; 22.1.75/24.7.75)
14. FR 2 016 394 (3M Company; AE Raymond; US Prior. 26.8.68; 25.8.69/8.5.70)=OS 1 943 975
15. NE 6 906 886 (SVUK; CZ Prior. 6.5.68; 6.5.69/10.11.69)
16. US 3 663 472 (3M Company; AE Raymond; 15.3.71/16.5.72)
17. FR 1 378 064 (Soc. des Produits Tiffine; 18.7.63/5.10.64)
18. JA 50 93 373 (Nippon Burakah Kogyo KK; Nippon Miractoran KK; 26.9.91/16.4.93)
19. FR 2 482 144 (Toray Ind.; I Nakajima, S Sugawara, T Oishi, K Hamada; 7.5.80/13.11.81)=JA 80–59522
20. OS 2 324 960 (BASF; F Miksovsky et al.; 15.5.73/12.12.74)
21. FR 1 510 180 (M Gorce, M Moreau; 9.12.66/19.1.68)
22. US 3 137 664 (AD Little Inc.; J Shulman, JT Howarth; 21.6.60/16.6.64)
23. OE 221 280 (O. Zwoboda; 14.6.60/15.10.61)
24. JA 14 993/64 (T. Izeki; 5.2.62/29.7.64)
25. JA 13 640/66 (Hiroshima Kasei Co. Ltd.; 19.12.63/30.7.66)
26. JA 27 555 (A Tomikawa; 19.2.63/4.12.65)
27. OS 2 012 662 (Dainippon Ink and Chem.Inc.; H Shirota et al.; JA Prior. 17.3.69; JA 69–19 551 and 7.4.69; JA 69–26 173; 17.3.70/24.9.70)
28. US 3 660 218 (Nat. Patent Devel. Corp.; Th Shepherd, EJ Jacot; 15.10.68/2.5.72)
29. FR 1 453 978 (AJG Baudou; 16.8.65/22.8.66)
30. JA 1 357/67 (Nippon Rubber Co.; 16.8.63/23.1.67)
32. FR 1 461 755 (Union Carbide; CL Purcell, LG Imhoff; US Prior. 20.11.64; 18.11.65/2.11.66)=NE 6 515 055
34. JA 73–20 283 (Honey Kasei KK; 20.2.70/20.6.73)
36. JA 24 677/67 (Kurare Plastics Co.Ltd.; 20.4.64/27.11.67)

37. OS 2 253 926 (Tschech. Akademie; V Heidingsfeld et al.; CZ Prior. 5.11.71; 3.11.72/10.5.73)
38. DDR 116 070 (VEB Chemiefaserkombinat Schwarza; H Alsleben; 10.7.74/5.11.75)=(WP 179 829) and OS 2 521 211
39. GB 1 325 676 (AB Ehrnberg; JE Otterstedt; 3.9.69 and 19.9.69/8.8.73)
40. OS 1 811 593 (BASF; R Gäth et al.; 29.11.68/9.7.70)
45. OS 2 404 794 (BASF; R Fikentscher et al.; 1.2.74/ 7.8.75)
46. JA 06,330,472 (94,330,472) (Showa Denko KK; T Wada, Y Maeda, M Ookuba; 20.5.93/29.11.94)
47. WO 95 33 007 [WL Gore and Ass. Inc.; A Dutta; 26.5.94 (US94–249 912) and 24.6.94/7.12.95]
48. JA 070 70 600 (Showa Denko KK; 2.9.93/14.3.95)
49. JA 060 16 951 (Aroma Kagaku Kikai Kogyo KK; 2.7.92/25.1.95)
51. DDR 78 547 (Forschungsinst. f. Textiltechnologie; G Schmidt et al.; 6.11.69/20.12.70) (WP 143 539)

18.2
Hydrophilic Polyurethanes

See also: 8.1 [24], 8.2 [10, 23, 34, 35, 36], 9.3 [11, 12, 18, 29, 35, 36], 10 [19], 11 [13, 14, 64], 11.1 [13, 64], 15 [3, 5–9, 11], 16.1 [3, 18, 19, 23, 25], 18 [92], 18.3 [9–11, 16], 22 [37, 48], 24 [16], 27 [10, 11]

1. DOS 4 339 475 (Wolff Walsrode; D Schultze, N Hagartër, H-W Funk, R Kunold; 19.11.93/ 24.5.95)
2. Petrik S, Hadobas F, Simek L, Bohdanecky M (1992) Sorption of water vapour by segmented polyurethanes. Eur Polym J 28:15
3. OS 1 936 787 (Et.Pennel & Flipo; RH de Broutelles, L Duchene; FR Prior. 22.7.68; 19.7.69/ 10.9.70)=NE 6 911 229
4. FR 2 116 306 (Omnium de Prospective Industrielle; J Davidovits, M Lefebre; 7.12.70/13.7.72)
5. DOS 4 027 797 (H von Blücher, E de Ruiter; 1.9.90/5.3.92) addition to DOS 4 003 764
6. JA 32 60 181 (Hosokawa Micron KK; Daiichi Lace KK; 1.3.90/20.11.91)
7. BE 697 617 (Amicon Corp.; US Prior. 26.4.66; 26.4.67/2.10.67)
8. DAS 1 226 071 (Bayer; J Peter, E Müller; 6.4.62/6.10.66)=BE 630 578=FR 1 362 538
9. BE 626 803 (Bayer; J Peter, E Müller; D Prior. 5.1.62; 4.1.63/2.5.63)=FR 1 350 900
10. FR 1 348 959 (Bayer; J Peter, E Müller; D Prior. 3.3.62; 1.3.63/2.12.63)= DAS 1 220 384=GB 989 760
11. US 3 384 506 (Thiokol Chem. Corp.; HL Elkin; 18.5.64/21.5.68)
12. JA 15 754/70 (Sanyo Chem. Ind. Co. Ltd.; 21.6.65/2.6.70)
13. US 3 719 726 (Kuraray Co. Ltd.; K Hara, T Yoshitake; JA Prior. 27.3.70, JA 45/26 186+45/26 187; 18.3.71/6.3.73)=OS 2 113 865
14. Desai VM, Athawale VD (1995) Water resistant, breathable hydrophilic polyurethane coatings. J Coated Fabr 25:39
15. JA 07 149 869 (Dainippon Ink & Chem Inc.; 30.11.93/13.6.95)=JA 93–02 99 758
16. BE 639 578 (Bayer; J Peter, E Müller; D Prior. 6.4.62; 4.4.63/31.7.63)
17. EP 609 730 (Th Goldschmidt AG; E Yilgoer, I Yilgoer, EO Yilgoer; 5.2.93/15.9.94)=US 5 389 430=US 5 461 122
18. JA 60 162 872 (Unitika KK; 26.1.84/24.8.85)=JA 84–13 081
19. Yen M-S, Cheng K-L (1995) The effect on the physical properties of the chitin-added poly-urethane cast films and coated fabrics. J Coated Fabr 25:87
20. Krishnan K (1995) New applications for breathable hydrophilic and non-hydrophilic coatings. J Coated Fabr 25:103
21. FR 2 001 438 (AKU; NL Prior. 7.2.68; 3.2.69/26.9.69)
22. The reaction of hexamethylene diisocyanate with sodium bisulfite is described in Petersen S (1949) Niedermolekulare Umsetzungsprodukte aliphatischer Diisocyanate 5. Mitteilung über Polyurethane. Ann. 562:205
23. Phaneuf MD et al. (1997) Chemical and physical characterization of a novel poly(carbonate urea) urethane surface with protein crosslinker sites. J Biomater. Appl. 12(2):100
24. JA 73 20 283 (Honey Kasei KK; 20.2.70/20.6.73)
25. RU 2 023 098 (Film Materials Synth. Leather Res. Inst.; LF Golovkina, AA Kasyanova, AV Stapanova; 4.6.91/15.11.94)

26. EP 684 286 (BASF; U Licht, H Seibert, R Hummerich; 25.5.94/29.11.95)=DOS 4 418 157
27. JA 08059981 (Tonen Kagaku KK; H Ishizuka, H Toda, K Myasaka; 18.8.94/5.3.96) (=JA 94 215 210)
28. JA 51 06 173 (Hosokawa Micron KK; 11.10.91/27.4.93)
29. Belokurova AP, Vladychina SV, Makarov AS, Koifman OI (1996) Moisture permeability of thermoplastic polyurethanes and their blends with PVC. Plast Massy 3:9; Belokurova AP, Bratter AM, Koifman OI (1996) Moisture permeability of film materials based on photocured oligourethane acrylate compositions. Plast Massy 3:11
30. Träubel H, Schütze D-I, Pedain J (1994) Bessere Tragehygiene durch wasserdampfdurchlässige Polyurethane. Textilveredlung 29:171
31. EP 339 435 (Asahi Glass Co. Ltd., Y Matsumoto, N Kunii; 26.4.88/2.11.89)
32. Nepyshnevskii VM et al. (1979) Hydrophile Polyurethanharnstoffe für synthetisches Leder. Colloid J USSR 41(2):204
33. JA 61 072 032 and JA 61 072 033 (Dainichiseika Color Chem.; 17.9.84/14.4.86)
34. JA 61 086 250 [Dainichiseika Color Chem.; 5.10.84/1.9.86 (208 379)]; JA 61 086 249 [Dainichiseika Color Chem.; 5.10.84/1.5.86 (206 378)]
35. EP 0 052 915 (Shirley Institute; JR Holker et al.; 23.6.81/2.6.82; UK Prior. 22.11.80)
36. JA 57 005 [Dainichiseika Color Chem.; 16.8.80/12.1.82 (80 295)]
37. JA 58 036 [Dainippon Ink Chem. KK; 21.8.81/3.3.83 (130 124)]
38. Lomax GR (1985) Der Aufbau wasserdichter, wasserdampfdurchlässiger Gewebe. J Coated Fabr 15(7):40; Lomax GR (1990) Hydrophilic polyurethane coatings. J Coated Fabr Oct, pp 88–107
39. EP 0 165 345 [Toray Ind. Inc., T Nishikawa et al.; 16.6.84/27.12.85 (304 074)]
40. DE 3 815 720 (H von Blücher, E de Ruiter; 7.5.88/16.11.89)
41. DOS 3 903 704 (Frauenhofer-Ges.; St Meinhard, O v Stetten, H Hasenfratz-Schreier, H Schmidt; 8.2.89/9.8.90)
42. DOS 3 705 025 (Daiichi Kogyo Seiyaku Co. Ltd.; K Sato, M Komori; 17.2.86 + 26.2.86 + 26.2.86/10.9.87)=JA 86–33616 + JA 86–42645 + JA 86–42646
43. EP 313 951 (Bayer; 29.10.87/3.5.89)
44. DOS 2 244 231 (CPC Intern. Inc.; JP Mudde; 8.9.72/23.8.73; US Prior. 14.2.72, 14.4.72 and 26.6.72)
45. JA 61 086 250A (Dainichiseika Color Chem; 5.10.84/1.5.86)
46. DE 2 319 706 (WR Grace & Co.; LL Wood et al.; 18.4.73/15.11.73; US Prior. 3.5.72) + DE 2 506 576 + DE 2 621 126 + US 3 903 232
47. US 4 110 508 (WR Grace & Co.; 9.2.76/29.8.78)
48. US 4 182 649 (WR Grace & Co.; 9.2.76/8.1.80)
49. DE 1 943 975 (3M Corp; AE Raymond; 25.8.69/5.3.70; US Prior. 26.8.68)
50. GB 2 157 703 (Shirley Inst.; 19.4.84 + 18.4.85/30.10.85)
51. DE 2 020 153 (Teijin Ltd.; 24.4.69/27.12.70)
52. DE 2 544 068 [Teijin Cordley Ltd.; 3.10.74/8.4.76; JA Prior. 3.10.74) (113 292)] and JA 1 124 687 (Tejin KK; 28.9.84/17.5.89)
53. WO 8 505 322 (Thoratec Lab. Corp.; 21.5.84/5.10.85)
54. JA 57 155 241 [Toray Ind. Inc.; 23.3.81/25.9.82 (40 479)]; JA 86 195 117 (Toray Ind. Inc.; 26.2.85/29.8.86)
55. JA 58 222 840 (Toyo Rubber Ind. KK; 19.6.82/24.12.83)
56. JA 61 009 423 (Toyo Rubber Ind. KK; 22.6.84/17.1.86); JA 62 271 740 (Toyo Rubber Ind. KK; 21.5.86/26.11.87)
57. Nepyeshnevskii VM (1979) Hydrophile Polyetherurethanharnstoffe für synthetische Leder. Colloid J USSR 41:204
58. Hayashi S, Ishikawa N, Giordano C (1993) High moisture permeability polyurethane for textile applications. PUR World Congr. Proc, pp 400–404
59. JA 62 270 667 (Dainippon Ink and Chemikals Inc.; T Masuda, H Ozawa; 19.5.86/25.11.87) =JA 86 112 539
60. JA 59 064 620 (Achilles Corp; 6.10.86/12.4.84)=JA 82–175 481
61. US 4 182 649 (WR Grace & Co.; IE Isgur, AB Holmstrem, B Andrews, J Norman; 9.2.76/8.1.80); cip from US 4 110 508

62. US 5 238 732 (Surface Coatings Inc.; S Krishnan; 16.7.92; US 5 239 036 [Surface Coatings Inc.; S Krishnan; 16.7.92 (US 92–914 871) + 11.1.93 (US 93–2 640)/24.8.93 + 11.1.93 (US 93–2 640)/24.8.93)]; US 5 239 037 (Surface Coatings Inc.; S Krishnan; 16.7.92 (US 92–914871) + 11.1.93 (US 93–2747)/24.8.93)

63. JA 52 02 286 (Nitta Gelatine KK; 24.1.92/10.8.93)=JA 92–11 145; JA 52 02 298 (Nitta Gelatine KK.; 28.1.92/10.8.93)=JA 92–13 408

64. JA 52 22 293 (Ajinomoto KK; Ajinomoto Takara Corp KK; 11.3.91/31.8.93)= JA 91–125 660

65. JA 52 22 681 (Achilles Corp.; 13.2.92/31.8.93)=JA 92–59 392

66. JA 53 02 275 (Daicel Huels KK; 23.4.92/16.11.93)=JA 92–104 897; JA 53 02 276 (Daicel Huels KK; 24.4.92/16.11.93)=JA 92–107 025; JA 53 02 277 (Daicel Huels KK; 24.4.92/16.11.93)=JA 92–107 026

67. EP 494 381 (P Bocciardo, S Bocciardo; 9.12.91/15.7.92)=EP 01–121 074

68. JA 42 02 482 (Idemitsu Petrochem. Co.; 30.11.90/23.7.92)=JA 90-336 588; JA 95 047 701 (Idemitsu Petrochem. Co.)

69. JA 81–112 578 (Teijin; 7.2.80/4.9.81)=JA 80–13 034

70. JA 86–072032 (Dainichiseika Color Chem. Co.; 17.9.84/14.4.86)=JA 84–192 630

71. JA 81–11 787 (Kuraray; 7.5.73/17.3.81)=JA 73–51 030

72. JA 81–112 578 (Teijin; 7.2.80/4.9.81)=JA 80–13 034

73. US 4 192 928 (Teijin; A Tanaka, T Shinoda, M Mimura; 23.3.77/30.12.76)=JA 77–31 122

74. JA 80–82154 (Kanebo; 19.12.78/20.6.80)=JA 78–157 877; see also JA 80–81 135 and JA 80–82 155

75. JA 9228253 (Toray Industries Inc. and Daiichi Lace Mfg. Co. Ltd.; T Takahashi, T Nakano, M Uemoto, F Masanori, T Furuya; 14.2.96/2.9.97)=JA 96–50836

76. WO 97 38 854 (Gore & Ass. Inc.; A Dutta; 17.4.96/23.10.97)=US 96–633 714

77. Johnson L, Samms J (1997) Thermoplastic polyurethane technologies for the textile industry. J Coated Fabr 27(7):48

78. Schledjewski R, Schultze D, Imbach K-P (1997) Breathable protective clothings with hydrophilic thermoplastic elastomer membrane films. J Coated Fabr 27(10):105

79. CN 114 512 (Nanya Plastic Industry Stock-Sharing Co. Ltd.; Y Liu; 15.7.96/12.2.97)

80. PCT WO 98 11 854 (Tyndale Plains-Hunter Ltd.; MH Reich, K Nelson; 20.9.96/26.3.98) =US 96–717 356

18.3
Polyglutamic Acid Containing Compounds

See also: 8.2 [23]

1. JA 20 311 (Ajinomoto; 13.6.66/10.7.70)

2. GB 1 217 215 (Kyowa Hakko Kogyo Co. Ltd.; JA Prior. 20.10.67 and 4.6.68; 30.7.68/31.12.70)

3. OS 1 794 066 (Kyowa Hakko Kogyo Co. Ltd.; YM Fujimoto et al.; 2.9.68/13.5.71)

4. OS 2 018 189 (Ajinomoto Co. Ind.; A Akamatsu et al.; JA Prior. 16.4.69; 16.4.70/29.10.70) =JA 29 519–69

5. OS 2 152 538 (Kyowa Hakko Kogyo Co. Ltd.; Y Fujimoto et al.; JA Prior. 26.10.70; 21.10.71/ 27.4.72)

6. DOS 2 124 042 (Kyowa Hakko Kogyo Co. Ltd., Y Fujimoto, M Teranishi; JA Prior. 20.4.70; 14.5.71/2.12.71)=JA 42 410/70

7. US 3 594 351 (Hitachi Chemical; Sh Uchida, Y Sone; JA Prior. 10.4.67; 26.3.68/20.7.71)=FR 1 567 338=JA 67/22 419 and JA 67/22 422

8. US 3 729 366 (Kyowa Hakko Kogyo Co. Ltd.; Y Fujimoto et al.; JA Prior. 22.11.67; 15.1.71/24.4.73; cip 18.11.68)=JA 74 709/67=DOS 1 810 728

9. JA 73 11 923 (Dainichiseika Colour and Chem. Mfg. Co. Ltd.; 14.5.69/17.4.73)

10. JA 28 914 (Ajinomoto Inc.; 29.12.67/21.8.71)

11. JA 1 715/72 (Ajinomoto Inc.; 25.12.67/18.1.72)

12. JA 73–23 882 (Honey Chemicals; 6.11.68/17.7.73)

13. JA 19 714/72 (Honey Chemicals; 23.1.69/5.6.72); JA 19 713/72 (Honey Chemicals; 22.3.69/ 5.6.72)

14. OS 2 223 080 (Ajinomoto; K Toogoo et al.; JA Prior. 11.5.71)=JA 71/31 434; 12.5.72/16.11.72; describing mainly a finish for artificial PVC leather

15. OS 1 469 572 (K Shibata; JA Prior. 12.8.64)=JA 64/46 637; 29.7.65/23.1.69=FR 1 455 985

16. JA 73–05 262 (Kyowa Leather Cloth Co. Ltd.; 14.12.70/15.2.73)

17. JA 49 074 798 (Daiichi Lace Co. Ltd.; 20.11.72/18.7.74)

18. JA 39 329/72 (Toyo Cloth Co. Ltd.; 27.9.68/4.10.72)

19. JA 73 500 1 (Ajinomoto Inc.; 25.12.67/13.2.73)

20. JA 75 034 081 (Ajinomoto Inc.; 4.9.64/6.11.75)

21. JA 75 034 082 (Kyowa Leather Cloth; 19.2.66/6.11.75)

22. JA 74 031 081 (Honey Chem. Co. Ltd.; 1.9.69/19.8.74)

23. JA 60 173175 (Unitika KK; 13.2.84/6.9.85)

24. JA 59 036 781 [Seiko Kasei KK; 23.8.82/29.2.84 (146 036)]

25. EP 0 104 049 [Kanebo Ltd., K Sato et al.; 14.9.83/28.3.84; JA Prior. 17.9.82 (162 995–82)]; JA 01 192 882 [Kanebo KK; 26.1.88/2.8.89 (15 303)]

26. JA 59 157 386 [Toyo Cloth KK; 22.2.83/6.9.84 (28 950)]

27. JA 59 140 219 [Mitsubishi Chem. Ind. KK; 31.1.83/11.8.84 (13 888)]; see also JA 59 140 217

28. JA 62 215 080 [Aichi Hikaku Kogyo KK; 10.3.86/21.9.87 (51 995)]

29. JA 62 162 518 [Mitsubishi Chem. Ind. KK; 23.1.86/18.7.87 (4 676)]

30. JA 5043792 (Ajinomoto KK; Seiko Kasei KK; 15.8.91/23.2.93)

39. DOS 3 816 648 [Ajinomoto; JA Prior. 15.5.87 (119 954)/24.11.88]

40. EP 225 060 (Toray Ind. Inc.; J Amano, M Shimada et al.; 31.10.85/10.6.87)=JA 85–242 960 (31.10.85) + JA 86–45 328 (4.3.86) + JA 86–112 773 (19.5.86)

41. EP 151 963 (Unitika KK; 23.1.84/21.8.85)=JA 84–10 853 (Prior. 23.1.84) + JA 84–1029 232 (Prior. 17.2.84) + JA 84–183 277 (Prior. 31.8.84) + JA 84–183 378 (Prior. 31.8.84) + JA 84–188 435 (Prior. 7.9.84) + JA 84–191 932 (Prior. 13.9.84)

42. JA 1 113 230 (Unitika KK; 27.10.87/1.5.89); JA 1 221 577 (Unitika KK; 29.2.88/5.9.89)

43. JA 2 026 735 (Unitika KK; 18.7.88/29.1.90)

44. US 4 145 266 (Honey Chemicals; K Hayasaka, Y Zamaoka; 21.12.76/20.3.79)=JA 76–152921 (JA Prior. 21.12.76)=JA 77–10719 (JA Prior. 4.2.77)=JA 77–10720 (JA Prior. 4.2.77)

45. JA 61 000 679 (Unitika KK; 12.6.84/6.1.86)=JA 84–121 067; JA 63 165 584 (Unitika KK; 25.12.86/8.7.88)=JA 86–314 237

18.4
Other Hydrophilic Products of Natural Origin

See also: 9.3 [39], 10.6 [3]

1. DE 2 124 042 [Kyowa Hakko Kogyo Co. Ltd.; Y Fujimoto, M Teranishi; 14.5.71/2.12.71; JA Prior. 20.5.70 (42 410–70)]

2. JA 072 47 373 (Ajinomoto Takara Corp.; 9.3.94/26.9.95)

3. JA 53 39 507 (Sumitomo Seika Chem. Co. Ltd.; M Yasui; 12.6.92/21.12.93)=JA 92–179 024

4. RU 2 004 553 (Film Material Res. Inst. Mosc. Light Ind. Techn. Inst.; MM Kukharchik, LV Moiseeva, IA Shalabanova; 30.4.91/15.12.93)=RU 91–4 938 273

5. EP 470 399 (Showa Denko KK; T Wada, T Yamaguchi; 8.8.90/12.2.92)=JA 90–211416

6. JA 41 53 378 (Japan Steel Works Ltd.; 15.10.90/26.5.92)=JA 90–275 667

7. JA 41 63 377 (Toyo Cloth Co.; 25.10.90/8.6.92)=JA 90–287 953

8. JA 4175369 (Ajinomoto KK; Seiko Kasei KK; 8.11.90/23.6.92)=JA 90–302 896

9. JA 92 28 249 (Kurashiki Spinning Corp and Shohikagaku Kenkyusho KK; M Yamada, K Tamahisa, T Hagiwara; 19.2.96/2.9.97)=JA 96–56 714

10. JA 09 310 286 (Toray Ind. Inc.; K Okamoto, A Muratsu; 24.5.96/2.12.97)=JA 96–129 750

11. DOS 19 704 651 (Fraunhofer Ges. e.V.; J Wissler et al.; 11.9.96/12.3.98)

12. Skarja GA, Woodhouse KA (1998) Synthesis and characterization of degradable polyurethane elastomers containing an amino acid based chain extender. J Biomater Sci Polym Ed 9(3):271

19
Production of Synthetic Suede

See also: 11.1 [51], 18.2 [28]
1. US 3 578 481 (DuPont; ChA Young; 18.6.68/11.5.71)
2. US 3 597 256 (DuPont; ChA Young; 18.6.68/3.8.71)
3. US 3 600 209 (DuPont; ChA Young; 18.6.68/17.8.71)
4. JA 75 016 401 (Teijin KK; 9.7.70/12.6.75)
5. OS 2 040 397 (Göppinger Kaliko; E Schwarzkopf, J Gorr; 13.8.70/17.2.72)
6. JA 73 00 925 (Toray Ind. Inc.; 21.2.70/12.1.73)
7. OS 2 125 443 (Degussa; H Neidhardt, W Bechthold; 22.5.71/30.11.72)
8. JA 48 103 874 (Kuraray Co. Ltd.; 11.4.72/26.12.73)
9. JA 23 902/72 (Kuraray Co. Ltd.; 4.10.67/3.7.72)
10. NE 6 805 359 (A.K.U. N.V.; 17.4.68/21.10.69)
11. OS 2 109 171 (Chem. Fabrik Stockhausen; R Peppmöller; 26.2.71/7.9.72)
12. BE 622 530 (United Merchants and Manufacturers Inc.; EA Smith; US Prior. 9.10.61; 17.9.62/16.1.63)
13. JA 5 195/64 (Koyo Kogyo Co. Ltd.; 27.11.62/21.4.62)
14. US 3 709752 (Pandel-Bradford Inc.; R Wistozky, RE Petersen; 20.1.71/9.1.73; cip 17.8.67)
15. JA 73 25 482 (Toray Ind. Inc.; 7.11.70/30.7.73)
16. US 3 861 937 (Hooker Chem. & Plastics Corp.; G Hanneken, C Wirth; 3.7.72/21.1.75)
17. OS 2 405 275 (Texon Inc.; GE Martel; US Prior. 8.2.73; 4.2.74/14.8.74)
18. JA 49 126 801 (Shiga Akiresu KK; 16.4.73/4.12.74)
19. JA 50 040 870 (Kurashiki Spinning KK; 15.8.73/14.4.75)
20. Trademark of Bayer AG, Leverkusen, Germany
21. JA 55 084 480 (Asahi Chem. Ind. KK; 22.12.78/25.6.80)
22. JA 55 098 975 (Kuraray KK; 17.1.79/28.7.80)
23. JA 58 008 174 (Mitsubishi Rayon KK; 6.7.81/18.1.83)
24. DOS 3 022 327 [Teijin Ltd.; N Minemura, T Kimura; 13.6.80/18.12.80; JA Prior. 15.6.79 (74 581)]
25. JA 57 021 576 (Toray Ind. Inc.; 9.7.80/4.2.82)
26. JA 06 264 370 (94,264,370) (Teijin Cordley Ltd.; N Ookawa, Y Suzuki, S Yamauchi; 10.3.93/ 20.9.94); JA 06 264 369 (94,264,369) (Teijin Cordley Ltd.; N Ookawa, Y Suzuki, S Yamauchi; 10.3.93/20.9.94)
27. EP 651 090 (Kuraray Co. Ltd.; T Ashida, H Yoneda, T Yamasaki; 29.10.93/3.5.95)
28. JA 07 126 986 (95,126,986) (Teijin Ltd.; H Kitawaki, M Mimura; 2.11.93/16.5.95); JA 07 126 985 (95,126,985) (Teijin Ltd.; H Kitawaki, M Mimura; 2.11.93/16.5.95)
29. JA 06 346 378 (94,346,378) (Kuraray Co., S Kawakami, T Yamazaki, 3.6.93/20.12.94)
30. JA 07 42 082 (95 42,082) (Achilles Corp.; K Oosawa; 27.7.93/10.2.95)
31. EP 674039 (Bayer AG; H Träubel, M Reiner, R Langel, HA Ehlert; 9.3.95/27.9.95)
32. JA 3260180 (Daiichi Lace KK; 1.3.90/20.11.92)
33. JA 062 20 774 (Kuraray Co. Ltd.; 22.1.93/9.8.94)
34. JA 08 13 339 (96 13 339) (Saito Yoshimitsu; Y Saito; 27.6.94/16.1.96)
35. DDR 234 886 (VEB Vogtländische Kunstlederfabrik; D Helbig, M Boehm et al.; 26.2.85/ 14.4.86)=DDR 85–273 533
36. JA 83 25 954 (Kanebo Ltd.; S. Kihara; 25.5.95/10.12.96)=JA 95–162931
37. DOS 2 905 185 (Toray Ind.; M Yagi, K Yagi, S Mizuguchi; 13.2.78/23.8.79)=JA 78–14047
38. EP 761 869 (Teijin Ltd.; N Ohkawa, Y Suzuki, K Sasaki; 7.9.95/12.3.97)=JA 95–230 398
39. DOS 2 726 569 (Toyo Cloth Co. Ltd.; T Sasaki et al.; 13.6.77/26.10.78; JA Prior. 20.4.77)=JA 77–46 093
40. JA 54005001 (Toyo Ink Mfg. Co.; 10.6.77/16.1.79)=JA 77–69 259
41. JA 79 133 692 (Unitika Ltd.; K Amamya, H Matszaka; 17.4.78/27.10.79)
42. DOS 3 041 088 (Toa Nenryo Kogyo KK; 31.10.80/14.5.81; JA Prior. 1.11.79)=JA 79–140 440
43. JA 57 171 772 (Toray Ind. Inc. 15.4.81/22.10.82)=JA 81–55 528
44. JA 61 207 675 (Kuraray KK; 8.3.85/16.9.86)=JA 85–46 858

45. OS 1 807 579 (Kurashiki Rayon; O Fukushima et al.; 4.11.68/3.7.69; JA Prior. 4.11.67)
46. OS 1 817 661 (Kurashiki Rayon; K Noda et al.; 30.12.68/18.9.69; JA Prior. 29.12.67)

20
Leather Board

1. DOS 1 802 129 (I Lorant, L Radnoti; HU Prior. 27.10.67; 9.10.68/29.5.69)
2. FR 1 494 094 (F Andrieu; 14.3.66/3.9.67)
3. Stather F (1959) Leder und Kunstleder. Berlin, pp 114–122; Herfeld H (1950) Die Qualität von Leder, Lederaustauschwerkstoffen und Lederbehandlungsmitteln. Berlin, pp 23–28, 162–167
5. Zemlicka P (1960) Untersuchungsergebnisse bei der Faserledertrocknung. Kozawstvi 10:72; C. 1961, 16.3.44
6. GB 930 928 (G Bertolaia; IT Prior. 27.1.61; 30.8.61/10.7.63)
7. JA 5193/64 (Teikoku Kasai Kogyo Co. Ltd.; 10.2.61/21.4.64)
8. RA 169 055 (RE Reizin et al.; 26.11.62/Nov. 65)
9. RA 152 937 (AE Rishin et al.; 27.4.62/March 63)
10. JA 5 258/67 (R Kimura; 20.12.63/3.3.66)
11. FR 1 567 432 (Collagen Corp.; 13.5.68/16.5.69)=IT 838 331
12. DDR 58 822 [A Fredl et al.; (WTZ Techn.Text.); 6.2.67/20.11.67]
13. GB 1 105 317 (Bonded Faber Fabric Ltd.; AES Fairfull; 28.5.65/6.3.68)
14. FR 2 077 236 (Ind. Fibre e Cartoni Speciale; IT Prior. 20.1.70; 20.1.71/22.10.71)
15. JA 50 039 364 (Mitsubishi Petroch. KK; 13.8.73/11.4.75)
16. US 3 236 923 (US Secr. of the Army; EF Degering; 8.11.62/22.2.66; div. from 8.11.62)
17. JA 16 378/63 (H Tatsuyama et al.; 3.4.63/15.12.69)
19. FR 2 482 144 [Tray Textiles Inc.; Kohoku Chem. Ind. Co. Ltd.; I Nakajima et al.; 7.5.81/13.11.81; JA Prior. 7.5.80 (59 533)]
20. DDR 298 298 (Forschungsinstitut Leder & Kunstleder, Lederfaserwerk Siebenlehn, Schuh Design; A Berger, U Beyer, T Feigel, C Menzel, G Reich, R Steinhardt, S Wuensch; 9.5.88/13.2.92)=DDR 88–315 515
21. DDR 298 297 (Forschungsinstitut Leder & Kunstleder, Lederfaserwerk Siebenlehn, Schuh Design; A Berger, U Beyer, T Feigel, C Menzel, G Reich, R Steinhardt, S Wuensch; 9.5.88/13.2.92)=DDR 88–315 516
22. DDR 298 669 (Forschungsinstitut Leder & Kunstleder, Lederfaserwerk Siebenlehn; A Berger, U Beyer, T Feigel, G Griessbach, C Menzel, G Reich, R Steinhardt, S Wuensch; 9.5.88/5.3.92) =DDR 88–315 517
23. Sykes G (1997) Leatherboard manufacture. Leather World, April, pp 83–87

20.1
Regenerated Collagen

See also: 8.2 [34–36, 63, 88], 8.6 [25], 15 [29], 22 [48]
1. FR 1 270 918 (United Shoe Mach. Corp.; FC Merriam, RA Whitmore; US Prior. 14.10.59; 14.10.60/24.7.61)
2. GB 960 830 (United Shoe Mach. Corp; US Prior. 23.11.60; 21.11.61/17.6.64)
3. RA 181 232 (MP Kotov; 26.10.64; published in Dec. 66)
4. JA 8 559/67 (H Sato; 20.2.64/15.4.67)
5. OS 1 470 977 (United Shoe Mach. Corp.; Shu Thung-Tu, RA Whitmore; US Prior. 3.5.60 and 27.3.61; 27.4.61/30.1.69)=FR 1 294 967=GB 979 310=US 3 071 483 and US 3 122 599
6. OS 1 494 740 (Fuji Spinning Co. Ltd.; H Sato, H Okamura; JA 30.4.64; 29.4.65/11.12.69) =JA 18 956/66
7. US 2 934 446 (USM Corp.; JH Highberger, RA Whitmore; 21.12.55/26.4.60)
8. US 3 542 910 (Collagen Corp.; M Barash, BS Anthony; 25.11.66/24.11.70)
9. US 3 619 275 (Collagen Corp.; M Barash, SA Benedict; 13.8.69/9.11.71; div. ex US 3 542 910 25.11.66)

10. FR 1 531 215 (Nihon Leather Kogyo KK; H Sato; JA Prior. 25.11.66; 17.7.67/28.6.68)= JA 77 216/1966
11. JA 16 265/70 (S Hisao; 24.8.67/5.6.70)
12. GB 1 147 657 (SVIT; L Bogdanovicz et al.; CZ Prior. 13.11.65; 1.11.66/2.4.69)=FR 1 506 366
13. FR 1 079 705 (ICI Ltd.; OB Edgar, HG White; GB Prior. 30.5.49; 10.5.50/2.12.54)=GB 671 398
14. FR 2 188 610 (Lincrusta; 2.6.72/18.1.74)
15. OS 1 470 891 (USM Corp.; Shu Tung-Tu; US Prior. 31.1.62; 20.12.62/7.8.69)=BE 627 838= FR 1 348 716 =GB 1 024 769=OE 256 321=SZ 412 325
16. FR 1 348 716 (USM Corp.; Shu Tung-Tu; US Prior. 14.7.64; 13.7.65/28.4.67)=US 3 223 551=BE 643 345 =NE 6 400 603
17. JA 73 3 899 (Honey Chem. Co. Ltd.; 6.9.68/3.2.73)
18. JA 9 062/65 (Japan Leather Co. Ltd.; 11.5.62/11.5.65)
19. Küntzel A, Leberfinger R (1965) Über die Herstellung einer künstlichen Haut aus löslichem Kollagen. Das Leder 16(5):97
20. JA 73-37 145 (Honey Chem. Co. Ltd.; 4.8.69/9.11.73)
21. GB 1 051 160 (USM Corp.; Shu Tung-Tu; US Prior. 25.9.61; 25.2.64/14.12.66)=US 3 136 682
22. DAS 1 254 817 (Nihon Hikaku KK; M Taniguchi; JA Prior. 12.11.63)=JA 60 768/63; 20.2.64/ 23.11.67)=FR 1 392 729=GB 1 018 911
23. Okamura H,. Ota H (1969) Der Einfluß von Kollagenfasern als Bindemittel auf die mechanischen Eigenschaften von Vliesen auf Basis von chromgegerbten Kollagenfasern. Hikaku Kagaku 15:14; (1970) JALCA 65:145
24. OS 1 945 244 (C.T.C.; GJ Pichon et al.; FR Prior. 27.11.68; 6.9.69/11.6.70)
25. FR 2 009 877 (Nichiko Co. Ltd.; S Kawamuro et al.; JA Prior. 13.5.68)=JA 32 047/68; 30.4.69/ 13.2.70=NE 6 907 086
26. BE 735 583 (SVUK; M Krems, V Hrabovsky; CZ Prior. 16.8.68; 3.7.69)=CZ 5 939/68
27. JA 16 262/70 (Fuji Spinning Co. Ltd.; 20.9.67/5.6.70)
28. JA 16 266/70 (H Sato; 20.9.67/5.6.70)
29. JA 15 824/70 (H. Sato; 15.7.67/2.6.70)
30. GB 1 196 308 (USM Corp.; US Prior. 2.8.66; 1.8.67/24.6.70)=US 3 483 016
31. OS 1 936 957 (H Schaller, G Schaller; 21.7.69/4.1.71)
32. JA 060 49 414 (Kyoeisha Kagaku KK; 31.7.92/22.2.94)
33. Johnston KP (1996) Science 271:624
34. Hebestreit G (1996) Textiltechnisch verarbeitbare Kollagenfasern aus Chromlederspaltabfällen. Das Leder 122
35. JA 50 65 457 (Maruhatchi Muramatsu KK; Nitta Gelatine KK; 1.3.91/19.3.93)
36. JA 50 97 900 (Otsuga Kagaku Yakuhin KK; 28.6.91/20.4.93)
37. JA 84-157383 (Asahi; 23.2.83/6.9.84)=JA 83-027651
38. DOS 2 705 669 (Chemiaro; 1.9.77)
39. JA 92 40 123 (Kako Seisakusho KK; 11.3.96/16.9.97)
40. US 3 651 210 (Ashland Oil Inc.; TH Shepler, DS Lord; 15.5.70/21.3.72)

20.2
Coating of Leather

See also: 11.1 [18, 19, 62], 15 [16], 17 [3]

1. Some publications dealing with split coating: Krum HF (1971) Veredlung von Spaltledern mit PVC- und PUR Folien. Das Leder 22:97; Herfeld H, Steinlein I (1972) Veredlung von Spaltledern durch Folienkaschierung. Das Leder 23:98; Zorn B (1971) Beschichtung von Leder mit einer neuartigen Polyurethanfolie. Das Leder 147

2. Schaefer P (1973) Spezialfolien und Verfahren zum Kaschieren von Spaltleder für die Herstellung von mit Hochfrequenz geprägten Schuhoberteilen. Das Leder 23:229; (1972) Schuhtechnik 66:12, 219

3. Träubel H (1974) Reaktive Polyurethane für die Spaltzurichtung. Das Leder 25:162; Träubel H (1975) The coating of split leather with high reactive polyurethanes. J Coated Fabr 5:114; US

3 245 827 (Phelan-Faust Paint Manuf. Co.; FC Weber; 2.1.64/12.4.66; FR 2 178 757 (Pechiney-Ugine-Kuhlmann; 5.4.72/16.11. 73) addition to FR 1 427 722 (28.10.64); GB 1 240 631 (Wharton Industries Inc.; US Prior. 9.5.68 and 10.4.69; 6.5.69/28.7.71)=US 3 539 424; FR 2 003 716 (Lankro Chem. Ltd.; A Gill et al.; GB Prior. 12.3.68 and 9.12.68; 14.11.69/11.3.69); GB 1 232 247 (Norwood Ind. Inc.; RG Sutton; 20.8.68/19.5.71)

4. OS 2 039 281 (Th Böhme KG; P Jacob, HD Graf; 7.8.70/10.2.72)
5. OS 2 146 889 (UCB; A Hansson; GB Prior. 21.9.70; 20.9.71/23.3.72)
6. OS 2 010 045 (Lederfabriek L. Mombers; PJ de Nijs; NL Prior. 10.3.69; 4.3.70/17.9.70)=BE 746 620
7. OS 2 003 861 (SAVA; RM Schmid et al.; JU Prior. 5.5.69; 28.1.70/19.11.70)
8. OS 2 046 318 (BASF; N Hoppe, F Förster; 19.9.70/13.4.72)
9. OS 2 205 388 (Rohm & Haas Co.; JG Brodnyan et al.; US Prior. 12.3.71; 4.2.72/19.10.72)= BE 780 487
10. Trademark of Bayer AG (Germany)
11. Träubel H, Weber KA, Pisaric KH (1986) The finishing of splits. J Coated Fabr April, pp 250–262
12. JA 07 150 479 (94–191 005; JA Prior. 21.7.94 and 93–276 224 (7.10.93) (Achilles Corp.; K Oosawa, K Mitsumura, K Sugaya; publ. 15.6.95); JA 07 242 900 (94–56 644) (Achilles Corp.; K Mitsumura, K Oosawa, M Sato; 2.3.94/19.9.95)
13. DOS 1 951 024 0 (Ph Schaefer; 29.3.94/5.10.95)
14. DOS 4 230 997 (Sandoz Ltd.; Sandoz Patent GmbH; R Helber, H Huber, H Knoch, T Reusch, W Walther; 16.9.92/24.3.94)=EP 662 985
15. JA 07 330 854 (95 330 854) (Hodogaya Chem. Co. Ltd.; A Ishii, S Katagiri; 10.6.94/19.12.95); JA 07 330 855 (95 330 855) (Hodogaya Chem. Co. Ltd.; A Ishii, S Katagiri; 10.6.94/19.12.95)
16. Uddin MK, Khan MA, Idris KM (1996) Development of polymer films and its application on leather surfaces. J Appl Polym Sci 60:887
17. JA 07 292 399 (Achilles Corp.; M Sato, K Mitsumura, K Oosawa; 24.4.94/7.11.95)
18. JA 53 45 384 (Achilles Corp.; 15.6.92/27.12.93)=JA 92–180 465
19. DOS 2 638 792 (Vyzkumny uytav kozedelny; E Mück et al.; 27.8.76/10.3.77; CZ Prior. 29.8.75) =CZ 75-5 908
20. Bailey AJ (1992) Collagen – nature's framework in the medical, food and leather industries. J Soc Leather Technol and Chem 76:111
21. Löbig W, Tardiello G (1996) Aktueller Stand und Möglichkeiten der Schaumzurichtung. Das Leder 9:182; Foam finish for upholstery leather: Wenzel W, Pisaric K-H, Schwaiger W (1995) Schaumzurichtung. Das Leder 46:217
22. DOS 19 632 925 (Bayer AG; S Groth, J Pedain, L Schmalstieg, D-I Schütze; 16.8.96/19.2.98)

21
Nonwovens

For wet manufactured nonwovens, see also: 18.2 [61] and for bonding of nonwovens, see also: 8.6 [35–45], 17.1 Surveys

1. Anon. (1977) Neue Vliesstoffprodukte durch verbesserte Verfahren und Maschinen. Allg.Vliesstoff Report 6, 8:238; Freitag HD (1983) Herstellung von synthetischem Leder – Vlieskunstleder. Coating 251; Harvey GB (1975) Pneumatisch gelegte Vliesstoffe. Formed Fabrics Industry 10; Hering E (1977) Beziehungen zwischen Faser- und Vliesstoffeigenschaften. Textiltechnik 27(2):99; Heintze EF (1978) Chemiefasern in porösen bis skelettartigen Artikeln: Spezialprodukte zwischen Kunststoff und Textil. Chemiefasern/textilindustrie 28/80(3):262; Jörder H (1967) Neuartige Erzeugnisse auf der Basis nicht gewebter, textiler Flächengebilde. Chemiefasern 730; Jörder H (1968) Nach neuen Technologien hergestellte nicht gewebte textile Flächengebilde. Melliand 771; Jörder H (1961) Vliesstoffe-Begriffsbestimmung und Untergliederung. Z. ges. Textil Ind. 63(11):914; Nottebohm C (1968) Vliesstoffe auf trockenem Weg. Chemiefasern 667; Loy W (1965) Eine Systematik der Methoden zur Vliesstofferzeugung. Melliand 46:933; Lünenschloß J, Albrecht W (1982) Vliesstoffe. Stuttgart; Smorada R Nonwoven Fabrics (Spunbonded) In: Kirk-Othmer Encyclopedia of Chemical Technology, 4th ed, Vol. 17, pp 336–368; Vaughin EA Nonwoven Fabrics (Staple

Fibers) In: Kirk-Othmer Encyclopedia of Chemical Technology, 4th edn, Vol. 17, pp 303–335

2. FR 1 501 816 (3M Comp.; AE Raymon et al.; US Prior. 16.9.65; 15.9.66/18.11.67)

3. Langenthal W v, Sinn G (1973) Coating base for synthetic leather manufacture. J Coated Fabr 3:111

4. Riess W (1975) Die Herstellung und die Eigenschaften von Poromerics auf Vliesbasis. Allg Vliesstoff Report 113

5. Peters W (1975) Syntheselederbasismaterial. Formed Fabrics Industry 24

6. Drösler H (1973) Vliese und Syntheseleder. Textil Praxis Int April, pp 214–217

7. Langenthal W v, Weber KA (1967) Über die mechanischen Eigenschaften von Vliesstoffen. Bayer Farben Revue, Sonderheft 9, pp 19–33; Bindung von Vliesstoffen; Reutlinger Kollo-quium 8–9 Feb 1973, publ. in (1973) Chemiefasern und Textilindustrie 23 Nr. 5–8

8. GB 1 312 969 [Vepa AG; D Prior. 31.7.69 (=1 938 966), 12.8.69 (=1 940 954) + 5.11.69 (=1 955 653) + 2.3.70 (=2 009 662); 23.7.70/11.4.73]

9. SZAS 1 233/70 [Vepa AG; H Fleissner; D Prior. 6.2.69 (=1 905 746); 28.1.70/31.1.73]=GB 1 273 311

10. OS 2 046 664 (Kalle AG; KD Hammer, H Porrmann; 22.9.70/13.4.72) addition to OS 1 951 977 =BE 757 409

11. OS 2 048 804 (Vepa AG; 5.10.70/6.4.72)

12. OS 2 146 059 (C Freudenberg; H Weber; 15.9.71/22.5.73) addition to 2 005 206

13. A survey of patent applications for the years 1909–1958 is given in Rupert W (1962) Vliesstoff-Fertigung. Z. ges. Textilind 64:581; Großsteinbeck R (1969) Die Entwicklung nicht gewebter Flächengebilde. Chemiefasern 19:698

14. OS 2 030 703 (DuPont; WA Hare, RK Smith; US Prior. 23.6.69; 22.6.70/14.1.71)

16. JA 4 720 473 (Nippon Cloth Ind. Co. Ltd.; 23.2.71/29.9.72)

17. FR 1 261 904 (Mölnycke Väfveriaktiebolag; 16.3.60/17.4.61)

18. GB 887 906 (British Nylon Spinners; 30.9.60/24.1.62)

19. GB 920 848 (Bonded Fibre Fabric Ltd.; AES Fairfull; 30.11.60/13.3.63)

20. US 3 097 127 (DuPont; BG Ostmann; 1.12.60/9.7.63)

21. DAS 1 148 521 (Rohm and Haas Co.; VJ Moser; 12.5.60/16.5.63)=BE 590 799=FR 1 264 661 =GB 890 149=SZAS 5512/60=US 3 012 911

22. OE 217 000 (O Zwoboda; 14.3.60/15.2.61)

23. OE 227 647 (O Zwoboda; 8.3.60/15.11.62)

24. FR 81 199 (AA Alegre; US Prior. 28.2.61; 23.2.62/1.7.63), see also BE 614 506; addition to FR 1 115 866

25. SZAS 1 036/61 (C Freudenberg; L Hartmann; 30.1.61/15.6.63)

26. SZAS 5 961/61 (C Freudenberg; L Hartmann, A Gräber; 23.5.61/30.8.63)

27. BE 620 089 (ICI; GB Prior. 11.7.61 and 20.6.62; 11.7.62/11.1.63)

28. BE 619 629 (Phrix-Werke AG; D Prior. 5.7.61;2.7.62/5.11.62)

29. OE 235 584 (Degussa; D Prior. 15.7.61; 26.6.62/15.1.64)=BE 620 057

30. SZAS 12 519/62 (C Freudenberg; D Prior. 16.5.62;24.10.62/15.6.64)

31. DAS 1 283 796 (C Freudenberg; R Schabert; 27.10.62/28.11.68)

32. BE 631 990 (ICI; GB Prior. 7.5.62/7.5.63)

33. GB 1 040 572 (Lantor Ltd.; GP Kiernan, BI Yones, FE Austin; 24.1.62/1.9.66)

34. BE 630 608 (Monsanto; JE Bromley, JM Green, R Horne, CM Irwin; US Prior. 4.4.62;/ 16.8.63)

35. JA 3 142/64 (Nippon Felt Kogyo Co. Ltd.; 12.1.62/26.3.64)

36. OE 227 224 (O. Zwoboda; 15.2.62/25.4.62)

37. US 3 125 462 (Hercules Powder Co.; MR Rachinsky; 10.4.62/17.3.64)

38. OS 1 444 068 (Bayer; G Kolb, KA Weber, B Zorn; 31.10.63/17.10.68)=BE 654 817=FR 1 413 373 =GB 1 020 142=US 3 294 580; vgl. SZ 441 211=SZAS 12 32/64

39. OE 250 897 (Bunse u. Biach; R Hemersan; 24.10.63/12.12.66)

40. GB 1 066 132 (C Freudenberg; H Fabricius et al.; D Prior. 21.5.63; 20.5.64/19.4.67)

41. GB 1 089 414 (C Freudenberg; D Prior. 29.11.63; 30.11.64/1.11.67) addition to GB 1 055 187

42. OS 1 635 684 (Glanzstoff; E Sommer et al.; 24.7.63/16.10.69)=BE 649 317

43. OS 1 546 476 (Glanzstoff; E Sommer, K Gerlach, K Boehme; 17.9.63/8.10.70)=BE 651 365 =FR 1 409 032=GB 1 036 041=NE 6 401 702=OE 266 573

44. US 3 304 220 (ICI; JE McIntyre; GB Prior. 7.5.62; 6.5.63/14.2.67)=BE 631 990=GB 967 350
45. GB 1 073 183 (ICI; St Davies, CR Sissons; 5.2.63/21.6.67)
46. GB 1 087 615 (ICI; CR Sissons; 29.11.63/18.10.67)
47. SZ 479 759 (ICI; HJ Marrinan, EI Riseley; GB Prior. 4.3., 4.7. and 15.7.63; 4.3.64/28.11.69)=BE 644 708=DDR 45 928=FR 1 392 744=GB 993 472=NE 6 402 166=US 3 317 335
48. GB 1 073 182 (ICI; S Davies; 1.3.63/21.6.67)
49. GB 1 073 181 (ICI; S Davies, CR Sissons; 5.2.63/21.6.67)
50. BE 644 708 (ICI; GB Prior. 4.3., 4.7. and 15.7.63; 4.3.64/3.9.64) see also reference 21[47]
51. JA 27 336/64 (Kurashiki Rayon Co. Ltd.; 20.9.63/30.11.64)
52. OS 1 469 546 (Kurashiki Rayon Co. Ltd.; O Fukushima; JA Prior. 19.3.63)=JA 12 803/63; 19.3.64/ 19.12.68=FR 1 386 044
53. SZAS 1 355/64 (British Nylon Spinners Ltd.; S Davies, CR Sissons; GB Prior. 5.2.63; 5.2.64/ 31.10.67); BE 643 421 and BE 643 422=FR 1 392 034 and 1 392 035=NE 6 400 929 and NE 6 400 932
54. OS 1 435 466 (C Freudenberg; L Hartmann; 24.10.64/20.3.69)=FR 1 450 274
55. JA 7 038/68 (S Yanagimachi; 14.3.64/15.3.68)
56. GB 1 130 571 (ICI; AJ Sanders; 19.8.64/16.1.68)
57. JA 27 271 (Kurashiki Rayon Co.Ltd.; 4.7.64/23.12.67)
59. OS 1 915 278 (Glanzstoff; K Boehme, E Sommer, K Gerlach; 26.3.69/15.10.70)
60. OS 2 013 912 (Monsanto Co.; P. Suskind; US Prior. 24.3.69; 23.3.70/1.10.70)
61. OS 1 934 835 (Dunlop; TE Jones, C Sutton; GB Prior. 10.7.68; 9.7.69/15.1.70)=NE 6 910 558
62. OS 1 918 674 (K. Narodni podnik; A Suchomel et al.; CZ Prior. 29.7.68; 12.4.69/30.4.70)
63. BE 719 796 (Glanzstoff; D Prior. 19.10.67; 22.8.68/3.2.69)=SZ 472 535=NE 6 814 772=OS 1 619 252
64. GB 1 212 750 (Kalle AG; D Prior. 30.11.67; 29.11.68/18.11.70)=NE 6 816 541=OS 1 704 777
65. GB 1 225 138 (Metallgesellschaft AG; D Prior. 13.12.67; 10.12.68/17.3.71=OS 1 704 854=NE 6 817740 addition to OS 1 214 509
66. US 3 520 834 (Mitsubishi Petrochemical; K Mizutani et al.; 18.8.67/21.7.70)
67. OS 1 560 801 (C Freudenberg; L Hartmann; 16.2.66/20.11.69)=SZ 444 107
68. NE 6 604 804 (AKU-Goodrich; 8.4.66/25.5.67)
69. US 3 533 902 (W. R. Grace Co.; RC Hoch; 29.5.69/13.10.70; cip 27.4.65)
70. NE 6 616 034 (Glanzstoff; D Prior. 14.12.65; 14.11.66/15.6.67)=BE 687 999=FR 1 504 370
71. US 3 436 303 (3 M Comp.; AE Raymond, JW Fraser, F Swedish; 16.9.65/1.4.69)
72. OS 2 053 497 (Kalle AG; KD Hammer et al.; 30.10.70/4.5.72)
73. JA 74–018 512 (Japan Vilene Co. Ltd.; 24.9.70/10.5.74)
74. JA 48–033 177 (Kuraray Co. Ltd.; 2.9.71/8.5.73)
75. OS 2 344 093 (ICI; KG Matthews et al.; GB Prior. 31.8.72; 31.8.73/7.3.74)
76. DBP 894 234 (Alkor-Werk; K Lissmann, GW Kühl; 10.2.43/29.1.53)
77. OS 1 560 768 (C Freudenberg; HA Flocke, E Demme; 22.6.62/9.10.69)=DDR 44 562=OE 233 512
78. FR 1 504 370 (Glanzstoff AG; D Prior. 14.12.65;13.12.66/1.12.67)
79. OS 1 619 105 (Kalle AG; B Kersting, W Seifried; 2.11.66/6.11.69)=BE 705 845=FR 1 543 638 =NE 6 713 625)
80. GB 1 072 445 (DuPont; US Prior. 8.3.65; 4.3.66/14.6.67)
81. OS 2 029 517 (DuPont; DCh Carbaugh, LR Harper; US Prior. 16.6.69; 15.6.70/23.12.70)
82. JA 74 030 864 (Kuraray Co. Ltd.; 7.12.70/16.8.74)
83. OE 250 289 (O. Zwoboda; D Prior. 14.7.62; 3.7.63/10.11.66)
85. JA 54 160704 (Asahi Chem. Ind.; 9.6.78/18.12.78)
86. JA 59 211644 (Mitsubishi Rayon KK; 11.5.83/30.11.84)
87. FR 2 217 456 [Mitsubishi Paper Mills; 28.11.72/6.9.74 (JA Appl. Nr. 72 42 233)]
88. Besso MM (1982) Die Beiträge des Binders und der Fasern zu den Eigenschaften von Nonwovens. Text Res. J 52(9):587
89. JA 06 240 582 (94 240 582) (Kuraray Co.; J Tani, T Akazawa, 15.2.93/30.8.94); JA 06 240 581 (94 240 581) (Kuraray Co.; J Tani, M Nakano, T Akazawa, 15.2.93/30.8.94)
90. FR 2 124 574 (Farbenfabriken Bayer; 8.2.71/27.10.72)=DOS 2 105 681
91. DDR 2 47714 (VEB Vowetex Plauen; D Helbig, U Hornung et al.; 12.3.86/15.7.87)=DDR 86–287783

92. US 4 397 892 (Es Cipoipari Kutato Ungar; I Lorant, G Murlasits, G Wlasitsch, F Farkas; Boer Muebor 26.4.82/9.8.83)=US 82–372039; GB 2 070 658 (Bor-Mu Bor Budapest; I Lorant, G Wlasitsch, Szomolanyi, G Murlasits 4.3.80/9.9.81)=GB 80–7258
93. DOS 2 444 782 (Artos; W Wittke; 8.4.76)
94. DOS 2 931 125 (DuPont; GM Parker; 31.7.79/20.3.80; US Prior. 7.9.78)=US 78–940 389)
95. EP 90 397 (Toray Ind. Inc.; H Kato, K Yagi; 28.3.83/5.10.83; JA Prior. 31.3.82+6.5.82)=JA 82–51119 + JA 82–74 582
96. JA 61 296 157 (Toray Ind. Inc.; 21.6.85/26.12.86)=JA 85–134 222
97. Anon (1990) Nonwoven production capacities on the increase. JTN, September, pp 31–34

21.1
Special Fibers

See also: 7.1 [26], 8 [21, 22], 8.6 [47], 23 [3]. Survey: Lauppe W (1976) Der Einfluß des Rohstoffs Fasern und seiner Eigenschaften auf Herstellung und Verwendung von Vliesstoffen. Melliand April, pp 290–300; Fourné F (1995) Synthetische Fasern. München, pp 551; Okamoto M, Kajiwara K Shingosen: Past, Present, Future. The Textile Dyer Vol. 27, No. 2: A survey about shingosen=new synthetic fibers, i. e. ultra fine fibers

1. JA 37 198/71 (Toray Ind. Inc.; 10.1.68/1.11.71)
2. US 3 589 956 (DuPont; W Kranz, WL Stump; 22.9.67/ 29.6.71; cip. 29.9.66)
3. JA 24 508/69 (Toyo Rayon; 19.3.66/16.10.69)
4. JA 73–20 282 (Toray Ind. Inc.; 2.2.70/20.6.73)
5. DDR 115 301 (ADW; Inst. f. Technologie d. Fasern; R Barthel, A Heger; 14.7.70/20.9.75) (WP 163 598) addition to DDR 112 060
6. JA 50 053 503 (Asahi Chem. Ind. KK; 17.9.73/12.5.75)
7. JA 50 042 004 (Kuraray KK; 21.8.73/16.4.75)
8. JA 17 478/65 (Kurashiki Rayon Co. Ltd.; 12.12.63/9.8.65)
9. JA 54 116 463 [Kanebo KK; 27.2.78/10.9.79 (22 567)]
10. JA 07 133 592 (95 133,592) (Kuraray Co.; M Makimura, T Hiramatsu; 9.11.93/29.5.95)
11. Schürer M (1996) Feinstfaservliese – ein Weg zu Kunstledern mit isotropen Eigenschaften. Coating August, pp 286–290
12. Riedel B, Knobelsdorf C, Mieck K-P, Seyfarth HE, Taeger E (1996) New ways to produce cellulosic microfiber nonwovens. Chem Fibers Int 46:454; Seyfarth HE (1987) Lederimitate aus superfeinen Fasern. Textiltechnik 118
13. Ishida T (1996) Manual of textile technology. JTN, Jan, pp 116–127
14. DOS 2 851 311 (Kuraray Co. Ltd.; O Fukushima et al.; 27.11.78/31.5.79; JA Prior. 28.11.77)=JA 77–143 062
15. JA 74 010 633 (Toray Ind. Inc.; 15.7.70/12.3.74)=JA 70–61 373
16. JA 9 228 222 (Unitika Ltd.; N Matsunaga, K Niikura; 29.2.96/2.7.97)=JA 96–42419
17. GB 1 120 781 (Kurashiki Rayon KK; JA Prior. 25.9.64)=JA 54 359/64; 16.9.65/24.7.68)
18. JA 48 093 796 (Kuraray Co. Ltd.; 22.3.72/4.12.73)=JA 72–28 703
19. FR 1 386 044 [Kurashiki Rayon Co. Ltd.; O Fukushima; JA Prior. 19.3.63=JA 12 803/63; (18.3.64/7.12.64)]
20. JA 34 923/72 (Toray Ind. Inc.; 26.4.68/2.9.72)
21. OS 1 955 673 (Toyo Rayon Co. Ltd.; T Hikota et al.; JA Prior. 5.11.68; 5.11.69/4.6.70)
22. JA 49–014 775 (Kuraray Co. Ltd.; 6.6.72/8.2.74); JA 49–014 776 (Kuraray Co. Ltd.; 6.6.72/ 8.2.74); JA 49–014 777 (Kuraray Co. Ltd.; 10.6.72/8.2.74)
23. JA 73 00 923 (Toray Ind. Inc.; 29.1.70/12.1.73)
24. JA 73 41 788 (Toray Ind. Inc.; 28.2.70/8.12.73)
25. Fukushima O, Kogame K (1976) Melliand 673; Hoashi K (1977) Japan Textile News No 4, 92; Oertel G (1993) Polyurethane Handbook, 2nd ed. Munich, p 581
26. FR 1 505 505 (DuPont; EG Parrish, J Farago; US Prior. 3.11.65; 2.11.66/15.12.67)=NE 6 615 458 =BE 689 190
27. OS 1 435 114 (DuPont; M Katz, M Makano; US Prior. 9.5.60; 9.5.61/17.10.68)=BE 608 645 =FR 1 297 784=GB 932 483; see also US 3 117 0566

28. JA 060 02 267 (Teijin Ltd.; 18.6.92/11.1.94); JA 061 58 496 (Teijin Ltd. 9.11.92/7.6.94)
29. Hengstberger M, Eiser R Schneider R, Oberhoffner S, Herlinger H (1995) Funktionale Flächengebilde aus Fibridmischungen. Technische Textilien 38(9):147
30. JA 52 30 772 (Kanebo Ltd.; 17.2.92/7.9.93)=JA 92–6 1542
31. JA 10 01 881 (Kuraray Co. Ltd.; H Nakashima, K Suetoshi, K Akamata, Y Komura, K Ouraya; 13.6.96/6.1.98)=JA 96–152 205; JA 10 08 382 (Kuraray Co. Ltd.; K Shikomata, H Yoneda, N Makiyama, H Uehara, Y Takahashi; 21.6.96/13.1.98)=JA 96–161 227
32. JA 10 53 957 (Kuraray Co. Ltd.; K Watanabe, H Hori, T Kowaka, S Sakane; 6.8.96/24.2.98)= JA 96 207 047

Part 4

Treatment of Man-Made Leather

Besides tanning and retanning, the optical and haptic effects of leather are determined by the dyeing and finishing operations. The dyeing and finishing of leather is described in detail in the literature (see 1 [1–5]); therefore, in the following chapter, only subjects covering the purpose of this book are discussed in detail.

CHAPTER 22

Finishing and Dyeing

Leather receives a surface treatment in the finish to influence its optical and haptic effects. Leather is dyed (Fig. 1-8, step 14) in the wet phase of its production. Nevertheless dyestuffs and pigments are also applied in the finish to adjust or intensify the color of the material (Fig. 1-8, steps 18 and 19). The use of dyestuffs and pigments is primarily done to get a colored product and secondly to create special effects like two-tone, glossy or mat effects and so on. Leather in most cases needs a finish. The finish is used to protect the surface of the leather and to unify the way the different pieces look. Being a natural product leather very often has different grain damages or grain structures which do not allow leather pieces to be easily matched in one article.

Consequently, to obtain similar effects, leather substitutes in many cases also receive an additional dyeing [32, 36] and/or finishing [1, 2, 11, 80) step (Fig. 22-1).

Leather substitutes in most cases are dyed with pigmented polymers. During their manufacture an additional coloring step is often not necessary. Sometimes, however, an additional dyeing at the end of the production is needed to get a

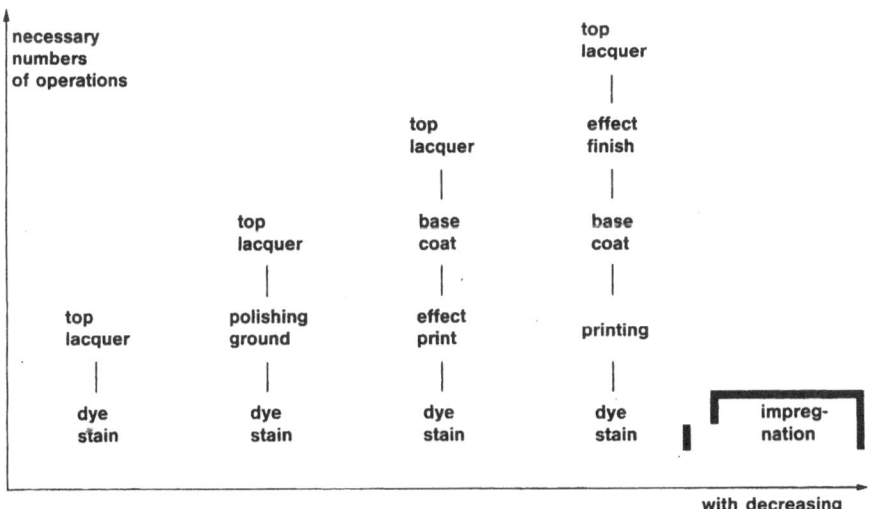

Fig. 22-1. Types of leather finishes depending on the quality of the leathers

special effect. This is normally done during the finishing operation where a colored finishing mixture to correct the final tone of the article or to improve or adjust the coverage of its surface is applied. There are only a few cases where dyeing is only carried out in a finishing operation. Therefore it does not make sense to differentiate between dyeing during manufacturing and coloring in the finishing.

The Dyeing. During the application of a polymer as a high solid, solution or dispersion, it is pigmented to match the shade which is needed. No special adjustment of the dyeing nuance is normally necessary. If necessary, 1 : 2 dyestuff complex [27, 28], i.e. dyestuffs with 1 molecule metal ion combined with two molecules of an azo compound [81], disperse dyes [75]; cationic and anionic ones, mordant or sulfur dyestuffs [4–7, 47] are recommended as suitable dyestuffs. The dyeing can be carried out in the usual devices for textile dyeing or sizing, like jig padding machines, spraying units [74], etc.

In textile dyeing a pre- or combined treatment with the dyestuff is usual. Microporous products are improved for the dyeing process if the surface of the artificial leather is pretreated with a methoxymethylated polyamide with tartaric acid as a crosslinker [19] or a polycationic component [68].

Metal-complex dyestuffs, insoluble in water, are dissolved in ethanol. An artificial leather is dyed with these solutions by flow coating (Fig. 22-2) or spraying (Fig. 22-3) [8]. A mixture of dyestuffs containing a metal-complex dyestuff with a cobalt atom can be used to obtain a green shade of a leather substitute having a polyamide fabric [50].

Leather substitutes may also be dyed in a dip coating machine [57], which is a device not common in leather dyeing (Fig. 22-4).

Surfaces of artificial leather may also be colored by a batik technique [20]. Water-soluble reactive dyestuffs, like Cibacron® brilliant yellow 3 GP (Ciba-

Fig. 22-2. A common flow or curtain coater as used in leather finishing

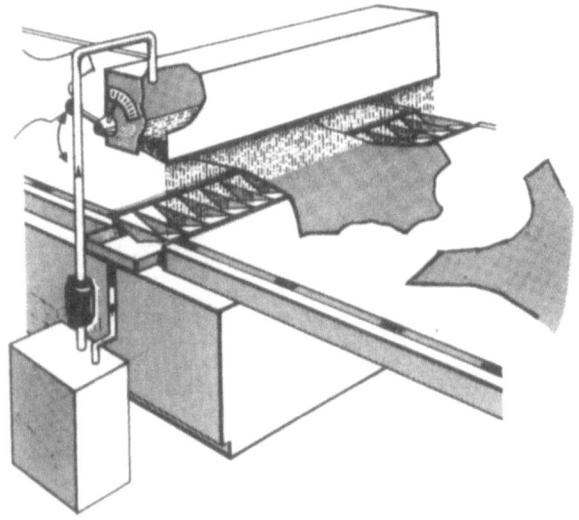

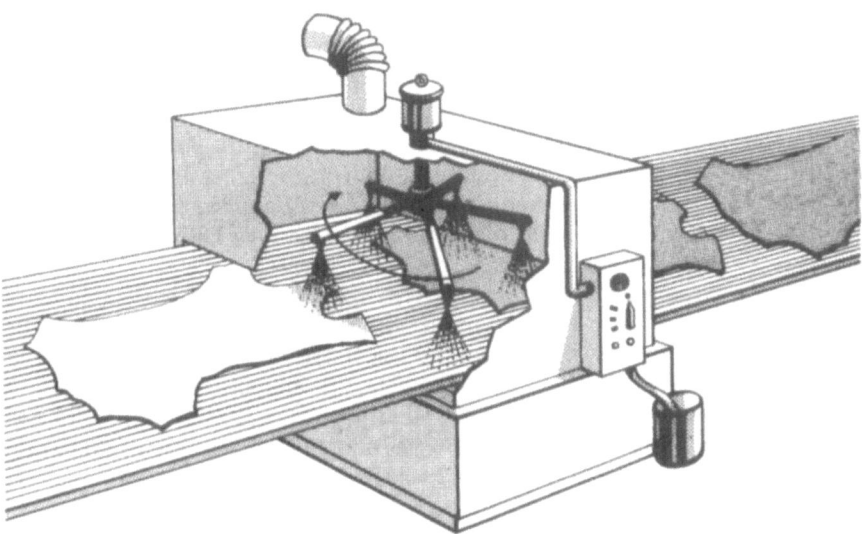

Fig. 22-3. A common spraying cabin as used in leather finishing

Fig. 22-4. A schematic view
of a dip coating machine

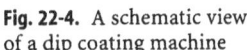

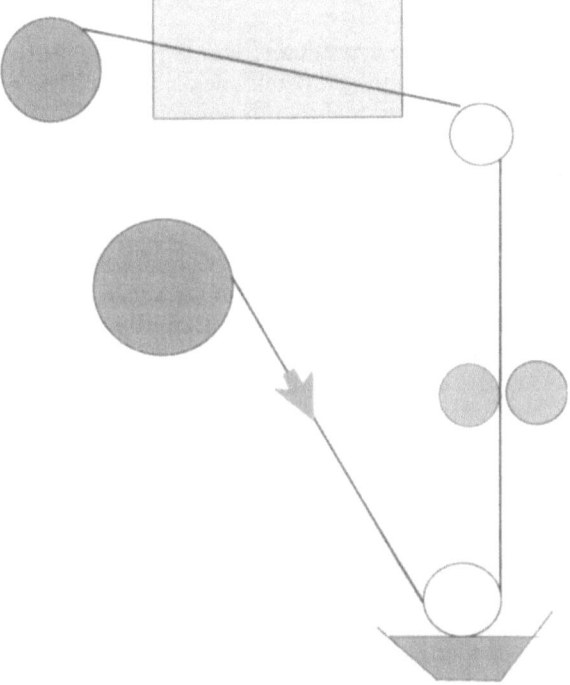

The substrate is dipped into the treating solution. After dipping it usually
passes two rollers where it is squeezed to eliminate a surplus of absorbed
material

Geigy) [45] may be added to a polyurethane solution in DMF to produce leather substitutes with a high degree of migration fastness [10].

It is also common to utilize a mixture of a binding agent, like polyurethane or polyamino acid, together with a pigment or a dyestuff [67].

Microporous materials are dyed by an impregnation of dispersions of pigments. These pigments are dispersed in a fluorocarbon, possibly containing a dispersed polymer, and an anionic surface active compound. After the removal of the fluorocarbon a treatment with an acid is necessary to fix and deactivate the surface active agent [23].

Leather substitutes of the Corfam® type may be colored with a padding dyeing or a printing ink [35]. For this type of dyeing the dyestuff should be soluble in DMF, amino alcohol or morpholine. A dyestuff solution is mixed with an alkoxylated phenol and a thickening agent, like alginate or starch. The substitute is treated for 10–30 s with this solution which may be split into two phases, and then heated at 200 °C [21]. Types of artificial leather with a different dyeing behavior may also be colored with such a system [46].

Leather substitutes may be dyed with pigments in butan-2-one and isocyanates if the polymer is treated with the dyeing slurry prior to coagulation [73].

If the surface of a leather substitute is difficult to dye then it should be treated with a dyestuff solution prior to heating for 10 s and up to 5 h at 40–150 °C. Then a solution is applied to the substitute containing a tensioactive substance. By this action the unfixed dye is able to penetrate the reverse of the substitute and color it [12].

Powders of polyurethanes with a particle size of 5–1000 μm can be used to color a water-based polyurethane dispersion. The powders are produced by prepolymers bonded to dyestuffs and reacted with diamines in water [31].

Azodyes are – after the dyeing of a leather substitute – metallized [33]. 2:1 metal-complex dyestuffs which have hydroxy or amino groups are applied together with isocyanates onto leather substitutes to become migration resistant dyeings [39]. Acid or mordant dyestuffs may be used to dye leather substitutes at a concentration of 0.002–8% and a pH of 1.5–11 at 50–100 °C for 2–20 s. The dyestuff solution is sucked through the substitute by reduced pressure at the opposite side. It is appropriate to use a chromium salt to fix the dyestuff [71].

Leather substitutes, produced by the wet coagulation process, are treated with a hot water based solution of a dyestuff having at least one of the following groups: sulfonate, carboxylate, sulfonamide, hydroxy, amine, urea, methoxy etc.; the dyestuffs should additionally have an equivalent weight of 200–600 [70]. Wet coagulated products can also be dyed by metal-complex dyestuffs [78]. 5–70% of a C_1–C_4 alkanol added to a dyestuff solution improves the dyeing [72].

A nonwoven consisting of ultrafine fibers is treated with a polyurethane emulsion to obtain a substitute which can be dyed easier [51].

Suede-like artificial leather is dyed by immersing the product consisting of very fine polyamide fibers and polyurethane in a dye dispersion [INDANTHREN® Red FBB (a trademark of BASF, Bayer, Cassella and Hoechst)] containing alkaline substances, adding a reducing agent, heating above 60 °C and oxidizing [79].

Undyed artificial leather is dyed by immersing it in a 0.01–5% solution of a disperse dye for 3–10 min in a solvent mixture of 80–95 parts of isopropanol and 5–20 parts of water [52].

A raised fabric consisting of ultrafine fibers is impregnated with a dissolved, colored polyurethane and coagulated. The coloring agent is a carbon black pigment able to reflect more than 60% IR rays with a wavelength of 900–1500 nm. The dyed artificial suede can be used in the automotive sector because it does not change color by heat or light [58].

1,4-Dihydroxy-2-phenoxyanthraquinone and aluminum tripropylate are added to a solution of a polyurethane. The solution is transformed into a film to be laminated by means of a polyurethane adhesive to a leather substitute which then is colored in a red shade [41].

Leather substitutes may be dyed with fat-soluble dyestuffs, such as Orasol® [65], which may be dissolved in an ethanol/water mixture [49, 61]. Dyeing with sublimating dyestuffs has also been investigated [53].

An artificial leather consisting of ultrafine polyester-fibers and a bonding agent containing polyester can be dyed with cationic dyestuffs as long as the polyester also contains sulfoisophthalate groups [55].

The surfaces of leather and leather substitutes treated with collagen powder may be dyed easily with acid dyestuffs [54].

Finishing. Artificial leathers produced by the indirect process do not normally need any finishing operation: They obtain their texture and appearance, their haptic and physical properties from the release paper and their top coat during the manufacture (see Chap. 2.1).

A top coat on a substrate can be regarded as a finish. *Laminated leathers* have already been discussed (see Chap. 20.2).

In leather finishing there are different types of finishes depending on the particular leather or the way in which it is finished [25, 43, 44]. The type and manner of finishes of leather depend on the surface structure of the material of origin. If the surface of the leathers is damaged than a lot of effort is needed to cover this damage. A survey of the finishing types depending on the quality of leather is given in Fig. 22-1. These types of finish also play a role in the finishing of substitutes and, therefore, they are discussed here.

Finishes containing a lot of – especially inorganic – pigments therein cover the surface of leather. Finishes which do not cover the surface of the substrate are called *Aniline Finish*, because usually only "aniline dyestuffs" are used for coloring. In some types of these aniline finishes only a minimal amount of – usually organic – pigments are used possibly together with a soluble dyestuff which does not cover as much the surface than a usual pigmented finish type. There are two steps in a normal aniline-type finish:

1. A spray coat with a dissolved dyestuff or an organic pigment and a binding agent such as a finely dispersed polyurethane dispersion is applied.
2. A topcoat consisting of a cellulose acetobutyrate-, a nitrocellulose- lacquer or – lacquer emulsion or a polyurethane solution or dispersion is applied. Usually only light-fast products are used. Cellulose acetobutyrate, polyamides and polyurethanes based on aliphatic isocyanates have a high light fastness.

Nitrocellulose and usual butadiene polymer products often used in leather finishes have only a medium light fastness.

Aniline-type leathers are the most expensive ones. Ergo an aniline type finish is a sign of an extremely good quality leather.

Leather is based on a natural product which is the animal hide or skin. Leather may have some superficial defects which consumers do not want. *Covering finishes* are applied to the surface of leather with a more or less damaged surface if needed. Normally 3–5 working steps are necessary for a covering finish:

1. If the surface is heavily damaged it is buffed.
2. To fix the rest of the finish and to guarantee a tight break of the grain the leather is impregnated with an acrylate or a polyurethane dispersion with a polymer of a small particle size distribution.
3. Then a base coat is applied consisting of an acrylate, butadiene and/or polyurethane dispersion [60] containing some pigment therein. The polymers used in the base coat are soft (Shore A hardness of less than 50). Acrylates have a rather high glass transition temperature (T_g), therefore they often need to be mixed with a polymer such as a butadiene copolymer with a very low T_g to obtain a good flexibility at a low temperature.
4. If desired an effect color can be applied; for example, a mixture of a dyestuff in a polyurethane dispersion which is printed to give a special effect.
5. Normally the leathers are embossed to obtain a grain pattern on their surface (which no longer exists due to the buffing operation).
6. Similar to the aniline finish a topcoat consisting of nitrocellulose, cellulose acetobutyrate or polyurethane is applied. The polymers used in the top coats are usually hard (Shore A hardness of more than 60). Nitrocellulose is sensitive to aminic compounds and plasticizers. Amines cause yellowing and degradation of the product. Therefore nitrocellulose products are rarely used with substitutes. The top coat products mostly used are polyurethanes.

To improve the resistance of the top coat, polyurethanes may be used in the form of their isocyanate prepolymers and mixed with a small amount of nitrocellulose [30].

The equipment used for the different finishing types are usually for an aniline finish, spraying guns [62] and/or printing machines [42]. In the impregnating and base coat steps for covering finishes which need more product per square foot more economic system applications are used with curtain coaters or reverse roll coaters (Fig. 22-5). In the past these product-intensive steps, like the base coat, were done by hand with a plush pad. The top coat is sprayed or applied by printing [42] or on a roll coater.

The water vapor permeability is reduced by the finishing operation. If the value of the water vapor permeability must be maintained either hydrophilic polyurethanes are used as bonding agents of the finish formulations [37] or the finish is applied discontinuously with a screen roller in a printing equipment. High priced leather substitutes are often finished in a similar way to genuine leather [59].

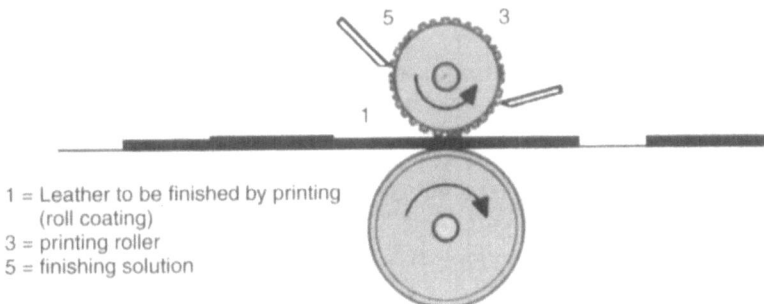

1 = Leather to be finished by printing
    (roll coating)
3 = printing roller
5 = finishing solution

Fig. 22-5. The principle of the roll coater printing finish

Finished leathers can also be characterized by their appearance. *Patent leather* is a leather with a highly glossy finish consisting of a two-component polyurethane [38] applied with a curtain coater.

Laminated layers of water vapor permeable foils on fabrics or nonwovens usually need a finishing because their surface is very uniform and no grain structure can be seen. It is easy to apply a finish onto these products. Usually only one [3] or two steps [64, 69] are necessary. In the first step a mixture of a bonding agent with a dyestuff and/or a pigment is applied. The second step is a top coat with a bonding agent, usually a polyurethane, and possibly some dyestuffs.

A two-step finish may consist of the following applications: A polyol, such as polypropylene glycol, is used for an impregnation then a polyurethane dissolved in DMF, consisting of polyethylene glycol, polypropylene glycol adipate, ethylene glycol as a chain-lengthening agent, and also MDI as well as Desmodur® L, an adduct of TDI on trimethylolpropane (Bayer AG), is applied [66].

An example of a finish for a leather substitute in several steps is as follows: A nonwoven bonded by a coagulated polyurethane is finished in several steps and finally embossed and plated by hot rollers [77]. Coagulated materials very often need a plating step prior to finishing with a plating roller [76].

A cationic polyurethane dispersion mixed with a methylol compound may be used as a finish for leather substitutes. With this finish the surface of the substitute becomes abrasion resistant and elastic [17].

Heat-resistant finishes for man-made leathers are created by finishes with a DMF solution of a polyamide. The polyamide is the condensation product of pyromellitic anhydride with 4,4'-diaminodiphenyl ether. Prior to application this amide is mixed with the solution of a polyamide-6 in methanol [24].

An artificial leather is treated with a solution of a polyamide containing a dyestuff, fixed by an acid then treated with a polyurethane [26].

Artificial PVC leathers may be finished with a polyurethane layer with a thickness of 0.1 – 0.01 mm then coated with a polyamide. The products are fixed onto the surface by heat [29].

Two latexes, the first consisting of a mixture of an acrylate with a vinyl chloride copolymer and a plasticizer and the second consisting of a polybutadiene with amino- or amido groups, are used as the finish to improve the abrasion and age resistance of the artificial leather [34].

A woven fabric consisting of polyester and polyurethane fibers is partially or completely impregnated with a polyurethane rubber to form an artificial leather [83].

A polyurethane with polyethylene glycol ether groups dissolved in DMF is printed onto a polyethylene foil and then coagulated. It is then printed with a pigment containing polyurethane. The polyurethane for printing contains polyethylene glycol ether groups as well [48].

An artificial leather containing polyamide, polyester and polyurethane fibers is treated with a colloidal solution of Au, Ag, Pd, Pt, etc. and a dispersing agent like naphthalene sulfonate. The result is an artificial leather with a metallic effect [56]. The surfaces of leathers or foams may be metallized with a coat with a mixture of a film-forming vinyl polymer mixed with a removable polyvinyl pyrrolidone and metallized without using electric devices [63].

Textiles coated with a polyurethane may be multicolored on the surface by a thermo-transfer printing process [22].

An artificial leather with a two-tone effect and good flexibility consists of a base sheet of an impregnated nonwoven, a surface layer of polyurethane, polyolefin, acrylate and/or nitrocellulose and inorganic particles, such as silicon dioxide, magnesium oxide, talcum etc. This layer has a thickness of 5 – 200 µm [14].

Finishing by Lamination. Thin foils of a certain thermoplasticity can be plated onto substrates. These thin foils may substitute a separate finishing operation. Foils may be plated directly onto leather or onto leather substitutes [18]. Another possibility is to use an adhesive prior to the lamination process. The lamination of thin foils is suited for coagulated substitutes [40]. The foil may be applied directly after production. So, for example, a thin polyurethane foil is produced on a release paper. By a heat setting resin this foil is transferred onto an artificial leather [13]. This method is claimed to have the advantage of having a tight break.

A solution of a polymer, e.g. a solution of polyurethane in THF, is applied onto a nonwoven by a polished roller prior to a lamination. By this action the nonwoven obtains a glossy surface [9]. The same can be done with a leather substitute [16].

An artificial leather with good chemical and abrasion resistance is prepared by lamination of a non-porous polyester elastomer, such as 1,4-butanediol–dimethyl terephthalate–polytetramethyleneglycol copolymer or adipic acid–1,4-butanediol–dimethylterephthalate–hexanediol copolymer, on a fibrous base such as a polyurethane impregnated polyester nonwoven. A porous intermediate layer can be applied if desired. The layers are pressed to form embossed or glossy surfaces [15].

A polyurethane soluble in DMF is made by polyadding 3-methyl-1,5-pentanediol adipate, MDI and butanediol. A second, slightly softer polyurethane is produced from the same ingredients with another molar ratio of the components. A nonwoven which has been impregnated by a polyurethane is coated by melt extruding of a polyurethane containing a blowing agent. Then it is again coated by melt extrusion by a polyurethane containing pigments. The material is finally embossed. The pigmented polyurethane is harder than the initial one. The result of this procedure is a three-layered artificial leather [82].

An impregnated polyester nonwoven is laminated with a silicone- and a fluorine-containing polyurethane of a Shore A hardness of 85 and embossed simultaneously [84].

CHAPTER 23

Modification of Physical Properties by Chemical Methods

The surface properties of leather substitutes often need to be improved in their water, solvent or abrasion resistance. Crosslinking is one of operations mostly used to modify a coating or an impregnation.

An application of a suitable solvent onto the surface of a microporous sheet usually leads to a local collapse of micropores on the outer parts of the surface. The result is a reduction in water vapor permeability and a smooth, uniform surface being more abrasion resistant than the original one.

Onto a surface of a microporous coating on a nonwoven DMF or cyclohexanone is sprayed. Simultaneously hot air with a temperature of 40–199 °C is blown over the surface. By this treatment the surface becomes smooth and leather-like [1].

When a microporous material is treated under heat and pressure with a solution of 20–50 parts of DMF, 15–30 parts of cyclohexanone and 0–50 parts of acetone then the material obtains a glossy surface [2].

In order to obtain a smooth surface, a nonwoven with fibers consisting of polystyrene and polyamide is impregnated with a polyurethane dissolved in DMF. Then the polyurethane is coagulated and the nonwoven is coated with a DMF solution of a polyurethane which is also coagulated. Polystyrene is removed by the action of toluene and the resulting material is treated with a solvent mixture of 1 part DMF and 9 parts of THF and pressed at 100 °C [3].

In order to create a surface that is similar to calf leather, microporous thermoplastic polymers, such as polyurethanes, are spot sprayed or printed with DMF which results in a slightly reduced water vapor permeability [4]. Similarly, a two-layer leather substitute without any textile substrate is treated with a solvent, such as DMF, cyclohexanone or acetone, possibly some dissolved or dispersed polyurethane can be added and then the material is heated to 160 °C under pressure. The thickness of the material is reduced by 20–30 % and its surface is solidified. If during this process the surface is also covered by a glossy, releasable foil the leather substitute obtains a glossy surface [5].

Microporous suede-like products are sprayed with a solution of an isocyanate to produce a heat-resistant article [6].

Special effects, such as how to obtain a stronger matting effect or how to obtain a glossy structure, are also a kind of finishing. It is possible, however, to alter the surface of a composed material just by the action of heat and/or solvent.

Coagulated surfaces become glossy if they are treated with a solvent and – additionally in a mechanical action – by a glossy or embossed roller [14].

Nonwovens with a coagulated polyurethane layer coming into contact with a hot roller acquire a solidified surface by a partial melting of the polyurethane [15, 16]. Onto a coagulated sheet a thin superficial layer of a polyurethane can be obtained if this polyurethane is applied in a melted form [17]. A polyurethane surface layer being too hard may be softened by adding a water-based dispersion of dibutyl phthalate [7] or silicone [11].

To obtain a softer substrate, a nonwoven containing polyester fibers is impregnated by polyurethanes to be coagulated afterwards. This nonwoven is then treated by alkali hydroxide. The alkali treatment hydrolyzes the polyester partially and decreases the contact of polyurethane to the fiber thereby softening the article [12]. A nonwoven impregnated with a coagulated polyurethane may be coated by a polyurethane dispersion to form an artificial leather [10].

Coating of silk, coagulating, drying and buffing leads to an imitation of a suede leather [13].

Microporous foils may be laminated to suedes by hot melt adhesive [9]. Coagulated man-made leathers may be bonded by an adhesive based on a dispersion of polychloroprene in water in the shoe factory. Prior to the bonding step the surfaces of the man-made leathers should be treated with a < 10 % solution of a chloride of Zn, Cr, Fe or Ca [8].

23.1
Modification of Physical Properties: Improving Water Resistance, Stain Resistance and Flame Retardation

In this chapter the treatment or a processing of finished leather substitutes is predominantly discussed. Water repellency or hydrophobing, oleophobic effects or flame resistance can be achieved by including the appropriate ingredients into a polymer prior to processing. This process of refining, e.g. by introducing functional groups into a polymer, is discussed in Chap. 25.

Many of the desired properties may be achieved by mixing a polymer blend with an active material. Flame resistance is usually achieved by adding to the polymer bromine- and/or chlorine-containing products or products containing a lot of nitrogen or phosphorus [19]. An undesired oxidation of polymers caused by free radicals can be avoided by an addition of a radical quencher like tri-*tert*-butylphenol [14].

Tri-*tert*-butylphenol

The same products may also help to improve light fastness [15]. Anti-static products are used against electrostatic charges on the surface of the coated materials [16]. Gliding agents are used to avoid stickiness when the materials come into contact with each other [13, 31].

Water Repellency and Oleophobic Effects [25, 29]. Paraffin wax or zirconium or aluminum salts of a wax, pyridinium salts, substituted ureas, chlorides of fatty acids, methyl compounds and silicones are suitable products to produce a water repellent surface [18].

Due to their large internal surface it is easy to impregnate microporous membranes to obtain an oleophobic effect (16 [13]).

A dyed fabric consisting of polyamide and a coagulated polyurethane coating is immersed in Asahiguard®, a fluorocarbon (trademark of Asahi Chemicals, Japan), to make it waterproof [1].

Microporous membranes containing a polar filling agent are made waterproof by a treatment with trimethylchlorosilane [17].

Hydrophilic polyurethanes are coated on one side with a polymer containing perfluoroalkyl groups [27].

Trifunctional isocyanates are reacted with a (meth)acroyl, allyl or a perfluoroalkyl compound containing hydroxylic groups. The resulting product containing urethane-groups may be applied to artificial leathers to give them waterproofness of a high permanence [4].

To get a water-repellent surface, coagulated products can be treated with paraffin waxes, substituted urea compounds, fatty acid chlorides or silicones [18].

Flame Retardance [28]. For PVC, flame retarding properties are not as important as for polyurethanes. It is more difficult to burn PVC – but if it is burned, hydrochloric acid is produced which may be toxic to the surrounding area or damage installations. Polyurethanes, especially those with aromatic isocyanates, may burn easily. So a flame-retardation process should be carried out. Polyurethanes become flame retardant by the addition of the following products, either alone or in mixture: decabromo- or pentabromodiphenyl oxide, hexachloro- or hexabromobenzene, pentabromoethylbenzene, antimony trioxide or pentoxide, magnesium hydroxide, magnesium carbonate, molybdenum oxide, aluminum hydroxide, red phosphorus, calcium carbonate, $CaOAl_2O_3 \cdot 6H_2O$, CuO and Cu_2O, and the additives are ZnO, ZrO_2, TiO_2, SiO_2, SnO, BaB_2O_3. The polyurethane should be ground with the additives, then it can be used for coating or coagulation [3].

For the coating in an indirect process, a hydrophilic polyurethane is mixed with 5% of an antimony compound, 10% of magnesium hydroxide, 10% of aluminum hydroxide and 6% of molybdenum oxide (based on the weight of the polyurethane) and used in this composition as a top coat. For the adhesive coat more or less the same composition is used [5].

Suede-like leather substitutes become flame resistant if the polyurethane used is mixed with red phosphorus, a halogen-containing compound, and a metallic oxide or hydroxide. For instance, a fabric consisting of polyester is impregnated with a polyurethane, such as Crisvon® (trademark of Dai Nippon Ink, Japan), mixed with decabromodiphenyl oxide and antimony trioxide. The polyurethane is coagulated to obtain a microporous product [2].

It is also possible to carry out a flame-resistance process on the reverse of a fabric coated by PVC, SBR or a microporous polyurethane. This is done by adding red phosphorus and a halogenated phosphoric acid ester to the polymer dispersed in water [8]. Polyurethanes modified by bromine compounds in combination with antimony trioxide can be used for flame-retardant coatings [24].

A leather substitute with a cotton substrate and a polyurethane coating applied by a DMF solution becomes flame retardant by impregnation with a water-based solution of Bischoffit, magnesium chloride hexahydrate, possibly in a mixture with other alkaline-earth compounds [11].

Additives possibly based on phosphorus- or halogen-containing triazinyl compounds transform polyurethanes into flame-retardant artificial leathers [22].

Protective clothing for firemen, in chemical plants, foundries etc. must be fire retardant. Clothes for firemen need to fulfill specific requirements [26].

Electric Conductivity. Movements and especially rubbings of non-conductive materials create electrostatic charges on the surfaces of the materials [30]. Just rolling up a coated textile, the surface of it obtains a surface charge because the coating polymers usually are non-conductive. As long as non-conductive materials contain ions or a slight amount of water they are able to conduct electric charges. Hygroscopic materials, such as polyethylene glycol, transform non-conductive materials into conductive ones.

Coatings produced by conductive compounds based on Cu, Ag, Cd, Sn, Pb or MnO, together with titanium dioxide are mixed with polyurethanes to achieve electric conductivity. Accidents caused by electrical charges in the presence of flammable materials may be avoided by this treatment [9].

Bactericidal Additives (2.2 [28], 10 [89]). Many textile substrates must be protected against biological attack. There is a need to protect the user of the textile for hygienic or medical purposes., the textile itself against bio-deterioration caused by mold, etc. and to protect the textile against insects [32]. Coated and impregnated materials have the same needs for protection as the textile itself.

Artificial suedes consisting of ultrafine polyterephthalate fibers containing a zeolite with a bactericidal metal, like Ag, Cu, Zn, become resistant to microbiological attack. These suedes may be used in the production of golf gloves [7]. Fibers of Ag, Cu, Zn, Au or Pb phosphates can be added in an amount of 0.1–15% to an artificial suede to produce a bactericidal product [12]. Ag, Li or K containing Zr or Ti phosphates are used to produce bactericidal leather substitutes [20, 21].

Resistance to $(NO)_x$ or Light. Nitrogen oxides are created by combustion of oil-based products. They may cause a yellowing effect on white or pastel shade materials or even react with polymers under deterioration. Leather and its substitutes become light and/or heat resistant by treating the articles with a water-based emulsion of an emulsifying agent, a sterically hindered amine which is insoluble in water and an UV absorber which is also insoluble in water [6]. Polyurethanes are mixed with phenols which are sterically hindered, benzophenone or piperidine derivatives. Leather substitutes containing these polyurethanes are resistant to light and $(NO)_x$ [23].

Polyurethanes consisting of polyalkylene oxide, diphenylmethane diisocyanate and hydrazine are resistant to $(NO)_x$ [10].

23.2
Modification of Physical Properties by Physical Methods

Directly after production the surface of man-made leathers often needs a physical treatment: Embossing and plating (see Chap. 22) are the usual operations with calenders or embossed rollers. As previously mentioned, this operation reduces the porosity of the material. A presupposition for a good embossing or plating effect is a thermoplasticity of the polymer. By heating thermoplastic, microporous layers lose part of their water vapor permeability. For instance a microporous polyurethane foil with a water vapor permeability of 10 mg/hcm^2 after lamination to the substrate will have a decrease in water vapor permeability to 4–6 mg/hcm^2, after embossing to 2–4 mg/hcm^2, and at the end of the finishing to 1–2 mg/hcm^2 comparable to finished genuine heather.

Increasing water vapor permeability reduces the tensile strength of the film (see Fig. 23-1). The same effect occurs with the tear propagation strength. The pore structure also has an influence on the water and air permeability of microporous sheets [6].

To improve the physical properties of microporous sheets a treatment with high frequency [8] or IR rays at 120 – 150 °C has been discussed. The IR treatment causes a superficial melting of pores which decreases water vapor permeability and eliminates stress in the polymer [5].

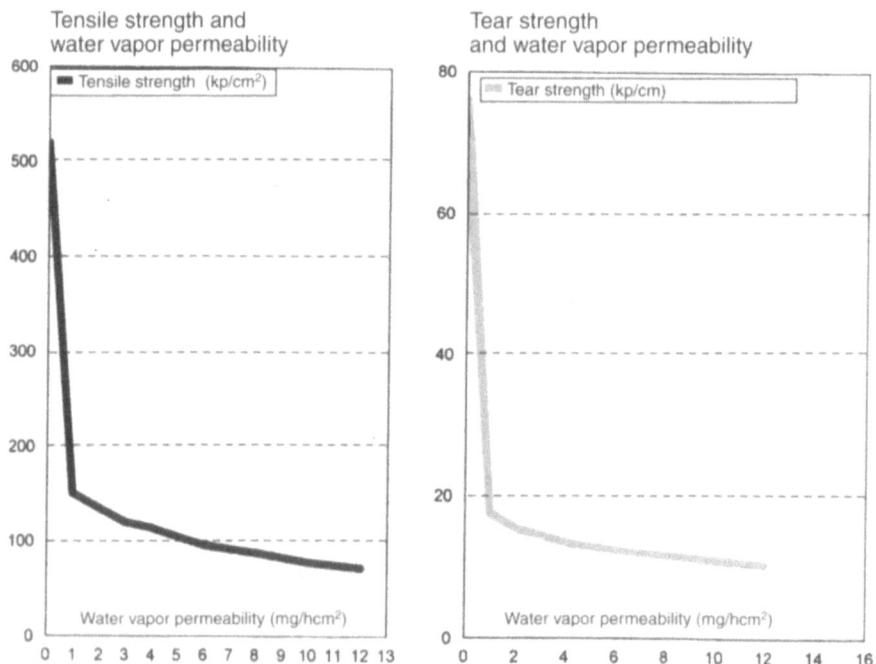

Fig. 23-1. The tensile strength and the tear strength of different microporous films with the same composition decreases with increasing water vapor permeability

The density of a material is increased if the material is heated under pressure at a temperature below the melting point of the polymer. This heating may be carried out together with a plating or embossing action [2].

A layer of a microporous polymer with a pore volume of 56–70% is pressed between two rollers at 150–190°C and 1–2.5 bar in such a way that its pore volume decreases to 40–50%. The size of the pores, whereby 50% are between 0.1–5 µm and the other 50% at 5–100 µm. Now 53–70% of the pores have a size of 0.1–5 µm. The polymer consists of a mixture of PVC and polyurethane [3].

Embossing of a coagulated man-made leather can be carried out by a silicone rubber mold containing copper powder [4]. Embossing may also be done under low pressure [7] (see also 20.2 [13]).

Microporous leather substitutes are immersed in water and then treated in a tumbler to get a crinkle effect [9].

Sheets consisting of a nonwoven which contain polyamide fibers are impregnated with a polyurethane. The sheets are folded and fixed by threads. Then they are treated with warm water to relax them. By this process an entanglement is prevented [1].

CHAPTER 24

Conversion of Water Vapor Permeable Products

There are processing techniques specific for leather substitutes. In contrast to leather, man-made leather can be processed in a continuous manner. Thermoplastic conditions are often used to form the surface to a grain pattern or gloss in a continuous process [14].

The lamination of a lining material to a man-made leather is also done in a continuous process. Therefore microporous foils may be adhered to textile substrates by means of a polyurethane adhesive coat containing polyethylene oxide groups and applied by a printing technique [16]. A proposition for an easier sewing of clothing with water vapor permeable textile composite materials impermeable to liquid water concern the part time adhesion of a lining material to a textile which after sewing is warmed to lose its adhering property. Sewing of the cloth is easier by this technique [18].

A disadvantage of coated materials is that it is not possible to rip out a seam without leaving visible holes. Another difficulty with artificial leathers with thermoplastic top coats is that the heat from the needles during sewing may damage the material [19].

Forming of PVC-coated fabrics under the influence of high frequency energy (HF) has been published several times [9,10]. A shoe upper, a part of a pocket or appliques are formed in a silicone rubber mold or in a metal mold. A coated textile or another thermoplastic material is put into a mold and exposed to HF and then the surface of the mold is transferred to the thermoplastic coating [11, 17]. It is possible to cut costs for sewing operations because the seams are formed by the surface of the silicone rubber mold. (see also Chap. 20.2). Using this technique it is possible to form seams or special structures, e.g. with silicone rubber molds [1]. It is also possible to adhere dyed PVC foils to a coated material to achieve special effects [7].

So it is also possible to produce a complete shoe by injection molding: A sole is placed on a form, the upper produced with plasticized PVC is laid on it and adhered by pressure and temperature [2]. Boots with a textile inlay can be produced in the following way: a textile lining is posed over an inner form and shrunk by hot air (220 °C) onto the form. Then the textile is coated with rubber which can be vulcanized. The vulcanization is carried out in a press to finish the boot [3].

A silicone rubber mold of an upper is treated with a swelling agent for a polymer, like acetone or trichloroethylene. Then a poromer is put into the form and a vacuum is applied. The swelling agent softens the poromer. Using this process it is easier to transform the shape of the mold [4].

A mold with a releasing surface is dipped into a thermoplastic powder or a thermo-coagulating latex. After heating the finished formed article can be stripped from the mold [5]. Similarly, a porous mold of a shoe upper can be dipped into a slurry of fibers. Then a vacuum is applied inside the mold. The fibers are then deposited on the surface of the mold where they are bonded by a latex. The resulting product may be used as a lining for a shoe [6].

Artificial leathers may be cut easily with knifes and HF [8].

Special advice is given for the manufacture of articles using Xylee® [12, 15] and Clarino® [13] as man-made leathers.

References to Part 4

22
Finishing and Dyeing

See also: 8.5[28, 30], 8.6[20], 18.3 (14), 23[9–12]
1. Träubel H (1975) Die Oberflächenbehandlung von Leder. Metalloberfläche 29:358; Eitel K (1959) In: Ullmanns Encyolopädie der technischen Chemie, Vol. 11. München, pp 577–585; Fischer A et al. (1980) Moderne Lederzurichtungen. Leder und Häutemarkt, August
2. Eitel K (1970) Über das Färben von nassen Ledern. Das Leder 21:281; Otto G (1962) Das Färben des Leders. Darmstadt ; Eitel K (1959) In: Ullmanns Encyolopädie der technischen Chemie, Vol. 11. München, pp 517–577
3. OS 1 941 412 (Kalle; H Lind et al.; 14.8.69/4.3.71)
4. JA 48 092 484 (Kuraray Co. Ltd.; 10.3.72/30.11.73)
5. JA 73 280 42 (Kuraray Co. Ltd.; 13.5.70/29.8.73)
6. OS 2 120 953 (Kuraray Co. Ltd.; S Ohnishi et al.; JA Prior. 30.4.70 (45–37 350/70); 28.4.71/ 18.11.71)
7. OS 2 334 316 (Hodogaya Chem. Co. Ltd.; S Maeda et al.; JA Prior. 26.7.72) (JA 72/74 162)
8. BE 723 367 (Glanzstoff AG; D Prior. 23.12.67; 5.11.68/16.4.69)
9. OS 1 769 706 [Kurashiki Rayon Co. Ltd.; S Nakajo; JA Prior. 30.6.67 (=JA 42 707–67); 1.7.68/ 1.7.71]
10. JA 49 081 452 (Dainippon Ink and Chem.; 8.12.72/6.8.74)
11. NE 6 700 490 (DuPont; US Prior. 12.1.66 and 29.12.66; 12.1.67/13.7.67)=FR 1 513 598
12. JA 27 277/67 (Kurashiki Rayon; 7.9.64/23.12.67)
13. JA 74 024 642 (Daiichi Lace Co. Ltd.; 29.10.69/25.6.74)
14. JA 09 241 978 (Kuraray Co. Ltd.; K Akamata, T Kawakami; 1.3.96/16.9.97)
15. JA 09 267 456 (Kuraray Co. Ltd.; T Oshita, K Hirai; 29.3.96/14.10.97)
16. OS 2 047 675 [Kuraray Co. Ltd.; Sh Ohnishi et al.; JA Prior. 29.9.69 (=JA 44–77555); 28.9.70/ 3.4.71]
17. NE 6 906 684 (BAYER; B Zorn et al.; D Prior. 3.5.68; 1.5.69/5.11.69)=OS 1 769 302
18. OE 286 489 (Kalle AG; D Prior. 14.3.68; 11.3.69/10.12.70)=FR 2 003 909=OS 1 660 088
19. JA 73–18 804 (Kokoku Chem. Ind.; 5.12.68/8.6.73)
20. JA 49 681/72 (Latsuraya Fine Goods KK; 23.5.68/13.12.72)
21. GB 1 097 461 (Sandoz AG; W Furrer et al.; SZ Prior. 13.2.64+2.3.64+11.12.64; 3.2.65/3.1.68)= BE 659 179=FR 1 424 350=SZ 426 725
22. FR 2 084 162 (Ciba Geigy; CH Prior. 4.3.70; 4.3.71/17.12.71)=BE 763 777
23. US 3 716 397 (FP Civardi; HG Kuenstler; 21.12.70/13.2.73)
24. JA 75–002 003 (Honey Chem.Co. Ltd.; 20.2.70/23.1.75)
25. O'Flaherty, Roddy WT, Lollar RM (1962) The chemistry and technology of leather, vol 3I. New York, p 263–295
26. JA 73 5015 (Kuraray Co. Ltd.; 24.3.70/13.2.73)
27. DDR 87 559 (R. Steinhardt et al.; 29.1.71/5.2.72)=(WP 152 773)
28. JA 47 20 303 (Tokai Plastics Kogyo KK; 19.2.71/28.9.72)
29. JA 48 099 306 (Kawaguchi Rubber Ind. Co.; 7.4.72/15.12.73)
30. US 3 816 168 (Rohm and Haas Co.; SN Lewis, MR Yunaska; 2.4.73/11.6.74)

31. OS 2 425 810 (Bayer; KA Weber et al.; 28.5.74/18.12.75)
32. JA 49–092 196 (Daiichi Lace KK; 6.10.72/3.9.74)
33. JA 49 081 502 (Daito Chem. Ind. KK; 14.12.72/6.8.74)
34. SU 443 858 (Film Mater Synth.; 10.3.72/18.4.75)
35. JA 49 133 466 and 49 133 467 (Kuraray KK; 25.4.73/21.12.74)
36. JA 49 132 201 (Kuraray KK; 20.4.73/18.12.74)
37. JA 50 005 503 (Kuraray KK; 17.5.73/21.1.75)
38. Fischer W, Leukroth G (1976) Untersuchung über den Zusammenhang zwischen Aufbau und Filmeigenschaften von PU-Lacken. Leder und Häutemarkt 28:89
39. JA 49 101 507 (Hodogaya Chem. Ind.KK; 5.2.73/25.9.74)
40. JA 75 034 083 (Kyowa Leather Cloth; 18.12.67/6.11.75)
41. JA 50 049 403 (Toppan Printing KK; 10.9.73/2.5.75)
42. US 3 821 012 (Immont Corp.; CJ Lattarulo, FP Civardi; 3.9.71 /28.6.74; cip 16.6.69)
43. Tork L, Träubel H (1976) Moderne Lederzurichtungen unter Berücksichtigung der verschiedenen Anforderungen an das Leder. Das Leder 27:142
44. Ullmanns Encyclopädie der technischen Chemie, 4th edn, vol 16
45. Reactive dyestuff of Ciba-Geigy
46. JA 55 098 973; JA 55 098 974; JA 55 098 975 (Kuraray KK; 17.1.79/28.7.80)
47. JA 55 001 365 [Kuraray KK; 21.6.78/8.1.80 (75 708)]
48. DE 2 406 126 [Kuraray KK; O Fukushima; 8.2.74/22.8.74; JA Prior. 17.2.73 (19 374–73]= GB 1 455 374
49. JA 06 200 483 (94 200 483) (Dainichiseika Color Chem; E Sugawara, M Iwasaki; 28.12.92/ 19.7.94)
50. JA 07207170 (Taoka Kagaku Kogyo KK; 21.1.94/8.8.95)
51. JA 072 29 071 (Asahi Kasei Kogyo KK; 14.2.94/29.8.95)
52. JA 061 23 083 (Itakura Shoji KK; K. Shibata, 13.10.92/6.5.94)
53. JA 061 92 867 (Achilles Corp KK; 24.12.92/12.7.94)
54. JA 062 35 177 (Showa Denko KK, 8.2.93/23.8.94)
55. JA 063 06 734 (Unitika Ltd.; 22.4.93/1.11.94)
56. JA 514778 (Toda Kogyo Group Corp.; 22.11.91/15.6.93)
57. TW 213 499 (High Cedar Enterprises Co. Ltd.; M Wang, 27.11.91/21.9.93)=TW 88–100 862
58. JA 53 21 159 (Seiren Co. Ltd.; 14.5.92/7.12.93)=JA 92–165 224
59. EP 504 701 (Lorica SPA.; G Poletto; 19.3.91/23.9.92)=US 5 290 593
60. JA 80–1 304 (Teijin; 8.5.78/8.1.80)=JA 78–53 787
61. GB 2 026 903 (Kuraray; H Ito et al.; 13.2.80/1.8.78)=JA 78–94910; specially claimed are (R)-ORAZOL dyestuffs of Ciba
62. JA 79–157802 (Teijin; 2.6.78/13.12.79)=JA 78–65678
63. US 4 244 789 (Stauffer; M Coll-Palagos; 24.1.79/13.1.81)=US 79–6 141
64. JA 80–1 304 (Teijin; 8.5.78/8.1.80)=JA 78–53 787
65. Trademark of Ciba-Geigy (Switzerland)
66. SU 560 466 (Film Mat. Synth. Leather; 13.6.75/7.11.80)=SU 75–146 112
67. JA 78–130 403 (Toyo Cloth; 20.4.77/14.11.78)
68. JA 76–130 503 (Yoshimura Abura; 12.11.76)
69. US 3 958 057 (Kuraray; T Nishimura; 18.5.76)
70. BE 657 721 (DuPont; JA Adams; 29.12.64/29.6.65; US Prior. 30.12.63)
71. GB 1 246 507 (DuPont; 18.11.68/15.9.71; US Prior 29.11.67)
72. JA 73–22 354 (Kuraray; 25.12.69/5.7.73)
73. JA 49/047 502 (Nippon Cloth Ind.; 8.9.72/8.5.74)
74. US 4 190 572 (Kuraray Co. Ltd.; T Nishimura et al.; 15.9.78/26.2.80; JA Prior. 14.6.76)= JA 76–70 061
75. JA 54 117 002 (Kanebo KK; 2.3.78/11.9.79=JA 78–24 859); JA 54 105 201 (Kanebo KK; 30.1.78/ 18.8.79)=JA 78–9 450
76. DOS 2 836 307 (Kuraray Co. Ltd.; T Nishimura et al.; 18.8.78/22.3.79; JA Prior. 5.9.77)= JA 77–107 116 and JA 77–107 117
77. JA 54 157 802 (Teijin KK; 2.6.78/13.12.79)=JA 78–65 678; JA 54 147 901 (Teijin KK; 8.5.78/ 19.11.79)=JA 78–53 786; JA 55 001 304 (Teijin KK; 8.5.78/8.1.80)=JA 78–53 787

78. JA 55 107 585 (Unitika KK; 13.2.79/18.8.80)=JA 79–15 883
79. JA 92 41 980 (Kuraray Co. Ltd.; H Nakajima, H Yoneda; 6.3.96/16.9.97)=JA 96–48 692
80. Yoshihiro T (1997) Surface finishing of man-made leather. Hikaku Kagaku (Sci.) 43(2):72
81. Schweitzer HR (1964) Künstliche organische Farbstoffe und ihre Zwischenprodukte. Berlin, p 470
82. JA 09 239 886 (Kuraray Co. Ltd.; T Oshita, K Hirai; 11.3.96/16.9.97); JA 09 254 322 (Kuraray Co. Ltd.; H Adachi, T Oshita, K Hirai; 19.3.96/30.9.97)
83. PCT WO 97 37 073 (DuPont; DP Zafiroglu; 29.3.96/9.10.97)=US 96–625 058
84. JA 09 300 546 (Kuraray Co. Ltd.; H Adachi, T Oshita, H Koji; 17.5.96/25.11.97)=JA 96–123 246

23
Modification of Physical Properties by Chemical Methods

See also: 8.1 [18, 23]

1. BE 698 352 (Porous Plastics Ltd.; 11.5.67/16.10.67)
2. OS 2 112 056 (Porvair Ltd.; GSt Hathorn et al.; GB Prior. 12.3.70; 12.3.71/30.9.71)
3. JA 50 063 103 (Kuraray KK; 2.10.73/29.5.75)
4. OS 2 028 361 (Inmont Corp.; FP Civardi, J Lattarulo; US Prior. 9.6.69 and 16.6.69; 9.6.70/10.12.70)
5. OS 2 112 056 (Porvair Ltd.; GSt Hathorn et al.; GB Prior. 12.3.70; 12.3.71/30.9.71)
6. OS 2 054 312 (Nairn-Williamson Ltd.; K Norcross; GB Prior. 13.11.69; 4.11.70/19.5.71)
7. JA 26 998/70 (Kurashiki Rayon; 26.10.67/4.9.70)
8. GB 1 253 677 (Glanzstoff AG; 21.11.68/17.11.71; D Prior. 23.1.68)=BE 726 311
9. GB 1 493 823 (Inmont Corp.; KDN Kearney; 19.10.74/30.11.77)=GB 74–25 641
10. JA 9031861 (Dainichiseika Color Chem.; I Kondo, H Sato, M Hirose; 12.7.95/4.2.97)= JA 95–197 935
11. JA 54 073 101 (Kanebo KK; 18.11.77/12.6.79)
12. JA 54 101 403 (Toyo Prod. KK; 27.1.78/10.8.79)=JA 78–7 348
13. DDR 148 245 (WTZ des VEB Kombinats Wolle und Seide; R Grünwald, A Meixner; 12.12.79/13.5.81)=DDR 218 023 (WP)
14. US 3 764 363 (Inmont Corp.; FP Civardi et al.; 21.7.70/9.10.73; cip 22.7.69+20.10.69)= DE 2 036 448=GB 1 325 928=BE 753 768
15. JA 73–19 924 (Kuraray; 7.6.69/18.6.73)
16. OS 1 810 733 (Kurashiki Rayon; S Nakajo et al.; 21.11.68/28.8.69; JA Prior. 21.11.67 + 23.1.68)
17. OS 1 960 266 (Inmont Corp.; FP Civardi; 1.12.69/9.7.70; US Prior. 2.12.68)

23.1
Modification of Physical Properties: Improving Water Resistance, Stain Resistance and Flame Retardation

See also: 16 [13] and 25.4 [24]

1. JA 06 272 168 (94 272 168) (Unitika Ltd.; T Furuta, K Kamemaru, H Nakamura; 22.3.93/27.9.94)
2. JA 07 18 584 (95 18 584) (Toray Ind., Daikyo Chem.; Daiichi Lace KK; M Saito et al.; 30.6.93/20.1.95)
3. US 5 393 569 (China Textile Inst.; SC Yao et al.; 14.2.94/28.2.95)
4. JA 32 65 700 (Asahi Glass KK; 16.3.90/26.11.91)
5. CN 94–102 420 (Chinese Textile Industry Research Center; Z Wu, K Zheng, Q Cai; 7.2.94/23.8.95)
6. EP 665 294 (Ciba Geigy AG; V Arnold, H Dbaly, G Püntener. R Rembold, F Wyss, A Püntener; 19.1.94/2.8.95)
7. JA 061 16 872 (Kuraray Co. Ltd.; 30.9.92/26.4.94); JA 061 42 268 (Kuraray Co. Ltd.; 6.11.92/24.5.94); JA 061 46 175 (Kuraray Co.Ltd.; 30.10.92/27.5.94)
8. JA 061 46 174 (Achilles Corp KK; 30.10.92/27.5.94)
9. JA 061 92 969 (Kanebo Ltd.; 25.12.92/12.7.94)
10. JA 052 71 761 (Kuraray Co. Ltd.; 22.3.93/27.9.94)

11. RU 2 010 899 (Volg. Poly.; VE Debrisher, RA Mikailovskaya, VD Vasileva; 9.1.92/15.4.94)
12. JA 063 46 376 (Kuraray Co. Ltd.; 4.6.93/20.12.94)
13. Riedel T (1996) Gleitmittel. Kunststoffe 86:982
14. Pauquet J-R (1996) Antioxydantien. Kunststoffe 86:940
15. Kramer E (1996) Lichtschutzmittel. Kunststoffe 86:948
16. Lichtblau A (1996) Antistatika. Kunststoffe 86:955
17. US 3 931 067 (Amerace Corp.; BS Goldberg, DE Johnson; 16.8.74/6.1.76)=US 74–498 154
18. JA 73–4 474 (Kuraray; 26.4.69/8.2.73)
19. Troitzsch J (1996) Flammschutzmittel. Kunststoffe 86:960
20. JA 51 71 572 (Toa Gosei Chem Ind. Ltd.; 20.12.91/9.7.93)=JA 91–355 302
21. JA 81 70 275 (Toray Ind.; S Nagura, M Ikeda, I Tanaka; 14.12.94/2.7.96)=JA 94–310 645
22. JA 53 042 295 (Asahi Chemical Ind. KK; 30.9.76/17.4.78)=JA 76–116 572
23. JA 53 20 499 (Kuraray Co. Ltd.; 19.5.92/3.12.93)=JA 92–150 108; JA 53 20 500 (Kuraray Co. Ltd.; 19.5.92/3.12.93)=JA 92–150 109
24. JA 63 012 769 (Kokoku Chem. Ind. KK; 1.7.86/20.1.88)=JA 86–154 408; JA 63 085 185 (Kokoku Chem. Ind. KK; 22.9.86/15.4.88)=JA 86–224 212
25. EP 087 537 (Toray Ind.; A Morioka, Y Naka, A Uchida, K Kawakami; 22.2.82/7.9.83) =JA 82–26149
26. Abbott NJ, Schulman S (1976) Schutz vor Feuer: Nichtbrennbare Gewebe und Beschichtungen. J Coated Fabr 48
27. PCT WO 97 36 951 (W. L. Gore & Ass. Inc.; YX Shen; 14.1.97/9.10.97)=US 97–512
28. Lewin M (1984) Flame Retardance of Fabrics. In: Lewin M, Sello StB (eds) Handbook of fiber science and technology, vol 2. New York and Basel, pp 1–141
29. Kissa E (1984) Repellent finishes. In: Lewin M, Sello StB (eds) Handbook of fiber science and technology, vol 2. New York and Basel, pp 143–210
30. Sello StB, Stevens CV (1984) Antistatic treatment. In: Lewin M, Sello StB (eds) Handbook of fiber science and technology, vol 2. New York and Basel, pp 291–315
31. JA 09 307 991 (INOAC Corp.; N Fujita, M Suzuki, K Karo; 9.5.96/28.11.97)=JA 96–140 849
32. Vigo TL (1984) Protection of textiles from biological attack. In: Lewin M, Sello StB (eds) Handbook of fiber science and technology, vol 2. New York and Basel, pp367–426

23.2
Modification of Physical Properties by Physical Methods

1. JA 10 204 781 (Kuraray Co. Ltd.; N Makiyama, 10.1.97/4.8.98)=JA 97-2623
2. FR 2 030 806 (Inmont Corp.; FP Civardi, CJ Lattarulo; US Prior. 9.12.68; 9.12.69/13.11.70) =OS 1 961 732=GB 1 296 060
3. US 3 565 981 (Du Pont; MR Lauro; 9.5.67/23.2.71)
4. JA 50 030 953 (Kuraray KK; 19.7.73/27.3.75)
5. OS 1 939 053 (Satra; LG Hole et al.; GB Prior. 1.7.68 and 10.12.68; 31.7.69/5.2.70)
6. Barthau R (1990) Untersuchung der Porenstruktur und deren Auswirkungen auf die physikalischen Eigenschaften mikroporöser Polyurethan-Beschichtungen. Dissertation, Deutsche Institute für Textil- und Faserforschung, Stuttgart, Germany
7. TW 228 491 (Nan Ya Plastics Corp.; W Wang; 30.7.93/21.8.94)
8. JA 53 18 592 (Yamazen Sangyo KK; 8.6.91/3.12.93)=JA 91–163 668
9. JA 76–42 161 (Mitsubishi Rayon; 13.11.76)

24
Conversion of Water Vapor Permeable Products

1. Pape N (1971) Das Hochfrequenz-Schaftprägeverfahren. Schuh-Technik 65:690; Rische UW (1971) Das Hochfrequenz-Schweißen von Kunststoffen, Kautschuk und Gummi. Kunststoffe 24:411
2. GB 1 021 731 (Pirelli; IT Prior. 6.3.62; 2.1.63/9.3.66)
3. US 3 324 220 (Dunlop Rubber Co.; RS Goy; GB Prior. 6.11.63; 28.10.64/6.6.67)

4. OS 2 030 071 (Porvair Ltd.; MW Denton, KDN Kearney; GB Prior. 18.6.69; 18.6.70/23.12.70)

5. OS 1 912 892 (Plastic Coating Ltd.; DEK Elliott, MSt Dance; GB Prior. 13.3.68; 13.3.69/16.10.69)=NE 6 903 870

6. FR 1 551 084 (Dunlop Co. Ltd.; GB Prior. 13.10.66; 11.10.67/27.12.68)=NE 6 713 725

7. GB 1 237 573 [Premier Footwear (Fleetwood) Ltd.; M Hodgkinson; 20.7.68/30.6.71]

8. JA 39 330/72 (Kuraray Co. Ltd.; 30.9.68/4.10.72)

9. OS 2 112 619 (Chaussures Andre S.A.; FR Prior. 17. 3.70+15.5.70+23.7.70+19.1.71+19.11.70; 16.3.71/14.10.71)

10. FR 2 109 415 (LC Saltel; 15.10.70/26.5.72)

11. BE 612 576 (Ferguson, Shies Ltd.; GB Prior. 14.1.61; 12.1.62/2.5.62)

12. Hubert A et al. (1972) Die Thermofixierung von Schuhen aus Xylee. Schuh-Technik 66:226; Anon. (1972) Xylee nun generell HF-schweißfähig. Schuh-Technik 66:23

13. Fujinomoto K (1974) Das Prägen von Clarino. Schuh-Technik 68:695

14. US 4 341 581 (Immont Corp.; FP Civardi, MJ Getting; 23.5.79/27.7.82)

15. Riess W, Stiegele D (1975) Die Arbeit mit poromerischen Schaftmaterialien. Schuh-Technik No 7, pp 434–488 and No 4, pp 270–275

16. JA 81/11 787 (Kuraray; 7.5.73/17.3.81)=JA 73–51 030

17. DOS 195 10 240 (Ph Schaefer; 21.3.95/5.10.95)

18. DOS 4 423 592 (Kleber Textil; K Kleber; 6.7.94/11.1.96)

19. Gilke U (1997) Sewing leather and artificial leather. Tex Decor 24

Part 5

Chemistry, Testing Methods, Other Industrial Applications, Ecology

In Part 5, polyurethane chemistry is discussed as far as this chemistry is related to water vapor permeability.

Many testing methods are described in the literature. Some of these testing methods claim to be very similar to real wearing situations but most of the testing is rather complicated. Other methods are simple and do not need much time to be carried out. Their adaptation to reality, however, is probably not very comparable with reality.

Spin-offs from man-made leathers are described in the chapter with other industrial applications. Ecology, as far as man-made leathers are concerned, is also discussed briefly.

CHAPTER 25

The Chemistry of Polyurethanes –
Especially for Water Vapor Permeable Products

The author will not attempt to explain all of the literature about the chemistry of polyurethanes existing today (for summarized articles see [1]).

Polyurethanes are not produced by a polymerization or a polycondensation but are built by a polyaddition reaction. Therefore they are called *polyadducts*.

As seen in the chapters dealing with coagulation and selective evaporation (see Chap. 8 and Sect. 9.2), the structure of a polymer plays the decisive role whether a polymer can be transferred into a microporous form or not. The structure of a polyurethane determines its hardness, lipophility, solubility, tensile strength, etc. The chain length depends on the mixing ratios of the starting materials. Crosslinking increases the chain length and influences the characteristics of the polymer further.

Polyurethanes consist of *soft segments*, the polyols, like polyester, polyether, polycaprolactone, and *hard segments*, the urethane and urea groups (see Fig. 25-1). Hard segments are able to build hydrogen bridges between themselves.

Fig. 25-1. Schematic view of a polyurethane with hard and soft segments

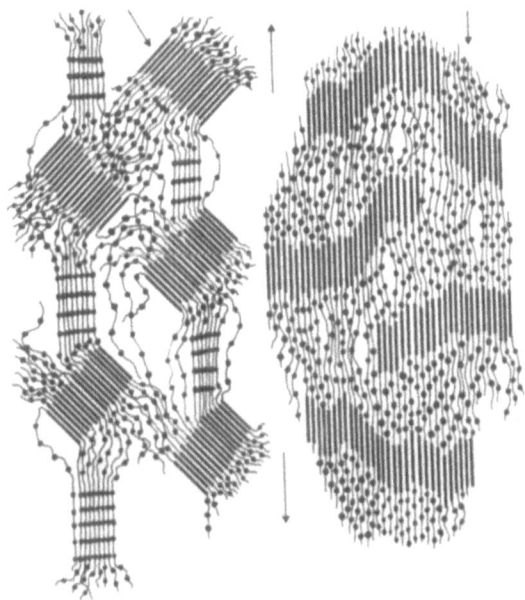

The hardness and modulus increase with the more hard segments a polyurethane contains. These hydrogen bridges are in a crystalline form. They are responsible for a high melting point, the tear strength and the tensile strength and
influence the glass transition temperature. Hard segments influence the solubility in water, alcohol and other polar solvents.

The soft segments are responsible for elasticity and solubility in unpolar,
organic solvents. They are responsible for a rather low glass transition temperature as long as the polyesters themselves do not melt at a high melting range. The
right selection of hard and soft segments in a polyurethane allows it to obtain a
product of the desired shape.

One essential property of polymers for coatings is the glass transition temperature (T_g). PVC, for instance, has a T_g of −75 °C, polyethylene gylcol terephthalate a T_g of ca. 80 °C and dimethyl polysiloxane a T_g of −123 °C [5]. Polyurethanes
with polyether soft segments have a low T_g, polyurethanes with polyester soft
segments have a high T_g. In comparison, ethylene glycol containing polyadipate
esters have a T_g of above 0 °C which even rises if the chain length of the glycol,
such as butanediol, increases. Diethylene glycol, or mixtures of glycol ethers,
decreases the T_g of a polyurethane.

The tensile strength and the elongation at break measured depend on the
temperature during testing. A film of a polyurethane with polydiethylene
glycol adipate (I) and a polyhexamethylene glycol carbonate (II) soft segments
decreases from 140 (I), respectively, 260 kp/cm² (II) at 20 °C to 15 kp/cm² at 120 °C
(Fig. 25-2). This property may cause problems for instance in shoe production

Fig. 25-2. Decrease in the
tensile strength with rising
temperature

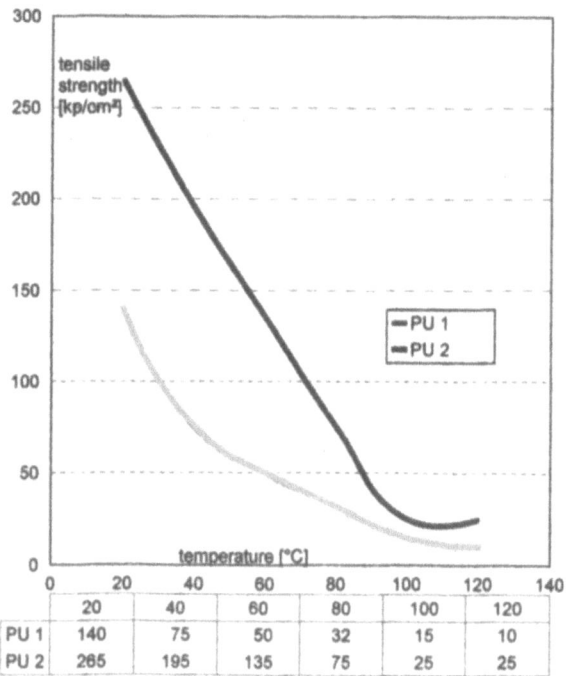

	20	40	60	80	100	120
PU 1	140	75	50	32	15	10
PU 2	265	195	135	75	25	25

when a lasted shoe upper passes the so-called thermo setting. Thermosetting is a treatment of a shoe upper by vapor and temperature at around 110–120 °C. Thermosetting is needed for an upper to keep its shape. It may be that the polymer of a top layer of a coated material is stretched for instance in the toe zone of a shoe more than its elongation at break at that temperature. The result are small cracks in the polymer film. The resulting shoe would be rejected.

The molecular weight, the physical properties and the solubility of solvents are determined by the molar amounts of the components applied. The polymers may be characterized by the molar proportion of the hydroxy, amino or isocyanate group containing products. The NCO/OH, the NCO/NH or the NCO/(OH + NH) proportions determine if the resulting product will be linear, a prepolymer or a crosslinked product. A linear polymer with the highest molecular weight is produced at an NCO/OH ratio of 1. At an isocyanate/hydroxy ratio of 1.03 up to 1.2 the resulting polymer still contains isocyanate groups. These isocyanate groups react over 24 h (at room temperature) with existing urethane groups to allophanate or with the humidity of the air to urea groups. If the NCO/OH proportion is much higher than 1.5 or 2 an isocyanate group containing prepolymer is the result. This prepolymer still contains reactive NCO groups which may be reacted with an amine (see Chap. 25.2). At an NCO/OH ratio of less than 0.8 a prepolymer with hydroxy end groups results [4] which can be terminated, for example, with a trifunctional isocyanate. Prepolymers with hydroxy groups can be dissolved rather easily, e.g. in ethyl acetate (see Fig. 25-3). The

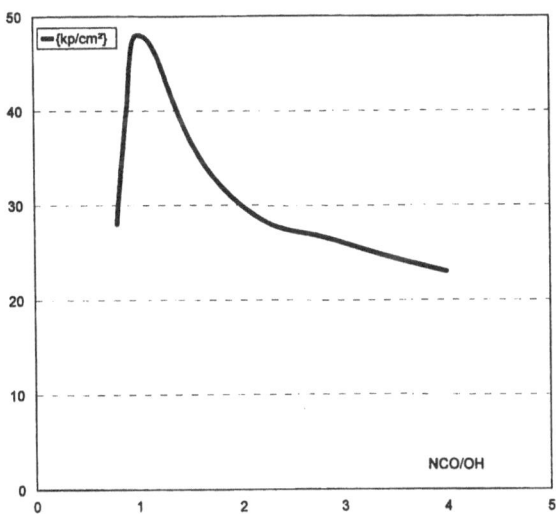

Fig. 25-3. The influence of the NCO/OH ratio on tear strength. There is a sharp decrease in the tear strength if the NCO/OH ratio is under 1; at NCO/OH ratios above 1 the tear strength also diminishes. The decrease in tear strength above an NCO/OH ratio above 1 is much slower due to the fact that in this case the free isocyanate groups are able to react further, e.g. with the humidity in the air to urea or with urethane groups to allophanate groups. This reaction increases the molecular weight of the polymer further. The same effect could be demonstrated with flexing properties or tear strength

first two-component systems were based on this chemistry (see Chap. 2.2): Impranil® C [3] is a prepolymer with hydroxyl end groups and Imprafix® TH a trifunctional isocyanate, the necessary chain extending product to obtain a high molecular polyurethane.

Systems carried out in practice based on isocyanate prepolymers are Adiprene® by DuPont [2] and Vulkollan® by Bayer.

Amino and hydroxy groups differ in reactivity. Therefore it is usually not possible to mix products containing hydroxy groups and amino groups and carry out a controlled reaction with isocyanates. Therefore two-step processes are necessary (see Chaps. 25.2 and 25.3).

25.1
Polyurethanes

The characteristic group of polyurethanes is the urethane group.

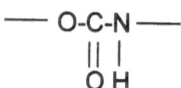

The urethane group

The word polyurethane is usually used in association with genuine polyurethanes, polyurethane-ureas and polyurethane-hydrazocarbonamides. The reason is that all of these compounds contain polyols and isocyanates which polyadd to urethane groups. The only difference between the different compounds named is the chain extension. When a glycol is used to extend the chain, the resulting product contains nothing but urethane groups. If an amine is used as a chain extender the resulting product will be a polyurethane-polyurea, etc. (for details, see the following sections).

Polyurethanes contain as polyols polyesters, polyethers, polycaprolactones, isocyanates and glycols [1–3, 5–7]. Because the reactivity of the hydroxy groups in these products does not differ much, polyurethanes may be produced in a one-step process, in one shot, whereby the reaction can be initiated by heat shock [Eq. (25-1)].

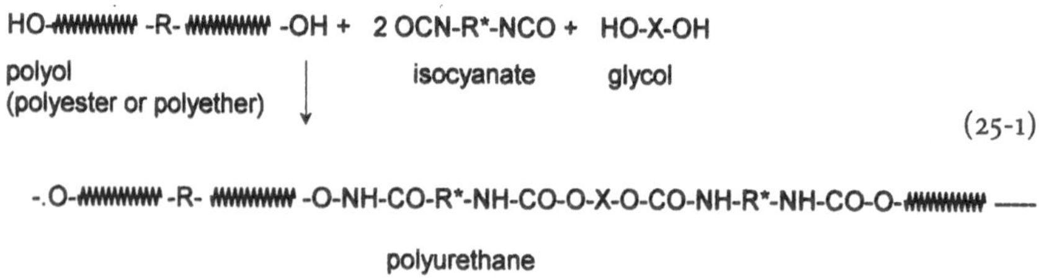

HO-⋀⋀⋀⋀ -R- ⋀⋀⋀⋀ -OH + 2 OCN-R*-NCO + HO-X-OH

polyol   isocyanate   glycol
(polyester or polyether) ↓

$$(25\text{-}1)$$

-.O-⋀⋀⋀⋀ -R- ⋀⋀⋀⋀ -O-NH-CO-R*-NH-CO-O-X-O-CO-NH-R*-NH-CO-O-⋀⋀⋀⋀ —

polyurethane

Polyaddition of a polyol, a glycol and a di-isocyanate to form a polyurethane (reaction in one step)

Polyurethane may be produced different ways; in solution [9], in situ, i.e. during their application, or by a melting process. They can be applied as solid parts, paints and lacquers, coatings, hard and flexible foams and so on.

Polyurethanes containing aliphatic isocyanates like hexamethylene diisocyanate (HDI), isophoron diisocyanate (IPDI), methylene dicyclohexyl-4,4′-diisocyanate (H12MDI) or their derivatives are light fast.

In general polyurethanes containing aromatic isocyanates are not light fast. Reasonable light fastness may be achieved by a mixture of an aromatic isocyanate such as MDI in combination with aliphatic isocyanates, such as methylene dicyclohexyl-4,4′-diisocyanate (H12MDI) [11].

Polyurethanes may contain unsaturated glycols or alkyl ethers of glycerol. They may then be vulcanized by sulfur. These vulcanizates can be calendered as foils or produced as solutions and applied to textiles [4]. Glycols containing sulfhydryl groups can be built into polyurethanes. These polyurethanes should contain (meth)acryloyl groups in the polyol and silicone groups as well. These products crosslink by themselves; they can be used in the manufacture of leather substitutes [10].

Hydrophilic polyurethanes whose soft segments contain 30–100% of polyethylene glycol ether and a polyester with monovinylic or α, β-unsaturated carbonic acids may be crosslinked with salts of the 1st, 2nd or 3rd main group of the periodic system. Vinylic group containing components may be graft polymerized onto these compounds. The resulting products are suitable for the production of leather substitutes [8].

25.2
Polyurethane-Polyureas

The characterizing group of polyurethane-ureas is the urea group.

The urea group

Polyethers or polyesters react with a diisocyanate and a diamine to form polyurethane-ureas. This reaction is usually carried out in two steps due to the fact – as previously explained – that OH and NH groups differ too much in their reactivity [Eq. (25-2)] [1–4, 7]). Amino groups react more than 100 times as fast as hydroxy groups. The reaction can only be carried out in one step when the reactivity of the amine is reduced, for instance, by steric hindrance of a chlorine atom in the ortho position to an aromatic amine [5].

In many cases the prepolymer is built in a melt. The chain extension is later done in a solvent. Usually DMF or DMAc is used. In the case of the aliphatic isophoron diisocyanate and or 4,4′-diisocyanato dicyclohexylmethane and isophoron diamine so-called *soft solvents*, i.e. a mixture of toluene and methanol or isopropylic alcohol may be used.

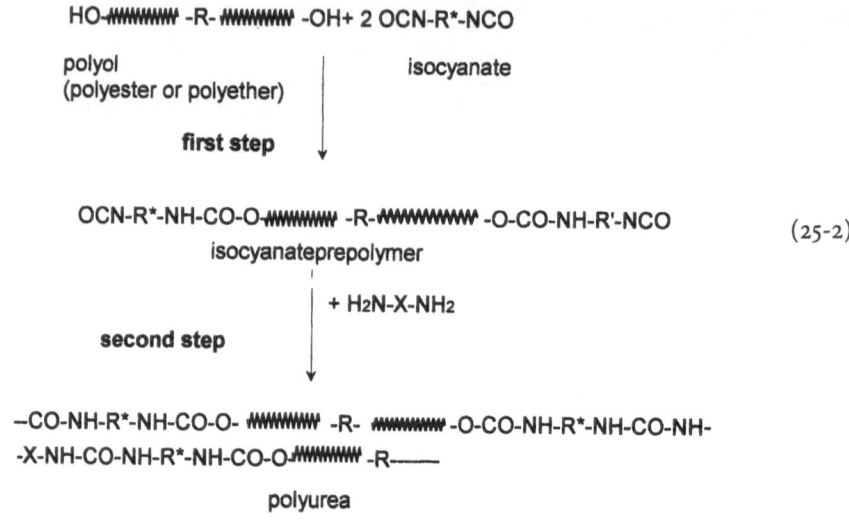

$$(25\text{-}2)$$

Polyaddition to form a polyurethane-polyurea. The reaction is carried out in two steps by forming a prepolymer consisting of a polyol and a di-isocyanate which – in the second step – reacts with the diamine.

As an amine terminal, amino groups at oligo amides having a molecular weight of 500–8000 can be used. This reaction is carried out in DMF which contains lithium or calcium chloride [9].

Polyesters dissolved in DMF can be prepolymerized with an isocyanate and the termination of the reaction can be carried out with water [8].

25.3
Polyurethane-Polyhydrazocarbonamides

Polyurethane-polyhydrazocarbonamides are products made out of polyesters or polyethers, isocyanates and hydrazine or hydrazine derivatives.

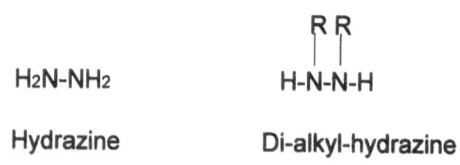

Hydrazine and substituted hydrazine derivatives

The resulting polymer is a polyurethane-polyhydrazocarbonamide.

A prepolymer (see Equation 25-2) is formed which reacts with hydrazine or a substituted hydrazine in a two step reaction to a polyhydrazocarbonamide.

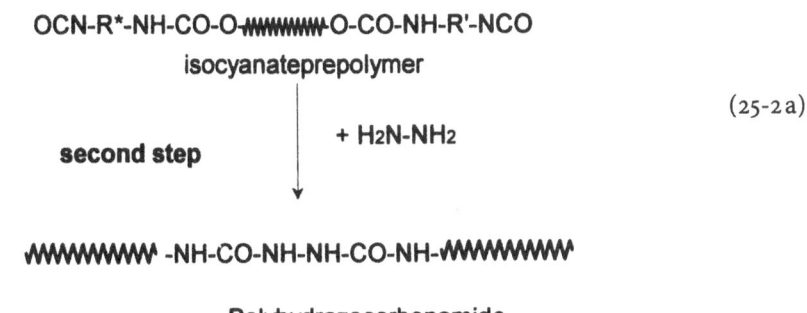

OCN-R*-NH-CO-O-wwwwwwww-O-CO-NH-R'-NCO

isocyanateprepolymer

(25-2a)

second step   + H2N-NH2

wwwwwwww -NH-CO-NH-NH-CO-NH-wwwwwwww

Polyhydrazocarbonamide

Polyurethane-polyhydrazocarbonamide

Hydrazine and its derivatives are as reactive as amines with isocyanates. They react too fast compared with hydroxyl groups. This reaction is also carried out in two steps.

The following chain lengthening agents can be used: hydrazine or alyklhydrazine [1], carbodihydrazide [2, 3], isophthalic hydrazide [8], terephthalic hydrazide [9] or aminocarbonic hydrazide [4, 5] as well as hydrazinocarbonic hydrazide [6].

The resulting polyurethane-polyhydrazocarbonamides are highly elastic polymers which can be used in the production of stretchable articles like swimsuits, etc. Polyhydrazocarbonamides are stable against attack by light and $(NO)_x$. Polyhydrazocarbonamides are mainly used in the production of stretchable fibers like Dorlastan® (trademark of Bayer AG, Germany) or Lycra® (trademark of DuPont, USA).

Since it has been determined by animal testing that hydrazine is a carcinogen [7], derivatives are used instead of hydrazine in the synthesis of polyurethane-polyhydrazocarbonamides.

25.4
Modifications in the Manufacture of Polyurethanes

In this section literature is given which relates to the production of polyurethanes especially suited for man-made leather.

The viscosity of the polyurethane solutions plays an important role [1]. The chain extension by means of a glycol in a two-step process, for example, should result in a solution with a better viscosity and, therefore, easier production [2]. A chain extension of an isocyanate prepolymer dissolved in DMF containing a glycol results in polyurethanes with a glass transition temperature which can be well destined [22].

Polyurethanes with an improved resistance to heat and light are produced by the reaction of an isocyanate prepolymer with an amino acid N-carboxyanhy-dride (see Chap. 18-3) such as, for instance, glutamic-γ-ethyl ester N-carboxyan-hydride, co-reacted with an amino acid [3 – 7].

Polyurethanes resistant to the action of oxygen and $(NO)_x$ contain phenols with steric hindrance ([30] see also 23.1 [10, 23]).

Polyurethanes resistant to hydrolysis are produced with polybutylene glycol adipate [8] or polypropylene glycol adipate [10], MDI, ethylene glycol and 0.001 – 10 % of anthranildiacetic or -dipropionic acid. Polycarbonates, possibly ethoxylated, in the soft segment [21] increases resistance to hydrolysis [20].

Polyurethanes containing uretdion groups can be mixed with products containing polyamine groups. The mixture crosslinks during or after forming on heating it up to 80 °C [9] [Eq. (25-3)].

(25-3)

A polyurethane containing a uretdion group

An amine attacks the uretdion group, opens the ring and forms a biuret. In the case of a diamine two polyurethane chains may be connected with each other by such a reaction.

Reaction of an uretdion group with an amine

Polyurethanes with polytetramethylene glycol ether or polypropylene glycol ether are resistant to hydrolysis and are easy to coagulate [33].

Polyurethanes well suited for the manufacture of artificial leathers contain mixtures of polyesters and polyethers [11, 12, 19].

If a surplus of amine is used for the chain extension, i.e. a NCO/NH ratio of less than 1, an amino prepolymer is created. This amino prepolymer may be chain extended in a second step by the addition of isocyanates. Polymers that are easy to coagulate can be produced by this method [13]. By using this kind of polyurethane, fading by light or $(NO)_x$ may be avoided [24].

Polyurethanes where starting materials are used in the molar ratios of 1 : 2 : 4 (for polyglycol/water and /difunctional alcohol) are easy to coagulate. To adjust the molecular weight and the viscosity the addition of a chain-terminating agent such as dibutylamine is used [14]. Dibutylamine as a monoamine stops a poly-addition reaction. Polyurethanes terminated with $C_1 – C_4$ dialkylethanolamines are easy to dye [16 – 18, 31].

Polyurethanes appropriate for bonding nonwovens consist of polyester, polycaprolactone, polyether, polycarbonate, etc., and a diol containing dimethyl-siloxane with a molecular weight of 1000 – 10,000. Substrates containing these polyurethanes obtain a high density and resilience [23].

Flexible polyurethanes resistant to heat are produced in a two-step process: The first step is the reaction of an isocyanate with a polyester, polycaprolactone or a polycarbonate where the molecular amount of the isocyanate is lower than the molecular amount of the hydroxyl groups (NCO/OH = <1). An addition of MDI then follows and finally a reaction with a diamine [28].

There are special polyols which are claimed to be very useful in the production of man-made leathers. Such a polyester contains octanediol-1.8, which may have additional methyl groups, or nonanediol. The resulting polyurethane needs additionally a diisocyanate and a chain extender and is highly flexible, resistant to heat and cold [29]. Polyesters with increasing chain length in the glycol have increasing melting points. The polyesters mentioned have a melting point of around 60–80 °C, which is – on the other hand – good for coagulation or other processes to create microporosity. Microporous products with a high glass transition temperature may even have a good cold flexibility. The same polymer as a homogeneous film would never have this.

Polyurethanes containing siloxane groups are able to crosslink in the presence of humidity. If these products are dissolved in DMF and coagulated they crosslink during coagulation [15].

Polyurethanes containing fluorinated polyoxyalkylene glycol ethers can be used as additives to polyurethane dispersions to improve the abrasion resistance and the gloss of the coatings or films produced [27].

Polyether-polyurethanes containing fluorine groups are very soluble and can be used as isocyanate prepolymers which are oleophobic and water-repelent products for leather substitutes [25, 26].

Abrasion resistant and well gliding polyurethanes for leather substitutes contain silicone and perfluoralkyl groups [32].

25.5
Polyurethane Dispersions

At the start of the production of leather substitutes only solutions of polymers in organic solvents were applied. Solvents are easy to eliminate because they do not consume much energy in vaporization (see Table 25-1). The values in the table show that it needs more than seven times the energy to evaporate water instead of toluene.

Table 1. Boiling points, specific heat and heat of vaporization of water and different solvents

	Boiling point (°C)	Specific heat ($J/°C \cdot g$)	Heat of vaporization (J/g)
Water	100	4.2	2480
Ethanol	78	2.5	870
Butanone	80		468
Ethyl acetate	77	1.9	420
Toluene	110	1.8	360

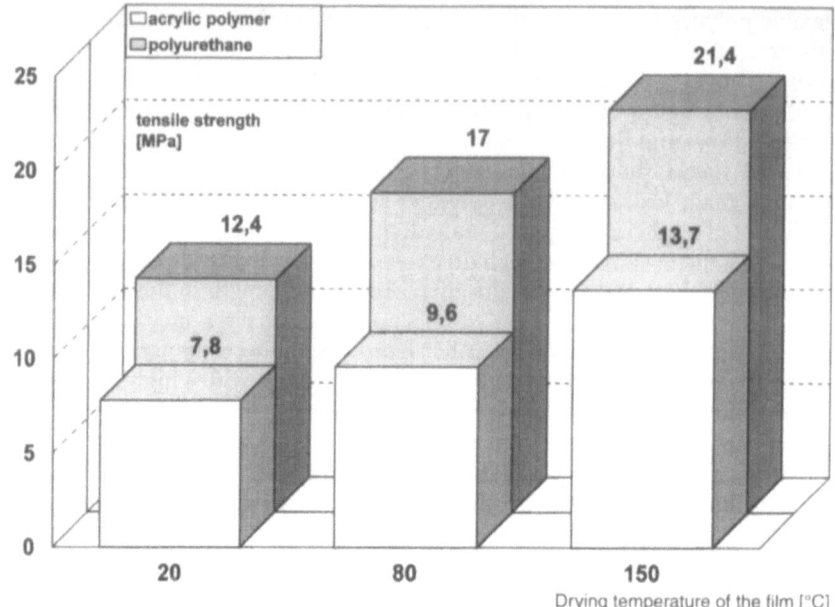

Fig. 25-4. Drying temperature and tensile strength of the resulting film

Film forming is easier with solutions of polymers than with dispersions, because dispersions usually contain the polymer in the form of fine distinct droplets which during drying need to flow into each other. So the drying temperature has an influence on the properties of the film. If the drying is only carried out at a low temperature the film forming does not take place accordingly and the resulting tensile strength and other physical properties are not optimal (Fig. 25-4). Spreading of water-based polymers on surfaces is more difficult than with solutions. The reasons for the late acceptance of water-based products by the industry were based on these difficulties [14].

Poyurethanes dispersed in water or water/solvent mixtures play an important role in the manufacture of leather substitutes.

Basically polyurethanes may be dispersed in water by three different methods:

– dispersion by means of an external emulsifying agent [6–8],
– dispersion with an emulsifying agent built into the polyurethane chain, or
– hydrophilic polyurethanes [1].

Often a combination of these three methods is applied. Polyurethanes with an internal emulsifying agent contain ionic groups [2–5], either anionic:

—X-NH-CO-O-wwwwww -O-CO-NH-X-NH-CO-O- wwwwww

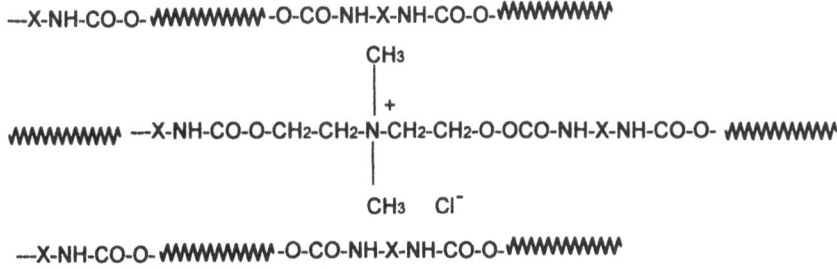

The polyurethane is dispersed in water by the anionic salt-like groups which act as an internal emulsifier.

Not every polyurethane needs to contain such an internal emulsifying group to get all the polyurethane chains dispersed. A cationic model of a polyurethane is:

—X-NH-CO-O- wwwwww -O-CO-NH-X-NH-CO-O-wwwwww

$$CH_3$$
$$|+$$
wwwwww —X-NH-CO-O-CH_2-CH_2-N-CH_2-CH_2-O-OCO-NH-X-NH-CO-O- wwwwww
$$|$$
$$CH_3 \quad Cl^-$$

—X-NH-CO-O- wwwwww -O-CO-NH-X-NH-CO-O-wwwwww

The polyurethane is dispersed in water by the cationic salt-like groups which act as an internal emulsifier.

Polyurethane dispersions containing internal emulsifiers are regarded as more stable than ones containing external emulsifiers. To produce a polyurethane dispersion by an external emulsifier usually a polyurethane dissolved in a water-miscible solvent like acetone, which has a rather low molecular weight, is dispersed under high shear forces by means of a non-ionic tensioactive substance in water. Originally ionic polyurethanes were produced more or less in the same manner. A polyurethane is produced in acetone then dispersed in water containing sodium, ammonium or chloride ions to transfer the polyurethane in the ionic form; afterwards the acetone is distilled off.

A special method for a polyurethane dispersion suitable for the production of leather substitutes is to produce a polycaprolactone ester in the presence of dimethylolpropionic acid [9].

Coatings with water-based polymers usually need a crosslinking reaction [13]. Polymers applied as water-based products in coating or film forming do not lose their emulsifying system. Due to their compatibility with water they remain sensitive against liquid water. Wet flexing, wet rubbing and tensile strength in the wet stage would be not good enough for clothing or shoes. This fact can be diminished by crosslinking of the polymer chains during or after application. As discussed previously (see Chap. 2.2) some reactive compounds – mostly with a functionality of three – are used on an industrial scale.

Isocyanates (2.2 [26]) [Eq. (25-4)], epoxides [Eq. (25-8)], N-methylol derivatives [Eq. (25-5)], aziridine (2.2 [25]) [Eq. 25-6)], and carbodiimide group [Eq. 25-7)] containing crosslinkers are used.

Tris-isocyanates crosslink polyurethanes via allophanate groups. The crosslinked polyurethanes do not swell as much in water or solvents as uncrosslinked ones.

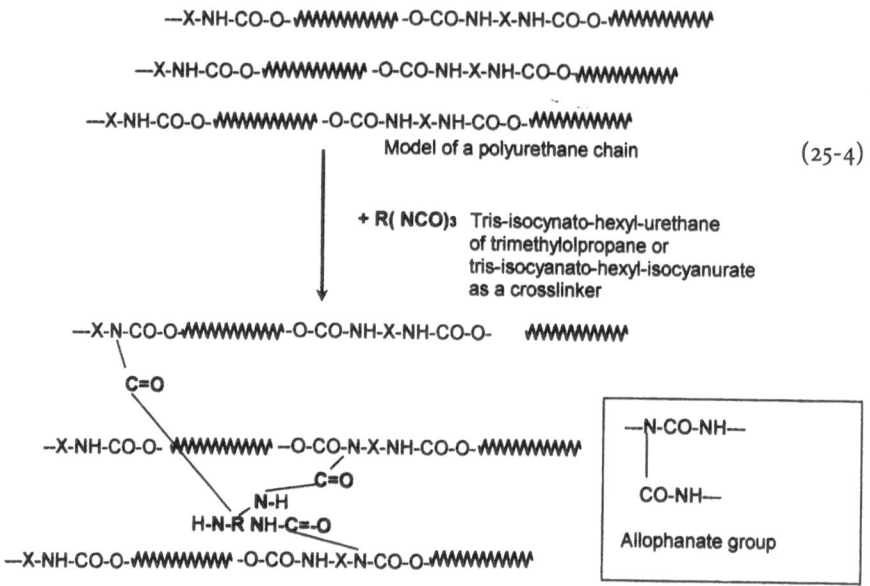

Crosslinking with a trisisocyanate

Isocyanate crosslinking is not ideal for polyurethanes having carboxylic groups as dispersing medium because the isocyanate carboxylate reaction is even slower than its competitive reaction with water. Aziridines, a product group with a certain toxicological potential, are better suited for the crosslinking of carboxylate groups. For polyurethanes with carboxylic groups, carbodiimide crosslinking can also be used but they are not as efficient as aziridines. Carboxylate groups also react with metal ions. With Zn or Ba, insoluble carboxylate salts are formed [Eq. (25-9)]. This reaction is usually carried out in the crosslinking of butadiene copolymer latex. For polyurethanes this reaction is not used due to the catalytic effect of these salts in the hydrolysis of polyesters.

This crosslinking reaction is only of theoretical value with polyurethanes because the ionic bonds are not very stable with water. Additionally polyester-polyurethanes suffer from metal salts because they reduce their resistance to hydrolysis remarkably. It is the usual crosslinking for butadiene copolymers containing carboxylic groups.

$$2 \; \text{-----} \; CO\overset{\ominus}{O} + Z\overset{\oplus\oplus}{n} \text{-----} \longrightarrow \; \text{-----} \text{-CO-O-Zn-O-CO-}$$

Zn, Ba and other carboxylate salts are usually insoluble in water; therefore their hydroxydes or oxydes can be used as crosslinkers. Only ZnO is used technically a s a crosslinker.   (25-9)

Crosslinking of carboxylate groups with bivalent metal salts.

An example where crosslinking by metal salts is used is to obtain an abrasion-resistant leather substitute which may be produced by treating a nonwoven consisting of ultrafine fibers with a polyurethane dispersion also containing an inorganic salt [10].

A polymer with terminated amino groups is treated with an epoxide then mixed with an isocyanate prepolymer which is blocked by an oxime and dispersed in water possibly by means of an emulsifier. If this mixture is heated, a stable dispersion is produced which may be used in the production of an artificial leather [11].

Polyurethanes exposed to water absorb it. Crosslinking diminishes the amount of the water which can be absorbed (see Fig. 25-5). The water absorbed is responsible for a decrease in certain properties.

The effect of crosslinking is demonstrated in the loss of tensile strength and tear strength of a wet film compared with a dry one. Rub fastness is also decreased if a polymer film gets wet (see Fig. 25-6).

To demonstrate the influence of water on the physical properties of polyurethane films, films were prepared from water-based polyurethane dispersions. The films were well dried. Two samples were taken from every film. Then tensile strength and elongation at break were measured. Parallel samples of the film were put in water for 24 h at 20 °C and the measurements were repeated. If the film was sensitive to attack by water, the tensile strength of the wet film would be much lower than that of the original dry one. Crosslinking reduces the sensitivity to water (Fig. 25-7). The decrease in tensile strength after a storage of a film in water is not specific for a polyurethane; acrylic polymers show a decrease in physical properties as well (Fig. 25-8).

Films prepared from polymers with an internal or external emulsifying agent absorb water after being stored in it. Water either is adsorped between the hard segments, or at the hydrophilic soft segments. This adsorption is responsible for the reduction of the internal crystalline bonds of the polymer chains which results in a loss in physical properties. The degree of water uptake is therefore also a measurement of the sensitivity of a polymer to water. If the reduction of the tensile strength in the wet stage – compared to the dry one – is less then 30 % of the original one, such a polyurethane should not be used for an outside coating.

A possible way to avoid crosslinking is to use polyurethanes containing groups which are able to react in the each other during drying so water-based

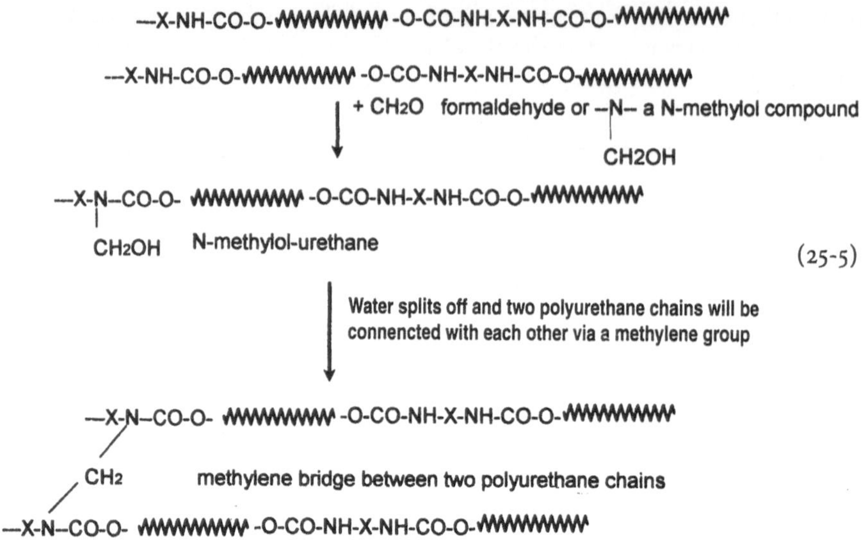

—X-NH-CO-O-〜〜〜〜〜 -O-CO-NH-X-NH-CO-O-〜〜〜〜〜

—X-NH-CO-O-〜〜〜〜〜 -O-CO-NH-X-NH-CO-O-〜〜〜〜〜

$$\downarrow \; + CH_2O \;\; \text{formaldehyde or} \; -\underset{|}{\overset{|}{N}}- \; \text{a N-methylol compound}$$

CH2OH

—X-N—CO-O- 〜〜〜〜〜 -O-CO-NH-X-NH-CO-O-〜〜〜〜〜
  |
  CH2OH   N-methylol-urethane

(25-5)

$$\downarrow \;\; \begin{array}{l}\text{Water splits off and two polyurethane chains will be}\\ \text{connected with each other via a methylene group}\end{array}$$

—X-N—CO-O- 〜〜〜〜〜 -O-CO-NH-X-NH-CO-O-〜〜〜〜〜
  /
 CH2   methylene bridge between two polyurethane chains
  /
—X-N—CO-O- 〜〜〜〜〜 -O-CO-NH-X-NH-CO-O-〜〜〜〜〜

Crosslinking with formaldehyde or products containing N-methylol-groups

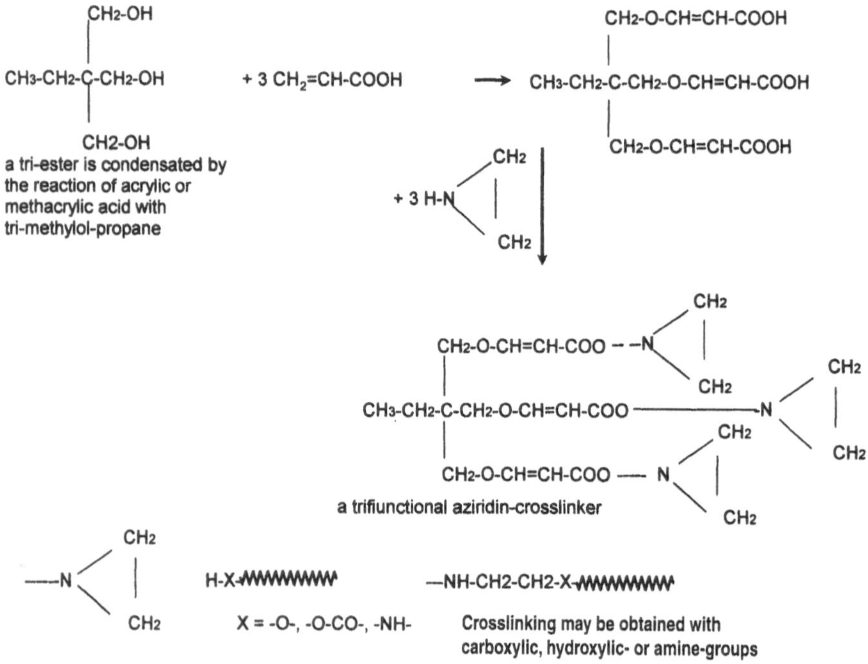

CH2-OH
  |
CH3-CH2-C-CH2-OH   + 3 CH₂=CH-COOH → CH3-CH2-C-CH2-O-CH=CH-COOH
  |
CH2-OH
a tri-ester is condensated by
the reaction of acrylic or
methacrylic acid with
tri-methylol-propane

CH2-O-CH=CH-COOH
  |
CH3-CH2-C-CH2-O-CH=CH-COOH
  |
CH2-O-CH=CH-COOH

$$+ 3 \; H-N\begin{array}{l}\diagup CH_2 \\ \quad | \\ \diagdown CH_2\end{array} \downarrow$$

CH2-O-CH=CH-COO - -N⟨CH2|CH2⟩

CH3-CH2-C-CH2-O-CH=CH-COO ——N⟨CH2|CH2⟩ with additional ⟨CH2|CH2⟩ N groups

CH2-O-CH=CH-COO —— N⟨CH2|CH2⟩

a trifiunctional aziridin-crosslinker

—N⟨CH2|CH2⟩   H-X-〜〜〜〜〜   —NH-CH2-CH2-X-〜〜〜〜〜

X = -O-, -O-CO-, -NH-   Crosslinking may be obtained with
                         carboxylic, hydroxylic- or amine-groups

Crosslinking with aziridine

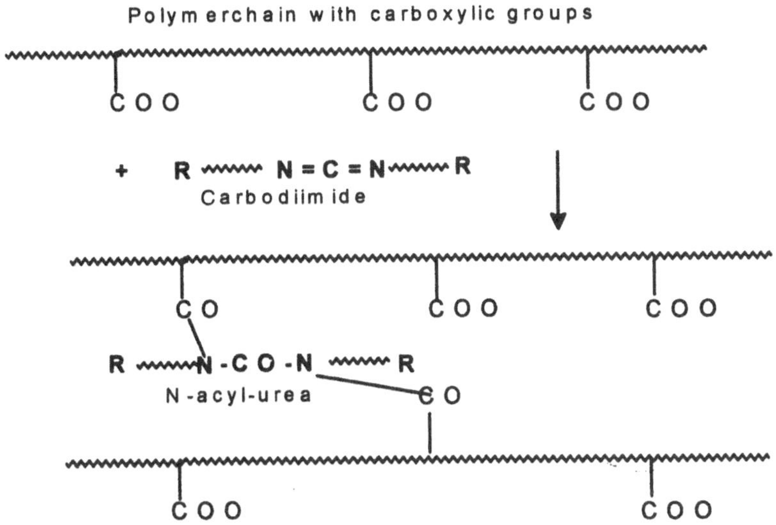

Polymerchain with carboxylic groups

Crosslinking with carbodiimide

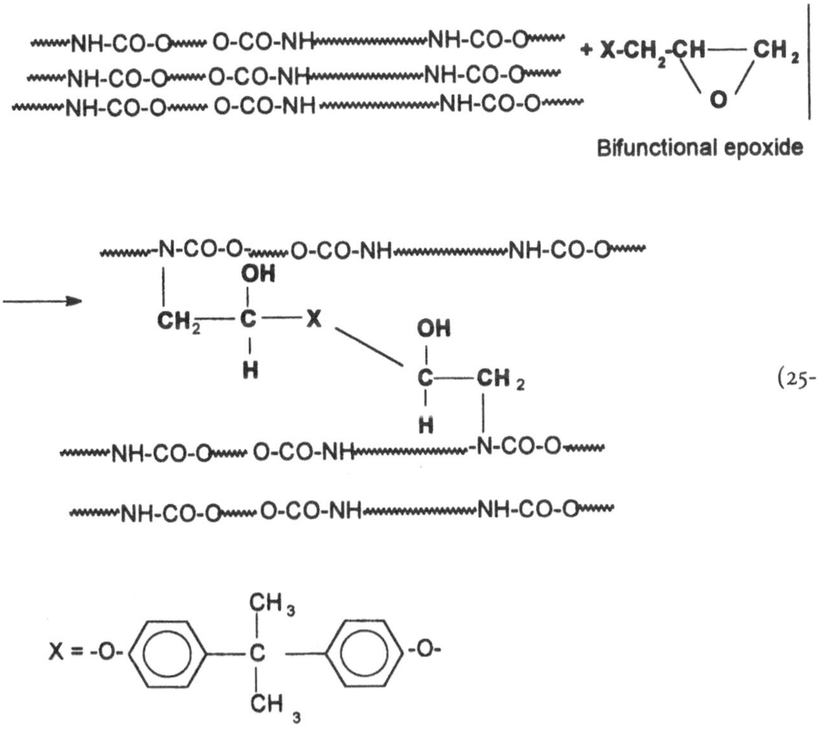

(25-7)

Bifunctional epoxide

(25-8)

$X = -O-\bigcirc-\overset{\overset{CH_3}{|}}{\underset{\underset{CH_3}{|}}{C}}-\bigcirc-O-$

Crosslinking with epoxides

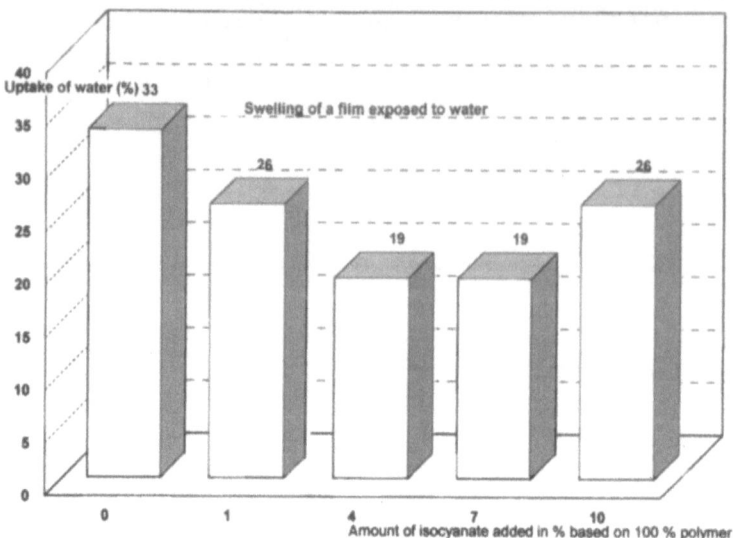

Fig. 25-5. Water uptake of different polyurethane films. The films were crosslinked by a hydrophilized isocyanate. Increasing amounts of isocyanate reduce the swelling of the film. With high amounts of crosslinking agent, however, the absorbency of water increases again due to the hydrophilizing agent in the isocyanate crosslinking agent

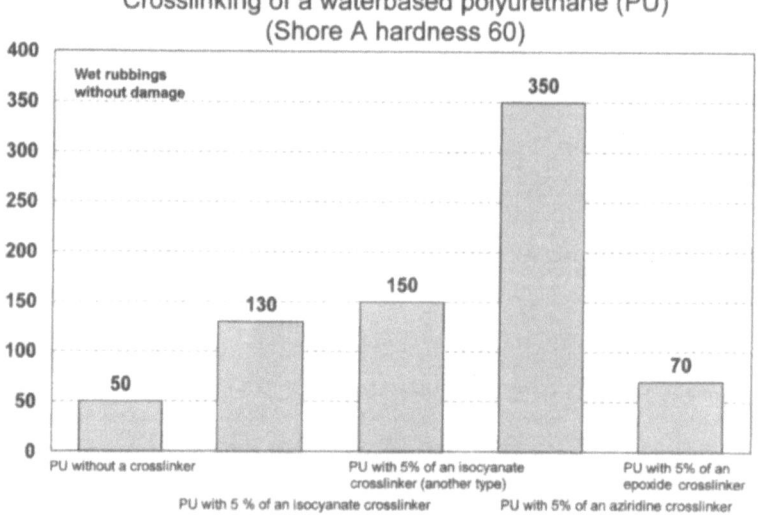

Fig. 25-6. Wet rub fastness of a coating using different crosslinking products for the polyurethane. The wet rub fastness is measured according to DIN 53339 by wetting a felt and rubbing over a coating without damage being seen. The wet rub fastness is an important property; this property should be as high as possible, because shoes, apparel etc. getting wet by rain should not be damaged by rubbing during use. In the test a high figure of rub movements should be reached. In the example demonstrated only the coatings crosslinked by aziridine had a sustainable good wet rub fastness

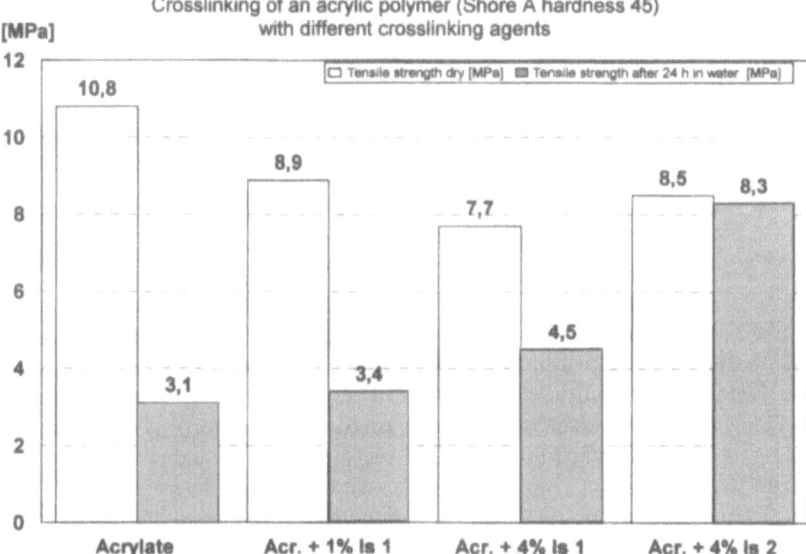

Fig. 25-7. Tensile strength of a polyurethane film in a dry state (*upper value*) and after storage in water for 24 h at 20°C (*lower value*). The films contained as crosslinkers: *Eph* Epoxide-highly reactive, *Epm* Epoxide-medium reactive, *Az* Aziridine crosslinker, *CD* Carbodiimide (18 [1])

polyurethanes can be used which contain cyanbiuret elements which under the influence of temperature or a change of pH split off their cation and react with one aother to form a high molecular network [Eq. (25-10)].

This chemistry has the advantage that the emulsifying group – after drying – no longer exists.

R-NCO + NH₂CN + Et₃N ⟶ [R-NH-CO-N-CN]⁻ [H-N-Et₃]⁺

an isocynate containing   by a change in
oligourethane reacts in presence   the pH or drying
of tri-ethyl-amine with cyanamide   triethylamine splits
to form a cyanbiuret   off and the biuret
    is reacting with itself   (25-10)

$$\text{R-NH-CO-}\overset{\displaystyle CN}{\underset{\displaystyle N\text{-H}}{N}}\text{-C-NH-CO-R}$$

a crosslinked polyurethane

CHAPTER 26

Testing Methods

When the development of leather substitutes was started the aim was to substitute leather in shoe production. In the USA, Japan and Western Germany the water vapor permeability was characterized by measurements according to DIN 53 332, respectively, IUP 15 or similar methods [1]. According to IUP 15 a circular testing piece is placed into a small vessel containing a desiccant like silica gel. The testing piece is conditioned at 23 °C and 60 % relative humidity overnight. The next morning the silica gel is removed and replaced by fresh silica gel. After weighing the vessel, the absorbency of water vapor from the standarzied surrounding air is measured by weighing it after 8 h, The weight gain is divided by 8 and divided by the area of the testing vessel which has a constant size. The resulting value is a water vapor permeability measured as an absorption of water vapor at the silica gel in (mg/h · cm²). In Japanese publications the result was transformed into mg/24 h · m². However, the results were comparable with the IUP 15 method.

According to this testing method the water vapor permeability for the following materials are:

- unfinished leather 6 – 12 mg/h · cm²
- suede leather 6 – 12 mg/h · cm²
- corrected grain leather heavily finished 1–1.5 mg/h · cm²
- patent leather 0.7 – 1.2 mg/h · cm²
- artificial PUR leather 0.8 – 1 mg/h · cm²
- artificial PVC leather 0.2 – 0.5 mg/h · cm²

The disadvantage of this method is the long testing time. An advantage is that the testing results correlate very well with wearing tests. Another advantage is the testing at 60 % relative humidity. The testing results correspond to wearing conditions similar to the closeness of the human body.

Newer testing methods are the Gore cup [7], a computerized method [13], the Lyssy vapor permeation tester [24] or the method which is called SST (Silica Schnell Test] [2], a method similar to the method of BTTG which also uses the absorption of water and water vapor with silica gel [3]. These methods do not need complicated testing equipment and the results are achieved in a shorter period of time.

"The most sophisticated method for measuring the water vapor transmission" (wvt) [4] is the skin simulation model at the Hohenheim Institute (Germany) according to DIN 54101. This method which is backed up by numerous wearing and action tests, e.g. working under defined climatic conditions and measuring the perspiration caused by work, is the basis of the new DIN 61536 and DIN 61539 describing the wearing comfort of clothing [5]. In these norms minimum standards for sports clothing and protective clothing against bad weather are defined. The lower the R_{et} value is the better is the water vapor permeability. Clothing is called water vapor permeable if it has an R_{et} value of lower than $200 \times 10^{-3} \, m^2 \, bar/W$.

A value lower than $60 \times 10^{-3} \, m^2 \, bar/W$ is regarded as excellent, up to $130 \times 10^{-3} \, m^2 \, bar/W$ is good and up to $200 \times 10^{-3} \, m^2 \, bar/W$ is fair. It is possible that in future these values will be regarded as the European standard.

If the measurement takes place under rain-like conditions then the water vapor transmission value is decreased by 30–70%. During these conditions, if the condensation of water occurs in the interior of the clothing, then the clothing is uncomfortable to wear [6]. The same occurs if perspiration is not eliminated fast enough through the textile and starts to condense inside. Therefore in Hohenheim a so-called "sweat arm" was developed to measure the possible condensation of water and the water vapor resistance according to the skin model [8]. These kinds of tests although complicated seem to be the best adaptation of real wearing conditions.

Some literature cites the comparability of the different testing methods [18, 19, 23]. According to one report [4] the values of the different testing methods cannot be compared. Woodbridge discusses negatively the variety of the different methods. All testing methods that exist today are discussed in this publication.

In another publication it is shown that the water vapor permeability of different composite materials can be compared if the values are correlated by correction factors. Then a consistent wearing property value of a special clothing can be determined [21].

If the transport of heat and sweat generated by human action of different parts of the clothing are considered it can be demonstrated that the parts covering the trunk should consist of products of a higher water vapor transmission than, for example, parts covering the arms of the person doing the test. In an examination it was found that fabrics with laminated PTFE membranes showed „outstanding performance in breathability and low thermal resistance while poromeric polyurethane laminated fabrics and hydrophilic laminated fabrics show moderate values in both; and polyurethane coated fabrics show poor breathability and high thermal resistance". Low thermal resistance of a fabric in the inside part of a piece of clothing helps to avoid condensation of water. Such condensation gives an uncomfortable feeling. The outside of the clothing should possess the lowest breathability and the highest thermal resistance [26].

It is possible to measure the temperature profile of a composite textile by IR thermography: This measurement correlates well with the wearing comfort as expressed by the test persons. According to a publication this measurement is easier to obtain than the skin simulation model [12].

The micro climate which is measured by interferometric tests is not determined by the material as such but by the pattern of the clothing [15]. In another published experiment carried out parallel to the interferometric tests it was found that an uncomfortable wearing property may be predicted by the water vapor absorbency of a material [16].

In a published automated testing method testing conditions were varied over a broad range. This method allowed the determination of a dynamic moisture permeability [22].

Underwear was also tested. Sixteen articles from different retail stores specializing in sportswear were tested. The result of these tests corresponds with the tests previously mentioned: Articles made of cotton are able to absorb a lot of humidity; modern materials, however, with almost no absorbency but which are fast drying have a better wearing comfort than cotton due to the fact that cotton keeps the absorbed humidity too long. Under a water vapor permeable layer near the skin a non-absorbing material is better for a fast transmission of perspiration than an absorbing layer. High wearing comfort depends on the moisture content of an article [17].

The wearing comfort of a shoe on a test person is determined by the temperature developing inside the shoe. This was measured by controlled tests and compared with the influence of the construction of the shoe, the material of the socks and the lining of the shoe [9].

There exists a computer simulation of a shoe whereby different amounts of perspiration were created under controlled temperature conditions. The perspiration is absorbed by the socks then the shoe and then transported outside. The faster the perspiration is transported outside the better the wearing comfort [10]. The wearing comfort of trekking, business, childrens and ladies shoes may be tested by a simulation apparatus called Cybor (cybernetic body regulation) [13].

The porosity of leather and its substitutes influences the water vapor permeability. Döring has published measurements of the gas adsorption, the mercury porosity and analysis of pictures of the materials produced by electro-scanning photography. In particular, with artificial PVC leathers the analysis of the photos gives the best results. Whether the porosity values measured correlate with the water vapor permeability was not published [14].

Comfort plays an important role in car seats. An examination of different materials showed that the upper layer of a car seat should not have a barrier against the transmission of water vapor. The upper layer should be hydrophilic. If the next layer is hygroscopic or hydrophilic the result is poor seat comfort. Polyester fabrics with a hydrophilic top layer or fabrics containing a mixture of wool and polyester are therefore better than wool alone. If the products are laminated the foil used in the lamination should be as thin as possible [20].

It is essential to know the water vapor permeability of a material in order to evaluate its protection against corrosion. A method and a definition of a water vapor permeability coefficient has been elaborated [11].

26.1
Test Results

High performance of products is usually more expensive than a low one. Industry needs to compromise between the costs of a product and its necessities. On the other hand the end articles are usually much more expensive than their raw materials. One square foot of a good leather costs around 2 $. One pair of shoes – for which two square feet of leather are needed – cost roughly 20 $ in the shoe factory. This shows that the value added is five times higher than the raw material. In clothing or automotives the value added can be much higher.

The qualities and properties of all artificial leathers and leather substitutes are orientated to fulfill the necessities and physical properties of genuine leather.

The first artificial PUR-based leathers were produced under the tradename of Vistram® (Bayer). Articles made of Vistram were only in use for a short period. The material was successful on the market. More and more chemical companies offered chemicals to produce these articles. Vistram was flexible and abrasion resistant and adhering dirt was easy to remove. Therefore, shoes, clothing for children and bags were produced from this material. A disadvantage was a certain amount of stiffness. In order to produce softer articles polyurethanes were developed based on diethylene glycol adipate ester. However this ester had a poor hydrolysis resistance. Then it was found that the material was also suited for the production of upholstery articles. After two to three years of use, because of the poor resistance to hydrolysis with this polyurethane, there was no further resistance to scratches and abrasion which resulted in claims from customers.

A positive result of these bad experiences was that all suppliers started to develop polyurethanes with a good resistance to hydrolysis. Polyurethanes containing polycaprolactone or hexandiol polycarbonate in the soft segment were developed which were able to be used for articles with long time use.

Minimal fastness properties besides resistance to hydrolysis which were necessary to take into consideration to guarantee the shape, form and utility of the articles were developed by different institutes for genuine leather, coated textiles [4] and leather substitutes. These fastness properties are essential if the articles are to resist during further production and for their later use. The fastness properties are orientated by the demands of the end product.

For testing leather and its substitutes the samples must be climatized according to DIN 50 014 (=ISO 2231 1989). The chemical analysis and the preparation of the samples must be done according to DIN 53 300 and 53 302/2 (ISO 4044) and the physical tests according to DIN 53 302/1 and 53 303/1(ISO 2419).

The German union of leather producers has formulated minimum requirements for the end use of articles to avoid claims (22 [43]).

Products for military use should have properties which fulfill the requirements of the standards as established by various countries. In Germany these properties are regulated by the "Technische Lieferbedingungen" (TL 8330 – 0001).

The most important property for functional clothing and articles for medical use is the ability to be cleaned easily. Uniforms for police, postmen, etc., should be waterproof and permeable to water vapor. Clothing which is heat resistant should have a high water vapor permeability at high temperatures. In a publication PTFE, for example, was compared with hydrophilic polyester materials and microporous polyurethanes under different degrees of heat. It was shown that microporous products lost their water vapor permeability at higher temperatures due to a certain melting of the surface which caused defects in the structure of the micropores [5]. Table 26-1 shows a selection of minimum requirements for different articles.

Table 26-1. Selection of minimum requirements for different articles

Test	Shoes [2]	Clothing	Upholstery [3]
Flexometer according to DIN 53351 (IUP 20, respectively, ISO DP 5402) leather		≥ 20,000 flexings (suede) ≥ 50,000 flexings (grain leather)	≥ 20,000 flexings
DIN 53340 dry wet	≥ 50,000 flexings ≥ 10,000 flexings		
patent leather dry wet	≥ 20,000 flexings ≥ 10,000 flexings		
coated split leather	≥ 150,000 flexings		
Adhesion of the finish dry (IUF 470E/ISO 11644) wet	3.0 (N/10 mm) 2.0 (N/10 mm)	≥ 2.0 N/cm (N)	1.5 N/10 mm
Rub fastness DIN 53 339 According to VESLIC. The testing fabric consisting of cotton should be according to DIN 53 339 dry cotton fabric and leather dry			50 rubs aniline-type leather, 500 rubs finished leathers
cotton testing fabric dry and	50 rubs		
leather wet	50 rubs		
Testing felt with alkaline sweat solution (pH 9)	50 rubs		
Ironing according to DIN 53 342 (IUF 458)	> 80°C		
Thermosetting with stitched patent leather	No cracks		
Distention and strength of the grain DIN 53 325 (IUP 9/ISO 3379)	Ball burst test ≥ 7.0 mm		
Lastometer distention to burst the grain (special leathers like clogs see DIN 53328)	≥ 9.0 mm and ≥ 35% until the grain bursts		

Table 26-1 (continued)

Test	Shoes [2]	Clothing	Upholstery [3]
Tear test DIN 53329	≥ 18 N for shoes with a lining ≥ 25 N for shoes without a lining	≥ 150 N/cm (suede) ≥ 200 N/CM (grain leather) ≥ 150 N/cm (lamb nappa)	≥ 20 N per 1 mm thickness of the leather
Substances soluble in dichloromethane DIN 53 306 (IUC 4/ISO 4048)	≤ 9% 1 comp. adhesive ≤ 14% 2 comp. adhesive	≤ 16% (box) ≤ 18% (kid)	
Water vapor permeability DIN 53 333 (IUP 15 resp. EN 344-pr EN 420)	1 mg/hcm^2 full grain	≥ 3.0 mg/hcm^2	
Water vapor adsorption DIN 53 332	0.9 mg/hcm^2 corrected grain ≥ 10 mg/cm^2		
Color fastness, resistance to water spotting (IUF 420/ISO 105 Bo1–Bo2)	≥ 5 min of surface penetration of a water droplet no stains after drying	≥ 5 min (suede) ≥ 10 min (grain leather)	
Tensile strength DIN 53 328 (IUP 6/ISO 3376)	≥ 150 N	≥ 1200 N/cm^2	
Hardness of polymers Shore A or Shore D, DIN 53 505			
Tensile strength, percentage of elongation at break with films of polymers DIN 53 504			
Dynamic water proofness DIN 53 338 (IUP 10) shoe uppers for boots	(Time with no penetration of water) ≥ 60 min (≤ 35 % water uptake)		
Other leathers Waterproof leather	≥ 30 min (≤ 35 %) ≥ 120 min (25 %)		
Flexing endurance at –20 °C DIN 53 351	≥ 30,000		
Rub fastness after exposure to light (IUF 450/ISO 11640) (ISO 3378/IUP 12)			Dry felt > 500 Wet felt > 50 (Color of the felt 3/4)

Table 26-1 (continued)

Test	Shoes [2]	Clothing	Upholstery [3]
Color fastness of leather to light (xenon arc) according to DIN 54004 (IUF 402/ISO 105 Bo2)	≥ 3 (international blue scale)	Suede ≥ 3 Grain ≥ 4	≥ 3 ≥ 4 (automotive)
Stability of the color DIN 53341	No visible yellowing		
Migration fastness of the finish and color DIN 53343 (test material plasticized PVC and/or TR-rubber) (EN/ISO/IUF 442)	No staining according to the grey scale of DIN 54002 and at the contact ≤ 3		
pH value of an aqueous extract DIN 53312 (IUC 11/ISO 4045)	≥ 3.5	≥ 3.5	≥ 3.5
Determination of water-soluble organic and inorganic substances in leather DIN 53307 (IUC 6/ISO DP 4098)	≤ 1.5%		
Color fastness of leather to washing IUF 423	–	For suedes no change of color 3 (grey scale DIN 54001), difference in the area max. ± 3%	
Color fastness by dry cleaning solution IUF 434	–	Suede no change in touch; suedes no change of color 3 – 4; difference in the area max. + 3%; Grain similar to suede no peeling of finish layers	
Content of pentachlorophenol (EN … in preparation)			
Content of formaldehyde (EN …)	≤ 150 ppm	≤ 150 ppm	≤ 150 ppm

Polymers are usually tested in film form. Polymer characteristics are determined by the thickness of the film (DIN 53353; ISO 2286 1986), tensile strength, elongation at break (DIN 53504), tear strength (DIN 53356; ISO 4674 1977), hardness in Shore A or D (DIN 53505) and the adsorption/swelling in water and organic solvents.

Resistance to hydrolysis or aging (ISO 1419 1977), abrasion resistance (DIN 53799; ISO 5470 1980) and light fastness (DIN 54004) are important for coatings. The light fastness is usually only tested as a change in color of a coating, etc. There is also a change in physical properties by the influence of light which is tested in the following way: The test specimens, as for the determination of the tensile strength according to DIN 53504, are exposed to light (xenon arc) up to 6 on the international blue scale. Then the tensile strength is tested again. A material is then regarded to be resistant to aging by light if the tensile strength and elongation at break do not differ by more than 10% with the unexposed film.

The resistance to hydrolysis is tested in a similar way: The tensile strength test specimens are placed in a closed vessel containing water and stored at 50 °C or even better at 70 °C for 4 or 6 weeks. The water on the bottom of the vessel creates a relative humidity of ca. 98% in the air above. At the end of every week three of the test specimens are tested to determine the tensile strength. A polymer is then regarded as being resistant to hydrolysis if after 6 weeks at 70 °C the tensile strength does not differ by more than 30% from the value of the specimen that was not aged (see Fig. 26-1). It has often been shown that a degradation occurs

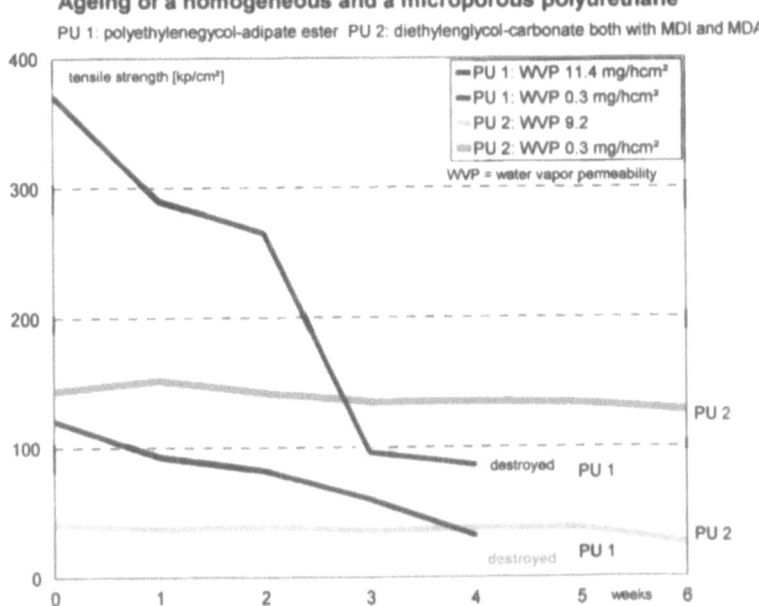

Fig. 26-1. Aging of two polyurethanes: Films were aged at 70 °C and ca. 98% relative humidity. The tensile strength and elongation at break according to DIN 53,528 were measured

with an increase of the elongation at break due to a punctual break of chemical (cross)linkages. For microporous products the fastness to hydrolysis is more important than for non-porous products. The porosity increases the surface area of a polymer so much so that it is much easier for water, perspiration, microorganisms, etc. to attack the polymer than with a homogeneous product. The aging in water vapor and 70 °C of a homogeneous product will decrease more or less in the same way as a microporous film if the tensile strength is measured. The homogeneous film starts at a much higher level than the microporous film but the percentage of decrease is more or less similar. For ester-polyurethanes the resistance to hydrolysis, and for ether-polyurethanes the aging by light can be a problem.

Besides DIN, IUC, IUF, IUP, CEN, ASTM, etc., other standards exist. These standards are not totally comparable with each other.

Leather and its substitutes also play an important role in the automotive industry. Every car producing company has its own catalogue of tests and minimum requirements for their products. In Tables 26-2 and 26-3, samples of some typical requirements and tests are shown.

- Resistance to low temperature (e.g. 1 h exposure to − 20 °C) no breaks or cracks after bending (Pirmasens method)
- Resistance to water; 72 h exposure at + 80 °C no breaks or cracks after bending
- Fastness to light according to DIN 75 202 and DIN 54 004 (international blue-scale 6) no change in color admitted after 1000 h exposure or a fastness to light of more than 6
- No fogging; i.e. fine droplets of organic material on the windows casued by volatile organic materials, like plasticizer in seat and/or wall covers
- No odor
- No flammability: materials should be according to US law 571.392

CHAPTER 27

Other Industrial Applications

According to an analysis carried out by the Freedonia Group [1] in 1994 the US market for coated textiles was 294 Mio/m². This had a value of $ 2.5 billion. Most was textiles coated with PVC for use in the field of transport. The sector of protective working clothing was 10 % of this quantity. This market segment had the biggest growth rate. In view of this, PVC will keep its dominant role in the market [1]. Polyester will increase further its share of the market as a textile substrate. The combination of PVC as a coating on polyester as a substrate is not expensive and fulfills all requirements.

Water vapor permeable materials are mostly used for functional textiles. Functional textiles (see Sects. 2.1, 26.1, and Chap. 18) are designed products adapted to the demands of their users and the required physical property. Examples are clothing for soldiers, protective clothing for workers in chemical companies or in plants working with hot metals or sharp-edged metals. Important requirements are flame resistance, resistance to chemicals, such as alkaline products, amines, solvents, carbohydrates, acids, etc. Further important physical properties are flexibility, abrasion resistance, bactericidal [89] or anti-static properties, easy to process into the finished article, washability and/or dry-cleanability and comfortable to wear [88].

Polyurethane-coated textiles are mostly used in the production of shoes and clothing [49]. The water vapor permeability is for all articles worn near the body a most important property (Chaps. 1 and 4). Bags (11 [13]) do not need water vapor permeability just as much but is nice to have this property. Here fashion-orientated effects, the grain pattern, touch, color, glossy or mat effects, besides abrasion resistance, and flexibility are regarded as being important.

Synthetic materials can be used in the normal production of shoes using similar production techniques to those used with genuine leather (1 [2], vol. 11). In fashion clothing almost no leather substitute has lasted for a long time apart from synthetic suede types (Alcantara®) or shoe uppers and gloves for (golf) sport. Materials showing a reasonable wear performance when used together with a lining have gained good market success in sportswear. They can be processed, e. g. by a hot melt adhesion of a microporous PUR membrane with a textile [13].

Functional Clothing. Clothing containing microporous materials are able to transport water vapor, created by perspiration of the wearer, out. Water should not be able to condense in the interior of such clothing even at external low temperatures [4]. Abrasion-resistant, water vapor permeable protective clothing may be produced from Aramid® fabric with a coagulated polyurethane containing quartzite [16]. Protective clothing for workers which can be worn near flammable liquids and are able to prevent electrostatic charges on its surface can be produced with leather substitutes (23.2 [9]).

Materials for clothing for people working in hospitals, especially in rooms where surgical operations are performed, besides being water vapor permeable should also be easy to bond together and resistant to electrostatic charges on the surface. However these kind of properties still have to be achieved [79]. Knitted polyester fabrics become microporous by a high temperature sealing process. Clothing with good wearing properties can be produced out of these fabrics [63].

Layers of a coagulated polymer containing active carbon and silicic acid on cotton, polyamide or polyester fabric can be used in the production of combat clothing because they protect against chemical weapons like gas or nuclear fallout. The same material can be used for packing weapons, ammunition, food and pharmaceuticals or to seal things against chemical weapons and nuclear fallout [41]. Good thermoinsulation as well as a high water vapor permeability is important for sleeping bags ([31], 16 [16]).

Protective gloves for industrial and military use are produced out of cotton fabrics with microporous layers on top of them. Another microporous layer is then applied which is a selective membrane consisting of cellulose, polyamide, polyurethane, etc., on top of the first microporous layer [9].

Sports articles are an increasing sector of water vapor permeable products; i.e. apparel, shoes, bags, gloves, bands for tennis rackets, grips etc. Membranes applied onto knitted or woven textiles or nonwovens which have elastic fibers are often considered suitable for outdoor sportswear (Fig. 27-1) (see also Chaps. 16 and 18.2 and [85]).

As previously mentioned (Chap. 18) one main use of water vapor permeable products is as a liner (Fig. 27-2; 3 [1]). Liners protect against wind and are water vapor permeable. The outside fabric protects the microporous layer inside against abrasion or the hydrophilic layer against the action of liquid water.

Materials used for making tents are based on water impermeable but water vapor permeable materials (29.1 [16], [6,7]).

Until now microporous and hydrophilic products were mainly used in the manufacture of shoes and clothing.

In a general sense, these kind of materials are membranes – allowing defined products to permeate a frontier between an inside and an outside part of a

Fig. 27-1. Sectors for water vapor permeable products (18.2 [2])

functional clothing
protective clothing
uniforms (post office, railway, police)
sportswear
workwear (medical, pharmaceutical, electronics)
military uses
filtration, perm-selective separating walls
packing, covers
slow release

 Coating
textile fabric

The coated textile can be worn
either with the coating outside
and the textile inside or
vice versa

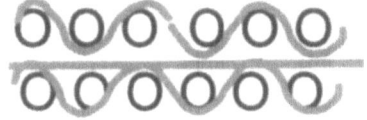

Lining inside two textile fabrics

Fig. 27-2. Construction of a liner

technical product. As found also in nature membranes are the most important part of a cell because they define its interior and its exterior. "Membranes play a central role in both the structure and function of all cells ... they define compartments, each membrane associated with an inside and outside ... they also define the nature of all communication between inside and outside" [90]. All water vapor permeable materials act as a membrane and allow a selective exchange of materials between the inside and outside.

Anti-static Properties (18.2 [68]). Hydrophilic polyurethanes are also anti-static [10,26]. This property may be increased by including metal powders in the polymer [11].

Coagulated, microporous products containing fluorine compounds can be easily charged electrostatically. This property enables the removal of dirt, residues of inks, etc. from printing plates [56].

Microporous materials can be used for insulation (10.1 [21]). High end cables for high fidelity use Gore-tex® as an insulating material to guarantee a low capacity of the highly inductive inside [107].

Office Use. It is easy to write on microporous surfaces (16.2 [4]). They can be used as typewriter ribbons (18.1 [18]) or as inkpads for stamping ink (16 [20]).

Polyethylene foils containing a filling agent may be used for printing sheets for ink jet printing [58, 66, 77].

To produce ink jet printing sheets an opaque layer, able to absorb organic solvents consisting of a polymer which is insoluble in water, containing silica particles as well as a microporous polymer is applied onto a paper surface [81].

Coated nonwovens may be used as typewriter ribbon. It is made out of a microporous coating on a nonwoven consisting of a polyurethane, 100 – 400 % of a printing ink containing a waxy mineral oil and carbon black [97].

Microporous dyestuff containing PVC sheets can be used for carbonless copying papers [9.3 [22]).

Cleaning and Filtration (7 [63], 12 [27], 16.2 [29]). Inorganic membranes used in ultra- and microfiltration are described in a monograph by Bhave [3].

It has been claimed by Staude that processes via membranes are very slow because they are regulated by diffusion. The market for membranes used for separation in the food sector, in medicine or in industry for the separation of materials is estimated at roughly 4000 Mio DM [2].

In a textile processing plant part of the residual liquors is purified via a membrane filtration. By this action the COD is decreased by 100 – 1000 times. If there are sticky wastes in the residual water this process cannot be applied [71].

Microporous membranes are suitable for filtration (10.3 [4], 10.5 [1], 12 [27], 18 [11], 18.1 [33]). Polyurethanes are especially suited for filtration due to their elasticity and their high tensile strength. Due to their physical strength and abrasion resistance polyurethanes are able to filtrate hard materials such as inorganic pigments.

Cell walls for batteries can be produced from microporous polyurethanes [15, 16, 18].

Products optionally in an electric field and under pressure can be filtered by reverse osmosis on microporous membranes. These membranes consist of acetyl cellulose, sulfonated 2,6-dimethylphenyl ether, polybenzimidazole, glass etc. Products to be separated are the residue of dyestuffs, proteins, sugar or lignin in the production of cellulose fibers [28]. Reverse osmosis at high pressure, e.g. 50 bar, can be carried out with membranes consisting of different polymer layers [50].

Semi-permeable membranes can be used as molecular sieves [27]. Ion-exchange materials are used increasingly for catalytic reactions; these materials may contain noble metals [59].

Kaolin particles can be separated from a suspension by rotating drum filters covered by coagulated, porous, impregnated fleeces [110].

In public water purification membranes may be used to remove bacterias, spores etc. [30]. Polyurethanes with an antibacterial appliance may also be used [42]. The surface of cellulose acetate membranes is treated with isocyanates, epoxides, phosgene or amines to improve their resistance to rotting [48].

Hydrophilic residues produced by the leather or textile industry may be mixed with cement. The resulting material may be used as filters or covers for dumps permeable to gases [45], bioreactors or to equalize pressure in storage tanks [84].

Impurities in the air can be removed with microporous membranes. This can be achieved with PTFE membranes [86] optionally covered by a hydrophilic polyurethane and laminated onto a textile fabric (Fig. 27-3). Microorganisms may be absorbed on the surface of this material [24].

A membrane can be used to increase the oxygen content in air. This is done with polybutadiene containing hydroxy groups. The hydroxy groups are reacted with H12MDI and N-methyldiethanolamine to form an ionic polyurethane membrane. This membrane also contains copper ions. The gas permeability of the membrane is correlated with the amount of copper(II) chloride [64].

High molecular weight substances or biological cells may be filtered by a composed material. This is achieved by placing a hydrogel on a microporous membrane or between two membranes. The flow of water or water-based solutions is regulated by the hydrogel [44].

Polypropylene foils biaxially stretched are microporous. They can be used in ultrafiltration, as membranes in gas-exchange reactions and as cell walls in batteries [62].

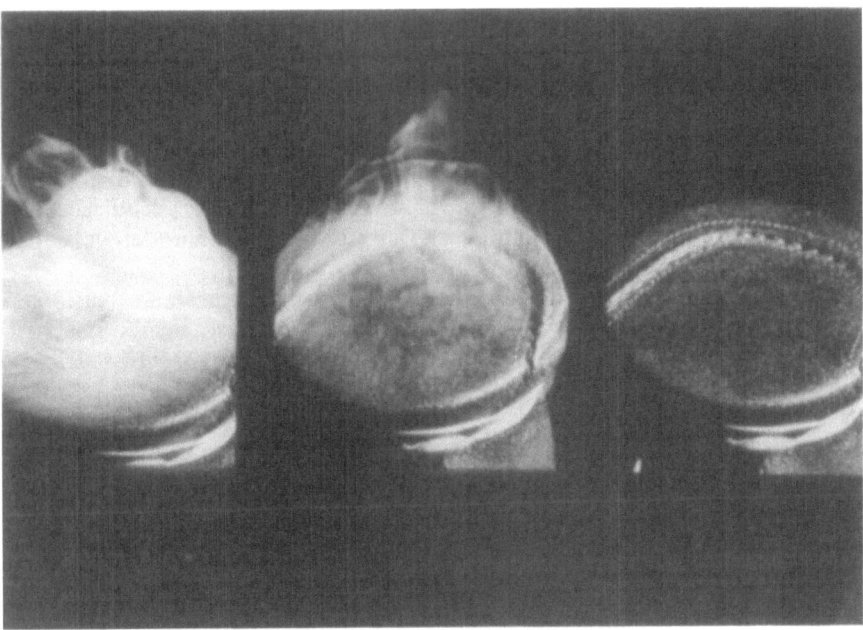

Fig. 27-3. To demonstrate a possible filtering effect of a microporous film smoke from a cigarette is blown through the film. Three stages are shown, the film as it is, a small quantity and a high quantity of smoke penetrating the film (from right to left)

Slow Release. Leather substitutes have a high internal surface. Humidity, temperature and the constituents of perspiration offer ideal living conditions to microorganisms. To avoid unpleasant smells, deodorant substances are often incorporated into leather substitutes [8, 32, 33, 73]. Mixtures of polycarboxylic acids, polyamines and hydrazine derivatives are used [102].

A textile substrate consisting of ultrafine cellulose fibers and filled by a polymer is treated with a cationic agent. The cellulose fibers are treated with a deodorant. A polymeric metal-porphyrin derivative is applied by padding. The deodorant agent adheres well to the fibers and is liberated slowly [54].

One or more polyurethane layers are laminated onto a knitted or nonwoven textile by an adhesive based on a polyurethane. To create pores the material is needled. At least one of the polyurethane layers contains a deodorant. The deodorant may be a mixture of 1 part manganese sulfate, 0.5 parts of ascorbic acid and 0.5 parts of citric acid. The artificial leathers produced in this manner deodorizes against ammonia, methylmercaptan, hydrogen sulfide and trimethylamine [82]. Natural or artificial perfume oils may be applied onto artificial leathers [68]. Water transfer with walls consisting of microporous PTFE containing chemicals release this active ingredients only slowly (16.1 [13]).

Medicine. The surface properties of biomedical polymers are important in the adsorption of proteins and in subsequent biological interactions [105]. Polyurethane seem to have a special compatibility with blood [106]. Membrane products produced by coagulation can be used in hemodialysis as an artificial kidney [107].

In medical or biological processes microporous membranes are of use in the separations of cells [23], or in hemodialysis [67]. Microporosity helps in the adaptation of implants with the human body [56]. Silicones with a microporous surface are more suitable as implants than non-porous ones [57]. Medical equipment (e.g. catheters) with a microporous surface are smooth under wet conditions and easier to apply [112].

A review of polymers, especially the thermoplastic polyurethane elastomers (TPU) in medical use, shows that polymers may be used in low risk non-invasive devices, blood bags and high risk applications like cardiovascular and orthopedic implants. The relationship between the chemistry of the products, processing parameters and the biostability of TPUs has been discussed [100]. Materials useful for the construction of implants have structural components that are subject to degradation in the body of the recipient [103]. By controlling the right amount of hydrophilic to hydrophobic polyethers in polyurethanes products are obtained which can be used as catheters, or substitutes for veins and arteries etc. [108].

Slow release of pharmaka is important in medicine [76] – microporous polyurethanes are able to release pharmaceuticals transdermally [21].

A broad use for microporous materials is as diaphragms (10.2 [14], 18.2 [4], [17]), in dialysis [69, 81] and for surgical clothing (18.2 [5], 6, 93]. Hydrophilic or microporous materials can also be used as synthetic skin ([14], see literature [83]), and as wound dressings (10.2 [14], 13 [1], 16 [12], 18.2 [4], [18, 19, 94, 109]). Hydrophilic polyurethanes may be used in the production of surgical gloves [5]. Membranes consisting of nitrocellulose with a pore size of 5 – 15 µm can filter selectively leukocytes from a blood-derived suspension [111].

Tubes, membranes, sheets etc. can be prepared by interpenetrating networks of polyurethanes with N-vinyl pyrrolidone and methyl-methacrylate. The permeation of glucose and insulin were found to vary with hydrophilicity of the membranes. Such membranes can be used as immunoisolation membranes [115].

The compatibility of blood cells with hydrophilic polyethylene glycol containing polyurethanes was examined. A molecular weight of 600 was found to be thrombogeneous [12].

Hydrophilic surfaces which are compatible with blood by passivation absorb in the first minutes of contact with proteins less than hydrophobic surfaces [34]. Composed materials consisting of polyvinylpyrrolidone and polyether sulfon may be used for equipment in which blood plasma is tested for HIV [72].

Cellulosic membranes may be modified in their blood compatibility as long as they have reactive hydroxy groups by the introduction of isocyanates, hydrophobic, hydrophilic or ionic functions [46].

Cellulose-acetate-polyurethane membranes were investigated by ultrafiltration techniques to separate proteins [116].

Leukocytes may be filtered out of blood with a nonwoven coated by a polyurethane which can be adhered onto PVC blood vessels by means of an adhesive based on a water-based polyurethane. The filter can be removed from the vessel easily [51].

Extremely purified polyurethane-ureas containing silicone groups can be used in medicine [22].

An absorbing layer for humidity consisting of a polyurethane foam, a second layer which is permeable for humidity and a third layer permeable for gas, can be used as wound dressings [29, 40].

Modified Lyocell® fibers can be used as wound dressings especially for chronically ill persons. These fibers absorb more water than alginate, which is usually used [87]. The material is sold under the name of Hydrocel®.

Collagen layers are suitable as wound dressings [78] or several layers of collagen fixed on each other [20].

A self-adhesive wound dressing is produced by applying a hydrogel layer to a vapor permeable bacterial barrier [96].

Porosity and compliance of microporous polyurethane based microarterial vessel has an effect on neoarterial wall regeneration [95].

Porous polyurethane films are coated with a layer of a water-absorbing material an epoxide crosslinked hyaluronate foam and a polyurethane top layer. The resulting wound dressing material is soft, resistant to bacteria and is biocompatible [36]. Gels of hydrophilic polyurethanes are also used as wound dressings [47].

Implants produced from aromatic isocyanates should be avoided because under reductive conditions carcinogenic aromatic amines may be formed which can be determined by chemical analysis. According to the author's opinion there is no miracle if, by using polymers based on aromatic isocyanates, carcinogenic aromatic amines are found [38].

Microporous membranes can be used to protect against noise, heat (11.1 [23]) and chemicals [35]. A publication is available which summarizes all types of protective clothing for the medical field [55]. All protection includes measures to avoid penetration by viruses and a good water vapor permeability. This publication also discusses the advantages and disadvantages of one-time use or disposable clothing, e.g. those of Kimberley-Clark and DuPont, as against reusable clothing e.g. of Gore.

Clothing and other uses of textile substrates in rooms for surgical operations are examined in a publication in regard to their cleanability and their ability to be sterilized by means of ethylene oxide, autoclave, heat etc. [75].

Incontinence and anti-decubitus articles can be produced with Lyocell® fibers and an optional coating [70].

Disposable diapers containing hydrophilic surface layers (12 [39, 40]) are claimed to have a good feel to human skin [81].

Cosmetic buffs may be produced with a coagulated polyurethane with pores of a size of 5 – 100 mμ [101].

Veterinary Use. Flea collars for dogs protection are produced from a polyurethane with a density of 0.3 – 0.45 g/cm³. The polyurethane is impregnated with an insecticide and laminated onto a textile substrate. As insecticides, pyrethroid, carbamates, organophosphorus compounds etc. are used [52].

Food. Hydrophilic foils can be used in food packaging (18.1 [32], [43]).

Other Industrial Applications. Microporous and mesoporous zeolites are used in chemical processes, in catalysis, in Diels–Alder reactions, oxidations etc. These products are usually inorganic and of a crystalline structure. They differ from membranes, usually polymeric and elastic, which are the subject of this book [104].

Hydrophilic polyurethanes when applied onto glass surfaces prevent dim surfaces (Fig. 27-4, [25]). Hydrophilic coatings applied on hydrophobic surfaces improve lubricity and wettability. Such coatings are useful in areas like medical devices, surgical gloves, lenses etc. [98].

Membranes used for ultrafiltration consisting of polyacrylonitrile are treated with enzymes. Then they are crosslinked by means of glutaric dialdehyde and activated with carbodiimide. These membranes now containing carboxylic groups combine biocatalytic effects and absorbency of special compounds [74].

Coagulated, microporous mixtures of PVC and polyurethane on a textile substrate can be used to polish silicon wafers [53].

Microporous products containing finely dispersed fillers which are insoluble in water can be used as abrasive materials [61].

Raised woven textiles coated with a coagulated polyurethane can be used for car interiors [60].

Construction Area. Microporous products may be used for insulation [10.1 [21]). For thermoinsulation of roofs, water vapor permeable foils are often used between the tile and the isolating material [37, 39, 65]. These foils are used to

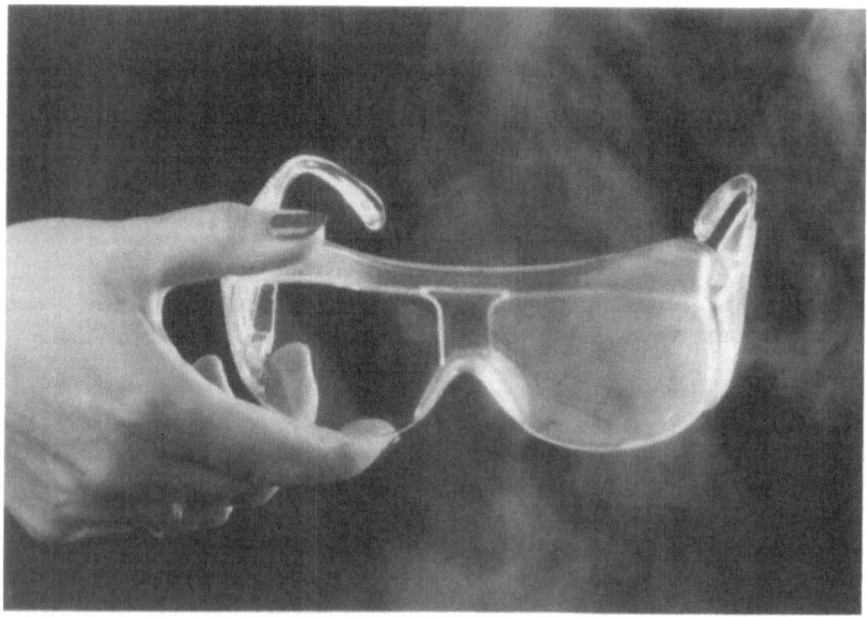

Fig. 27-4. Two glasses are shown whereby one is coated with a hydrophilic polyurethane in water vapor. The coated glass does not become dim

prevent the condensation of water vapor inside the insulation which would make the insulation ineffective. Panels with insulating layers are also used [10.2 [16]).

Coatings for panels consisting of aluminum which is corrosion resistant and used for heat exchange equipment are produced with a protective coat of a hydrophilic aluminum copolymer mixture (18 [8]).

To improve the properties of cement hydrophilic polyurethanes are added. Grouts containing hydrophilic polyurethanes improve concrete so that it may stop seepage and prevents cracks and consolidate foundations [92].

Roof panels protective against heat and cold consist of a polyester material coated on both sides with PVC. The PVC contains encapsulated solid hydrocarbons. By warming up the hydrocarbons they become liquid and use energy. Cooling causes crystallization and liberation of energy. These effects can be used to cool or heat room. The materials are called phase-exchange materials. To utilize the panels appropriately in the right thermo-interval they need a selection of hydrocarbons with the right melting points [90].

A water-based polyurethane with a glass transition temperature of $-30\,°C$ is mixed with cement and $CaCO_3$ and sprayed onto concrete to give a membrane as a sealant for use in the construction field [99].

CHAPTER 28

Ecology

As we have previously seen leather substitutes in most cases consist of a textile substrate with a polymer coating and or impregnations. Therefore, the ecological behavior of the textiles is as important as the polymeric materials used. There are several environmental influences:

(1) The impact of the chemicals used in processing during and after the production of the textile and the substitute.
(2) The chemicals needed to protect the article during usage, and
(3) The disposal of the article when no longer needed.

The following figures demonstrate the quantity of products involved: As an example, the textile industry in Germany uses 250,000 tons of water annually which is mainly recycled. 960,000 tons of used clothing are produced of which 300,000 tons are reused. 560,000 tons of used household textiles are disposed [26].

Besides the textiles, dyestuffs of textile substrates are a point of ecological discussion. Since certain dyestuffs are known to contain carcinogenic amines as components, all azo dyes are often regarded as dangerous. Only some dyestuffs contain carcinogenic components. The dyestuffs themselves do not need to be hazardous. Some years ago it was found that such dyestuffs in the human metabolism may resplit by azoreductase in the amines (see Fig. 28-1). Therefore, dyestuffs containing carcinogenic amines should not be used any longer. In most countries today their use is forbidden.

All major chemical suppliers test the products they are marketing for toxicological behavior. The safety data sheets [19, 23] which are delivered with the product include information about safe applications of the product so that hazards may be avoided for the workers and the environment [18].

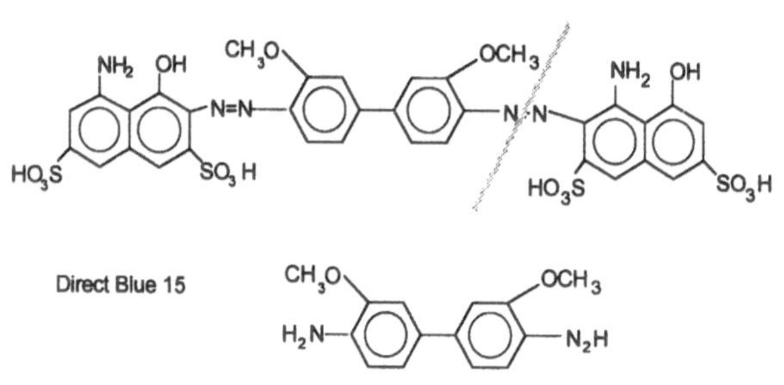

Fig. 28-1. Resplitting of an azo dyestuff

Direct Blue 15

One way to eliminate used textiles is *biodegradation*. Biodegradation of textiles is therefore of interest. In recent years textile biodegradation according to DIN 54900 has been examined and, due to research, this process has been improved. Most textiles of natural origin can be degraded by microorganisms. The chemistry of processing textiles, however, influences the biodegradation. The resistance to microorganisms according to DIN 53933, part 1 of cellulose was, for example, tested to discover whether it could be biodegraded [21]. It was seen that a cellulose having a crosslinking finish by formaldehyde to obtain a permanent press sizing was not biodegradable at all. A water-repellent treatment – contrary to such a permanent press sizing – did not influence the biodegradation. After use those textiles can be eliminated.

Polyurethanes which are important polymers for all kinds of leather substitutes can be built by using biodegradable components. In this case the final product may be biodegradable (13, 18.4 [12]).

Two amines, base products for industrially used isocyanates toluylene-diamine (TDA) and 4,4'-methylenedianiline (MDA), were investigated by measuring $^{14}CO_2$ as to whether they were biodegradable under aerobic as well as anaerobic conditions. It was found that they both decomposed under aerobic conditions. Under anaerobic conditions no $^{14}CO_2$ was liberated [31]

Another way to study polyurethane biodegradation is to look at older publications ([30], 7.1 [14]) about photo-decomposition [5] or about thermal or oxidative destruction of polyurethanes [6] to learn if these mechanism can also be used to eliminate polyurethanes. One way to eliminate polymers is to mix them with products of natural origin. For example, a mixture of polyurethanes with collagen powder or collagen fibers results in products which can be decomposed easily [27].

Recycling is a way of eliminating of used textiles. Textiles containing polyester can be recycled totally. With an alkaline treatment a polyester is resplit into dicarbonic acid and glycol which later can be polycondensated into a new polyester.

Recycling of polyurethanes, without a chemical treatment or a partial destruction, is difficult to achieve [32]. Ground polyurethanes can be used as a kind of filler in casting systems for microcellular rubbers [33].

Polyurethanes can also be destroyed by glycolysis [11]. The glycolysis results in amines and glycols. These components may be used for articles with minor physical properties [8]. Nonwovens can also be recycled by glycolysis and thermal decomposition – although this is a difficult process to handle [10].

If for instance a coated polyester substrate is recycled there should only be a minor problem doing so as long as a polyurethane is used for coating which consists of a polyester with groups easy to hydrolyze. After an alkaline hydrolysis the remaining HDI-bisurethane can be resplit also by high-temperature glycolysis back to the original starting materials of the HDI synthesis. As long as artificial leathers differ so much in their composition this is difficult to achieve. A suggestion to carry out such a process would be that every supplier of textiles and coating substances do the following:

– only offer a limited selection of products,
– manufacturers turning these products into articles do not change the performance of the articles too much from that of their competitors, and

– producers of the polymers and textile substrates as well as their users convince customers to accept products of lower quality and to collect the clothing after use.

Composed products similar to artificial leathers are carpets. It is reported that a joint venture will remove more than 90 000 tons of Nylon® from carpets and recycle them back into caprolactam [34]. That it is possible to treat artificial leathers to get low molecular weight raw materials is shown by the following example: Artificial leather consisting of fibers and coating mass can be heated to 110 – 170 °C at a pressure of 5 – 10 bar. Then the resulting mass is cooled with water to a temperature below 100 °C. The resulting agglomerate suspension contains 3 – 6 % solids. The agglomerate may be used as such or in a mixture in the production of films [14].

Reusing recycled products can be a problem. Recycling of used seats was examined in the automotive industry. The glycolysis of the polyurethane foams in seat upholstery results in polyols. These polyols can no longer be used for seat upholstery due to a lack of performance. They can only be used for sound insulation materials [12].

Heat, sound, vibration and ray absorbing materials may be produced by using wastes of the leather, textile and coating industry. The wastes are mixed with inorganic binding agents and heated to a temperature of 600 °C or to the melting temperature of iron. Skeleton-containing structures are produced by this method which contain hollow parts [17].

Polyurethanes are used more and more in medical applications (see Chap. 27). In a publication [16] the decomposition of hydrophilic polyurethanes is compared with hydrophobic ones. The polyurethanes examined contained azo groups. By the anaerobic flora of the intestinal tract these azo groups are easily reduced to hydrazo and – finally – to amine groups. Using hydrophobic polyurethanes this reduction did not result in amines, the reaction ended at the hydrazo level. This test demonstrates that a hydrophilic polyurethane may be degraded easier than a hydrophobic polyurethane.

The *burning* of used products is another way of elimination. Especially for composed articles which are difficult to recycle, burning may be the only economical alternative. In a study floor carpets consisting of polyamide, polypropylene, styrene-butadiene copolymer, sodium silicofluoride and ammonium acetate were burned in a way that, not regarding the fluorine components of the products, the smoke met the requirements of the 17. BimSchV (Bundes-Immissions-Schutzverordnung, Germany) and did not need to be additionally treated [9].

Textiles, leather and other products worn by people may suffer during their use due to attack by insects or bacterial or fungicidal or chemical (sweat) action. A problem arising when using biocide chemicals during processing is that the biodegradation is hindered. Therefore a dilemma exists. Either the biocide chemicals are used to protect the product and the product does not degrade properly – or the biocide chemicals are not used so the products biodegrade properly, but here there remains a risk that the customer will not be satisfied due to destruction by microorganisms [22].

There is public controversy about a potential negative effect of the processing chemicals used in the manufacture of leathers and textiles. Special concern deals with the chemicals used to improve the appliance of the textiles. Part of this concern may be based more on emotions rather than rational, factual reasoning and it may sometimes lead to public uncertainty and confusion. One specific concern was the presence of amines in products. This was discussed by looking at the reaction of isocyanates with water to an amine to find out whether an amine resulted and remained free in the product. It was not mentioned, however, that the amine instantaneously reacts with the isocyanates present to an urea which is physiologically inactive [28]. Therefore, according to the result of this work, it may be concluded that the negative discussions about certain chemicals used in the production of textiles and leathers may actually be a marketing ploy to enhance the sales of special products. In a publication it was reporten that dyed textiles consisting of wool-polyamide, viscose and cellulose acetate containing no formaldehyde, no carcinogenic amines and no heavy metals surprisingly may have a positive Ames test [24].

Eco labels should insure the consumer that a textile product does not contain hazardous components. [25,29].

Various manufacturers of fabrics with high wear comfort use environmental arguments which discriminate certain product groups. Polyurethanes are among the products affected by such emotional claims [1].

Polyurethanes are, in our opinion, advantageous in the manufacture of consumer goods. The properties of polyurethanes used to coat apparel fabrics are very similar to those of leather. We feel that this is an argument which favors the use of polyurethanes for consumer goods. They contain the same chemical elements as fresh animal hide, i.e. carbon, hydrogen, oxygen and nitrogen. Consequently, they do not release halogen-containing dioxins or furans (i.e. substances containing fluorine, chlorine or bromine) under combustion, as is sometimes incorrectly claimed.

The polyurethanes used in the apparel industry are usually based on aliphatic or aromatic isocyanates. Aliphatic isocyanates are generally somewhat more expensive but have the advantage of being fast to light.

These light-fast polyurethanes do not contain either toluylenediisocyanate (TDI) or 4,4'-diphenylmethane diisocynate (MDI): their structure is aliphatic. Consequently, aromatic amines cannot be released when they decompose, e.g. as a result of hydrolysis. In other words, a potential generation of aromatic amines can be prevented quite simply by using aliphatic isocyanates.

Isocyanates are produced industrially by the condensation of amines with phosgene. This process is carried out in closed plants which are subject to constant monitoring by the authorities and have high safety standards. Phosgene is only produced for immediate use, so there is absolutely no need for it to be stored or transported. Contrary to what is often claimed, the HCl gas generated in the production of isocyanates can be processed to chlorine gas (by electrolysis) and returned to the production cycle [2]. Diisocyanates produced in this way do not contain either phosgene or chlorine, nor do the polyurethanes produced from them.

Associating the chemical processes used to produce polyurethane [1, 3] with phosgene, which was used as a chemical weapon in the First World War, is quite unnecessary and only raises unjustified fears. It is tantamount to warning people not to use common salt, which we all know is about 50 % chlorine, because chlorine was also used as a chemical weapon in the First World War.

There is no justification for equating polyurethane with highly toxic isocyanates. At the end of the reaction process, polyurethanes do not contain any free isocyanates, let alone "highly toxic" isocyanates.

The author does not share the opinion that polyurethanes are problematic commodity plastics. On the contrary, because of their similarity with leather and natural skin, polyurethanes are the ideal starting materials for coating fabrics for use in the apparel industry. Polyurethanes have a wide range of properties, so products can be developed to meet all requirements in this area. They are therefore used in all materials used in the production of garments with high wear comfort.

Since they can be modified to a greater extent than other plastics, polyurethanes should not be lumped together. The polyurethanes used in textile coating are quite different from the rigid polyurethane foam used in refrigerators.

Like wool, leather and polyamide, polyurethanes contain $-CO-NH-$ groups. These substances release toxic gases on combustion in oxygen-depleted conditions (e.g. smoldering). The characteristic data for plastics containing such

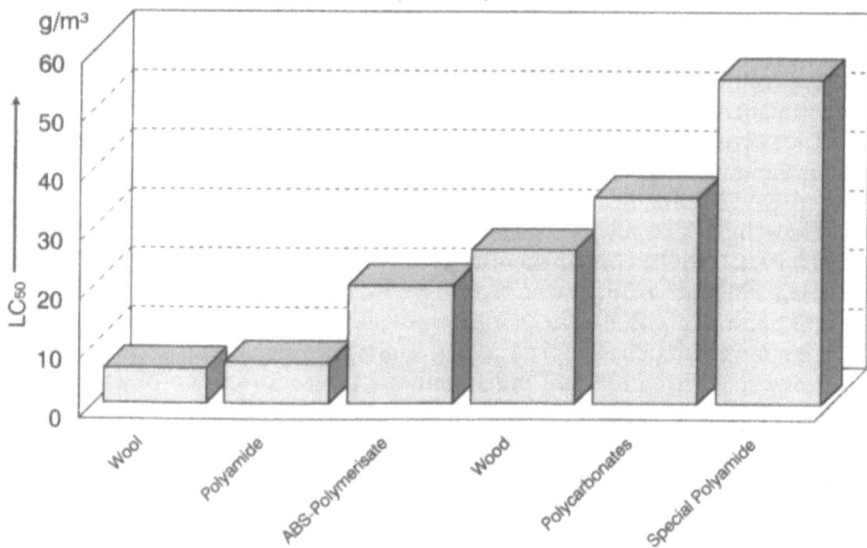

Smoke gas toxicity - material-specific data on the acute inhalation toxicity of thermal decomposition products

Fig. 28-2. Smoke gas toxicity of thermally decomposed polymers under oxygen depleted conditions. The values for polyurethanes are higher (i.e. better) than polyamide and a little lower than polycarbonates (LC_{50} = leathal conc. 50 %)

groups show that the acute inhalation toxicity of the decomposition products released when exposed to heat is lower, or at any rate not higher, with respect to the corresponding product data from natural substances such as wood and wool (see Fig. 28-2) [4].

For a better understanding of this, it is worthwhile looking at the transport classification of dusts and vapors, based on inhalation toxicity, as shown in Table 28-1. The data given are based on a 1-h exposure period [4].

There is not one substance, either natural or synthetic, which meets all the user's requirements as completely unproblematic substance, that when considering all the types of decomposition possible, i. e. hydrolysis, smoldering, etc, does not produce some sort of harmful substances in at least one of those reactions. Although the structure of polyurethanes resembles that of certain natural products which do not cause any particular problems. In any case, the debate about the environmental impact on products should always be based on facts.

Besides polyurethanes, polytetrafluoroethylene (PTFE) was regarded as hazardous. At first Prof. Dr. Höpner researched the burning of materials containing PTFE. It was found that when burning one piece of clothing containing ®Gore-Tex the amount of hydrogen fluoride (HF) liberated required 2000 m³ of air for dilution to render the HF liberated not hazardous for inhalation [7]. The manufacturer of Gore-Tex claims it is safe to dispose of their product with regular household waste. Pyrolized textiles containing PTFE liberate a complex mixture of inorganic compounds. These inorganic products are said to be not toxic [15]. The only toxic substance that may result is carbon monoxide if the burning takes place with a lack of oxygen.

Neither nature nor the synthesis carried out by man is able to produce products resistant to all kinds of end uses and is unproblematic to all kinds of possible decomposition reactions. Polyurethanes having a structure imitating natural products do not offer special problems.

Even if the statement is correct that we are living in a "throw-away society", this society delivers a high standard of living, a standard we like to have. It is important to consider every component and chemical used from the very beginning of the manufacture of a product to the end product with regard to recycling and biodegradability. Shoes or clothing are usually made out of a complex composite of materials and are therefore very difficult to recycle. Although this is true, the materials used in their production while fulfilling the demands of the articles containing them, should still be able to be recycled and/or biode-

Table 28-1. Threshold values (lethal conc. 50 %) for inhalation toxicity of dusts and vapor. Exposure time: 1 h

Transport	
Class	LC_{50} data (mg/l) ($= g/m^3$)
highly toxic	0.5
toxic	0.5 to 2.0
health hazard	2 to 10

graded. For the most part there has been ongoing research in recent years to develop new or improved biodegradable products [20]. Although until now specifically biodegradable products for the coating of textiles has been intensively researched.

References to Part 5

25
The Chemistry of Polyurethanes – Especially for Water Vapor Permeable Products

1. Bayer O (1941) Ann 549:286; Bayer O (1947) Das Di-Isocyanat-Polyadditionsverfahren. Angew. A 59:257; Bayer O et al. (1950) Über neuartige hochelastische Stoffe Vulkollan. Angew 62:57 and (1952) 64:523; Höchtlen A (1950) Kunststoffe aus Polyurethanen. Kunststoffe 40:3 and (1952) 42:403; Saunders JH, Frisch KC (1962) Polyurethanes. In: Chemistry and Technology, High Polymers, Vol XVI, part I, New York; part II (Technology), New York 1964; Müller E (1963) In: Houben-Weyl, Methoden der organischen Chemie, vol. 14/2 Makromolekulare Stoffe, Thieme, Stuttgart, pp 57–68; Rinke H, Istel E (1963) In: Houben-Weyl, Methoden der organischen Chemie, vol. 14/2 Makromolekulare Stoffe, Thieme, Stuttgart, pp 99–191; Dieterich D „Polyurethane, [Poly(carbaminsäure-ester)] Houben-Weyl, Methoden der organischen Chemie, vol E 20, pp 1561–1757
2. Trademark of DuPont
3. Trademark of Bayer AG
4. JA 50 049 402 (Dainichi Seiko Kogyo KK; 7.9.73/2.5.75)
5. Sorenson WR, Campbell TV (1962) Präparative Methoden der Polymerenchemie, Weinheim, p 49

25.1
Polyurethanes

For polyurethanes with fluorine groups, see: 8.1 [27], 25.4 [25, 26, 32]

1. JA 73 10 201 (Sanyo Chemical Ind. Co.; 21.12.66/2.4.73)
2. US 3 536 572 (Goodrich, WT Murphy, FD Stewart, 30.8.67/27.10.70)
3. JA 73 11 824 (Kuraray Co. Ltd.; 9.4.69/16.4.73)
4. GB 989 760 (Bayer; D Prior. 3.3.62; 4.3.63/22.4.65)=FR 1 348 959
5. US 3 823 111 (Inmont Corp.; FCh Loew, E Stone; 13.1.72/9.7.74; cip 25.4.69)
6. JA 73 25 436 (Kuraray Co. Ltd.; 11.11.69/28.7.73)
7. JA 73 11 825 (Kuraray; 11.4.69/16.4.73)
8. JA 75 018 917 and JA 75 018 916 (Kuraray KK; 22.12.70/2.7.75); JA 75 006 226 (Kuraray KK; 21.12.70/12.3.75)
9. OS 2 409 789 (Akzo GmbH; H Brahm; 1.3.74/11.9.75)
10. JA 51 63 682 (Dainippon Ink & Chem KK; 12.12.91/29.6.93)=JA 91–328 849; JA 52 62 847 (Dainippon Ink & Chem KK; 18.3.92/12.10.93)=JA 92–62 114; JA 41 78 417 (Dainippon Ink & Chem. KK; 13.11.90/25.6.92)=JA 90–306 679; JA 41 78 416 (Dainippon Ink & Chem. KK; 13.11.90/25.6.92)=JA 90–306 678
11. JA 60 047 038 (Dainippon Ink & Chem. KK; 24.8.83/14.3.85)=JA 83–153 093

25.2
Polyurethane-Polyureas

1. JA 72 00904 (Mitsubishi Rayon Co. Ltd.; 23.10.67/11.1.72)
2. JA 74 049 (Toyo Spinning; 3.4.65/25.12.74)
3. JA 72 00 906 (Mitsubishi Rayon Co. Ltd.; 31.10.67/11.1.72)
4. FR 2 014 588 (Glanzstoff; D Prior. 13.7. 68; 10.7.69/17.4.70)=NE 6 910 606=DOS 1 770 884
5. US 3 752 700 (Continental Tapes Inc.; R Dahl; 22.3.72/14.8.73; div. from 29.6.70)
6. JA 74 010 560 (Kao Soap; 17.6.69/11.3.74)
7. OS 2 355 073 (Chemie-Anlagenbau Bischofsheim GmbH; KH Hilterhaus; 3.11.73/7.5.75)
8. SU 1 825 504 (Polymersintez Res. Prodn. Assoc.; EL Korzyuk, FK Samigullin, AA Zaplatin; 22.2.91/30.7.94)
9. JA 79–46 295 (Toyo Cloth; 20.9.77/12.4.79)=JA 77–113799

25.3
Polyurethane-Polyhydrazocarbonamides

1. OS 2 115 419 [Asahi Kasei Kogyo KK; Y Nakahara, J Ohmura; JA Prior. 30.3.70 (=JA 45–25 912) 30.3.71/ 7.10.71]
2. JA 73 14198 (Kuraray Co. Ltd.; 16.5.69/4.5.73)
3. JA 73 29 559 (Kuraray Co. Ltd.; 16.5.69/11.9.73)
4. OS 1 952 394 (Bayer; W Thoma et al.; 17.10.69/29.4.71)
5. DAS 1 301 569 (Teijin Ltd.; E Negishi; JA Prior. 2.11.64; 2.11.65/21.8.69)
6. OS 1 900 515 (Bayer; H Träubel; 7.1.69/6.8.70)
7. DFG, „MAK- und BAT-Werte Liste 1997", Weinheim 1997, p 108 (hydrazine [302–01–2])
8. JA 85–130 616 (Kuraray; 13.12.83/12.7.85)=JA 83–238 093
9. DOS 2 905 185 (Toray Ind. Inc.; M Yagi et al.; 10.2.79/23.8.79; JA Prior. 13.2.78)=JA 78–14 047

25.4
Modifications in the Manufacture of Polyurethanes

See also: 23.1 [10, 23]
1. DDR 97 662 (Deutsches Lederinstitut; K Walter et al.; 16.3.71/12.5.73)
2. US 3 823 111 (Inmont Corp.; FCh Loew, E Stone; 13.1.72/9.7.74; cip. 25.4.69)
3. US 3 594 351 (Hitachi Chemical; Sh Uchida, Y Sone; JA Prior. 10.4.67; 26.3.68/20.7.71) =FR 1 567 338=JA 67/22 419 and JA 67/22 422
4. DOS 2 124 042 (Kyowa Hakko Kogyo Co. Ltd.; Y Fujinomoto, M Teranishi; JA Prior. 20.4.70; 14.5.71/2.12.71)=JA 42 41 070
5. US 3 729 366 (Kyowa Hakko Kogyo Co. Ltd.; Y Fujinomoto et al.; JA Prior. 22.11.67; 15.1.71/ 24.4.73; cip 18.11.68)=JA 74 70 967=DOS 1 810 72
6. OS 1 949 090 (Ajinomoto; A Akamatsu et al.; JA Prior. 30.9.68; 29.9.69/16.4.70)=JA 71 01 668
7. JA 49 080 205 (Tsunoda Kagaku; 7.12.72/2.8.74)
8. JA 48 034 246 (J Tanaka; 6.9.71/17.5.73)
9. OS 2 044 838 (Bayer; HD Winkelmann et al.; 10.9.70/16.3.72)
10. OS 2 364 938 (Kanebo Ltd.; K Asano; JA Prior. 28.12.72 (=JA 73 01 648); 28.12.73/11.7.74)
11. JA 73 5 892 (Toyo Spinning Co. Ltd.; 19.11.70/21.2.73)=JA 70–101 184
12. OS 1 964 820 (Kurashiki Rayon Co. Ltd.; M Tanomura, K Itoi; JA Prior 16.12.68, 19.12.69 and 15.3.69; 16.12.69/9.7.70)=JA 92 525 68, JA 93 598 68 and JA 19 748 69
13. DDR 60 146 (F Haobas et al.; CZ Prior. 17.3.66; 7.3.67/5.2.68)
14. JA 41 427 72 (Toyo Rubber Ind. Co. Ltd.; 25.8.67/19.10.72)
15. OS 2 243 628 (Bayer; K Wagner, HJ Müller; 6.9.72/14.3.74)
16. DOS 1 619 284 (DuPont; TV Peters; US Prior. 6.4.61; 27.3.62/2.10.69) addition to 1 520 488= BE 616 044=FR 1 322 442=GB 993 758=OE 250 016=SZ 440 706 and SZ 444 810=US 3 180 853
17. GB 1 035 310 (Du Pont; US Prior. 30.12.63; 16.12.64/6.7.66)=BE 657 722=NE 6 415 258
18. JA 48 075 704 (Hanny Chem. Co. Ltd.; 7.1.72/12.10.73)

19. GB 1 172 325 (Kurashiki Rayon KK; JA Prior. 10.2.67 (=JA 8 691/67); 5.2.68/29.11.69)
20. JA 59 100 778 [Kuraray KK; 30.11.82/11.6.84 (210 852)]
21. JA 59 066 577 [Kuraray KK; 30.9.82/16.4.84 (172 953)]
22. Hayashi S, Ishikawa N, Giordano C (1993) High moisture permeability polyurethane for textile application. J Coated Fabr July, pp 74–83
23. JA 071 50 478 (Kuraray Co. Ltd.; 1.12.93/13.6.95)
24. JA 071 65 856 (Sanyo Chem. Ind. Ltd.; 15.12.93/27.6.95)
25. JA 072 92 058 (Japan Energy Corp.; 28.4.94/7.11.95)
26. JA 072 92 239 (Sanyo Chem. Ind. Ltd., 28.4.94/7.11.95)
27. JA 061 45 598 (Dai Nippon Ink & Chem KK; 13.11.92/24.5.94)
28. JA 50 09 256 (Kuraray Co. Ltd.; 1.7.91/19.1.93); JA 50 65 324 (Kuraray Co. Ltd.; 4.9.91/19.3.93)
29. JA 51 12 713 (Kuraray Co. Ltd.; 18.10.91/7.5.93)
30. JA 51 56 151 (Kuraray Co. Ltd.; 6.12.91/22.6.93)=JA 91–0 348 544
31. DOS 2 912 864 (Sanyo Chemical; I Tanaka, TN Fujii; 31.3.78/15.11.79)= JA 78–38 640
32. JA 52 62 846 (Dainippon Ink & Chem. KK; 18.3.92/12.10.93)=JA 92–62 112
33. JA 56 026 081 (Kuraray KK; 3.8.79/13.3.81)=JA 79–99 768

25.5
Polyurethane Dispersions

1. DOS 2 314 512 (Bayer AG; K Noll; 23.3.73/17.10.74)
2. Dieterich D, Keberle W, Witt H Angew 82:53
3. Dieterich D, Reiff H (1972) Polyurethan-Dispersionen durch Schmelz-Dispergierverfahren. Angew Makrom Chem 26:85
4. BE 639 107 (Bayer AG; D Dieterich, O Bayer, J Peter; D Prior. 26.10.62; 24.10.63/17.2.64)
5. BE 653 223 (Bayer AG; D Dieterich, O Bayer; D Prior. 19.9.63; 18.9.64/18.1.65) and NE 6 410 928
6. US 2 698 575 (DuPont; JE Mallonee; 25.5.54/17.1.67; cip 30.6.53)
7. BE 660 518 (Wyandotte; SL Axelrod; US Prior. 2.3.64)
8. BE 663 102 (Wyandotte; RCh Dawn, NS Nichols; US Prior. 27. 4.64)
9. JA 063 13 024 (Daicel Chem. Ind. Ltd.; 4.3.93/8.11.94)
10. JA 063 16 877 (Asahi Kasei Kogyo KK; 30.4.93/15.11.94)
11. JA 4189813 (Dainippon Ink & Chem. KK; 22.11.90/8.7.92)=JA 90–316 024
12. DOS 3 441 934 (Bayer AG; W Schäfer et al.; 16.11.84/28.5.86); see (1986) Chem Abstr 105:154 824
13. Coogan RG (1997) Post-crosslinking of waterborne urethanes. Prog Org Coat 32(1–4), pp 51–63
14. Zorn B (1990) Aqueous polyurethane top coats. Leder und Häutemarkt No 8, p 21

26
Test Methods

1. IUP 15 published (1961) Das Leder 12:86
2. Schröer W, Schütze D-I, Thoma W (1992) Wasserdampfdurchlässige kompakte Textilbeschichtungen mit Polyurethanen. Coating No 9, p 290
3. BTTG (1993) Neue Testmethoden für Schutzkleidung. Textilveredlung 28, No 11, p 382
4. Woodbridge T (1993) Breathability – facts and fiction. The Nonwoven World, Fall, pp 59–68
5. Mecheels J (1986) Leistungsfähig und komfortabel in funktioneller Kleidung. Textilveredlung 21, No 10, pp 323–330; Umbach KH (1993) Wirkerei und Strickerei Technik 43:108; Umbach KH (1993) Feuchtetransport und Tragekomfort in Mikrofaser-Textilien. Melliand No 2, pp 174–178
6. Weder M (1990) Wasserdampfdurchgang an Regenschutzbekleidung auch unter Beregnung?. Textilveredlung 25:31
7. Kannekens A (1994) Breathable coatings. J Coated Fabr July, pp 51–59, Kannekens A (1994) Breathable coatings and laminates. 4th Int. Congr. on Textile Coating and Laminating, Zurich, 8–9 November

8. Weder M (1995) Schwitzarm. Melliand No 7–8, p 509–514
9. Kurz B (1992) Messung des Mikroklimas im Schuh zur quantitativen Beurteilung des Tragekomforts am Beispiel von Kinderstiefeln. Schuh-Technik 456
10. Anon. (1996) Footwear breathability and sweat management. World Leather October, p 67
11. Gibbesch B, Schedlitzki D (1996) Wasserdampfdurchlässigkeit von organischen Materialien für den Korrosionsschutz. Kautsch Gummi Kunstst 49(6):452
12. Sehm H, Teubner U, SpRÖßig P (1996) Erfassung der bekleidungshygienischen Eigenschaftenn von Textilien mit der Infrarot-Thermographie. Melliand No 10, pp 698–699
13. Anon. (1997) Cybor ein neues Testgerät von Gore. Leder und Häutemarkt March, p 12
14. Döring E (1997) Untersuchung zur Porosität von Leder und Kunstleder. Das Leder 48
15. Chu MS, Kato T, Kamata Y, Nakajima T (1994) Experimental and numerical analysis of clothing microclimate. ISF 94 Proc of the Int Symp on Fiber Sci and Technol Yokohama, 26–28 October
16. Ushioda H, Nakajima T (1994) The influence of water absorption characteristics of clothing fabrics for human soaking sense. ISF 94 Proc of the Int Symp on Fiber Sci and Technol Yokohama, 26–28 October
17. Bartscher R (1996) Funktionsunterwäsche – ein Vergleichstest. Outdoor No 6, pp 50–54, 54–63
18. Gretton JC, Brock DB, Dyson HM, Harlock SC (1996) A correlation between test methods used to measure vapour transmission through fabrics. J Coated Fabr 25:301
19. Pause B (1996) Measuring the water vapor permeability of coated fabrics and laminates. J Coated Fabr 25:311
20. Fung W, Parsons KC (1996) Some investigations into the relationship between car seat cover materials and thermal comfort using Human Subjects. J Coated Fabr 26:147
21. Gretton JC, Brook DB, Dyson HM, Harlock SC (1997) The Measurement of moisture vapour transmission through simulated clothing systems. J Coated Fabr 25:212
22. Gibson P, Kendrick C, Rivin D, Sicuranza L (1995) An automated water vapor diffusion test method for fabrics, laminates and films. J Coated Fabr 24:p 322
23. Pause B (1996) Measuring the water vapor permeability of coated fabrics and laminates. Ann Int Conf Text Coat Laminating, 6th paper (Leipzig); see (1998) Chem Abstr 127: 359 877 r
24. Mayer W, Mohr U, Schuirer M (1989) High-Tech Textilien, der Beitrag der Ausrüstung am Beispiel der funktionellen Sport- und Freizeitbekleidung. ITB Textilveredlung No 2, pp 16–32; Day M, Sturgeon PZ Water vapor transmission rates through textile materials as measured by differential scanning calorimetry. Textile Res J 56 (1986) No 3, pp 157–161
25. Gretton JC, Brook DB, Dyson HM, Harlock SC (1997) The measurements of moisture vapour transmission through simulated clothing systems. J Coated Fabr 25 (January), pp 212–220
26. Ruckman JE (1997) An analysis of simultaneous heat and water vapour transfer through waterproof breathable coatings. J Coated Fabr 26 (April), pp 293–307

26.1
Test Results

The compilation of the test values and methods was done by Mrs Ch Kaspar
1. Trademark of Bayer AG
2. Vereinbarung zwischen dem Verband der deutschen Lederindustrie (Leverkuser Str. 20, Frankfurt/M) und dem Hauptverband der deutschen Schuhindustrie (Waldstrasse 44, Offenbach/M)
3. Vereinbarung zwischen dem Verband der deutschen Lederindustrie (Leverkuser Str. 20, Frankfurt/M) und dem Verband der deutschen Polstermöbelindustrie (Enger Strasse 4b, Herford)
4. Bunten J (1975) Anforderungen an PU beschichtete Gewebe. J Coated Fabr 35; Bluestein C (1976) Polyurethan-Textilbeschichtungen. Advan Urethane Sci Technol 224
5. Rossi R (1997) Bekleidungsphysiologische und Schutzaspekte von Hitzeschutzkleidung nach thermischer Alterung. Textilveredlung 32:187

27
Other Industrial Applications

See also: 10.1 [21]; A survey of the use of membranes in reverse osmosis, separation of gases, dialysis etc. Staude E (1992) Membranen und Membranprozesse. Weinheim; Sedláček B, Kahovec J (1987) Synthetic polymeric membranes. Berlin; The use of membranes in medicine, see, e. g., (1985) Chiliellini E, Giusti P, Migliaresi C, Nicolais L (eds) Polymers in medicine II, New York, p 247; Piskin E Characterisation of membranes for artifial organs; The use of membranes as filters see, e.g., Matteson MJ, Orr C (eds) Filtration: principles and practices, New York, Basel, p 560 ff; Houghton J, Arnand SC, Purdy AT (1997) Characterisation of fabrics used for wet filtration. Textile Asia July, pp 59–66; Sirkar KK (1997) Membrane separation technologies: current developments. Chem Eng Commun Amsterdam 157, pp 145–184

1. NFA dated 19 July, 1995
2. Peinemann K-V (1995) Stofftrennung mit porenfreien Kunststoff-Filmen. Spektrum der Wissenschaft August, pp 88–92
3. Bhave, RR (1991) Inorganic membranes: synthesis, characteristics and applications. New York
4. JA 6 3109 046A (Daiichi Lace KK; 27.10.86/13.5.88)
5. JA 6 2006 903A (Diya Gum KK; 2.7.85/13.1.87)
6. EP 0 039 184 (WL Gore & Ass.; DJ Golke 16.4.81/4.11.81; US Pior. 22.4.80)
7. WO 9 000 643-A (WL Gore & Ass.; 6.7.88/25.1.90)
8. WO 9 008 218-A (WL Gore & Ass.; 17.1.89/26.7.90)
9. US 4943 475 (Membrane Techn.; 23.7.86/24.7.90)
10. JA 6 2206 083 (Teijin KK; 6.3.86/10.9.87) and JA 1 156 580 (86/20.6.89)
11. JA 2 035 842 (Teijin KK; 9.8.85/16.2.87)
12. Silver JA, Cooper SL (1994) Effekt der Polyolmolgewichte auf die Blutverträglichkeit von Polyurethanen mit Polyethylenoxidpolyol-Segmenten. Biomaterials 15:695
13. JA 06 70 804 (94–70,804) (Japan Gore Tex Inc.;29.8.92/15.3.94)
14. Hinrichs WLJ (1993) Artificial skin. Nat Tech 708; Chem Abstr 121:65349g
15. US 5 346 788 (WR Grace & Co.; VSC Wang et al.; 1.4.93/13.9.94)
16. JA 06 116 867 (94 116 867) (Teijin Ltd.; M Nakayama, A Matsunaga, K Koga; 19.8.92/26.4.94)
17. Kwon SK (1994) Studies on the polyurethane diagnostic membrane for diabetes 2: Effect of additives in membrane formulation of urine glucose. Pollima (Seoul) 18:1055; Chem Abstr 122:75652 k
18. Wu P, Gaylor JDS (1994) A model of water vapor transmission in hydrocolloid wound dressings. J Membr Sci 27; Chem Abstr 122:89311 k
19. DOS 4 241 479 (Bayer; H Hugl et al.; 19.12.92/16.6.94)
20. EP 637 452 [Y. Shimizu; 6.8.93/8.2.95 (=JA 93/195 755)]
21. WO 94 22 432 (Rexham Ind.; GD Gregory; 7.4.93/13.10.94)
22. JA 07 53 764 (95 53,764) (Asahi Chem. Ind.; H Yoneda; 17.8.93/28.2.95)
23. WO 95 13,860 (CJM van Rijn; 12.11.93/26.5.95)
24. EP 638 352 (WL Gore & Ass.; T Wiemer, F Gruber; 9.8.93/15.2.95)
25. JA 07 126 342 (95 126 342) (Saint Gobain Vitrage; Asahi Glass Co. Ltd.; S Kondo, H Watanabe, K Yoshida, H Shimoda; 21.10.93/16.5.95)
26. EP 633 277 (Takeda Badische Urethane Ind. Ltd.; T Hirono, S Higashi, Y Suzuki; 5.7.93/11.1.95)=JA 93/192 051
27. US 5 428 123 (The Polymer Technology Group; RS Ward, KA White, 24.4.92; cip 27.6.95)
28. JA 5 3 025–280 [Ebara Inflilco KK; 10.8.76/8.3.78 (JA 095 272)]
29. GB 2 290 031 (Seton Healthcare Group PLC, UK; GJ Collyer, PA Gray; 8.6.94/13.12.95)
30. Powell N Membrane microfiltration for municipal water treatment. Membrane Technology No 71, pp 7–9
31. Weder M (1996) Überprüfung einer Schlafsackevaluation durch Praxisversuche. Textilveredelung 31 No 1/2, pp 35–36
32. JA 062 28 431 (Dainichiseika Color & Chem. Mfg.; 23.3.87/16.8.94)
33. JA 080 13 352 (Unitika Ltd.; K Kamemaru, M Shinomya, 19.6.94/16.1.96)

34. Sevatianov VI, Drushilak IV, Eberhart RC, Kim SW Blood compatible biomaterials: hydro-philicity vs. hdrophobicity. Macromol Symp 1996, 103 (Polymers and Medicine), pp 1–4
35. Mastura Raheel (1996) Modern Textile Characterisation Methods p 395
36. JA 07 313 585 (Mitsubishi Kagaku KK; T Kuroyanagi, M Tsunoda; 24.5.94/5.12.95
37. DOS 4 428 304 (H von Borries; 10.8.94/15.2.96)
38. Sepai O, Henschler D, Czech S, Eckert P, Sabbioni G (1995) Exposure to TDA from PU covered breast implants. Toxicl Lett 77:371
39. EP 708 212 (E Doerken AG; J Fischer, K Urban, D Jablonka; 20.9.95/24.4.96)
40. PCT WO 96 09 165 (Exxon Chem. Pat. Inc.; LC Wadsworth, N Gosavi; 20.9.94/28.3.96)= US Appl. 309 841
41. DDR 301588 (Bundesamt für Wehrtechnik und Beschaffung; Forschungsinstitut für Leder und Kunstleder; S Fourier; D Harzer, M Herrmann, G Jurthe, G Reich, H Schoenfeld, J Semmer; 8.11.84/8.4.93); DDR 301 589 (Bundesamt für Wehrtechnik und Beschaffung; Forschungsinstitut für Leder und Kunstleder; S Fourier, D Harzer, M Herrmann, G Jurthe, G Reich, H Schoenfeld, J Semmer; 8.11.84/8.4.93); DDR 301 590 (Bundesamt für Wehrtechnik und Beschaffung; Forschungsinstitut für Leder und Kunstleder; S Fourier, D Harzer, M Herrmann, G Jurthe, G Reich, H Schoenfeld, J Semmer; 8.11.84/8.4.93)
42. JA 81 57 637 (Unitika Ltd.; Y Yabushita, H Yokoi, S Sakai; 6.12.94/18.6.96)
43. JA 81 64 590 (Matsushita Reiki KK; Mitsubishi Jukogyo KK; Komatsu Seiren KK; K Nagata, M Inatani, S Kayashi, S Kondo, M Yamaguchi, A Okuya; 13.12.94/25.6.96)
44. PCT WO 96 20 040 (Gore Hybrid Technologies Inc.; SL Mish, PD Drumheller; 23.12.94/ 4.7.96)
45. DOS 4 218 216 (Enviromental Technologies Europa Ltd.; H Fuchs, W Knaupe, F Markert; 11.2.92 + 3.6.92+24.8.92/17.6.93)=EP 625 962
46. Paul D, Malsch D, Bossin G, Wiese E, Thomaneck U, Brown GS, Heinz W, Falkenhagen D (1990) Chemische Modifizierung von Cellulose Membranen und ihre Blutverträglichkeit. Artif. Organs 14:122
47. WO 88 01 877 (CWG Ansell; UK Prior. 6.12.86+16.1.87+22.1.87/24.3.88)=GB 86–29 231 + GB 87–943+GB 97–944+87–1 434
48. Müller FJ, Krieger W, Kissing W, Reiner R (1981) Reduction of membrane fouling in reverse osmosis by means of surface modification of the membranes. Fundam Appl Surf. Phenom Ass Fouling Clean Food Process, Proc Int Workshop, pp 343–347
49. Gottwald L (1996) Wasserdampfdurchlässige PUR Membranen für wetterfeste Laminate. J Coated Fabr 25:168
50. JA 81 68 658 (Toray Ind.; M Kurihara, Y Fusaoka, T Ikeda; 20.10.95/2.7.96)=JA 95–273188
51. JA 82 15 521 (Kanebo Ltd.; Y Mizukami, T Tejima, K Agari, Y Fukumoto, Y Tanaka; 10.2.95/ 27.8.96)=JA 95–45 101
52. EP 576 267 (Sumitomo Chem. Co. Ltd.; S Sembo; JA Prior. JA 92–168897; 26.6.92/29.12.93)= JA 6141724; JA 53 36 858 (Sumitomo Chem. Co. Ltd.; 7.4.92/21.12.93)=JA 92–85350
53. DDR 299 520 (Forschungsinstitut für Leder & Kunstleder; A Bauch, G Feigel, G Hebestreit, U Loose, R Steinhardt; 22.12.87/23.4.92)=DDR 87–310 990
54. JA 42 45 980 (Daiwabo Create Co. Ltd.; 28.1.91/2.9.92)=JA 91–60 829
55. Greenwald E (1992) New fabrics for the medical market. The 1992 Industrial Fabric and Equipment Exposition, Medical Textiles, Phoenix, USA, pp 128–134
56. WO 94 18 242 (W Budinger; US Prior. 5.2.93/18.8.94)=US 93–14 018=US 5 384 337=EP 682 673
57. US 5 354 336 (WJ Ledergerber; 22.2.91/11.10.94)=US 91–660291
58. EP 289 859 [PPG Ind. Inc.; J Young, DD Leatherman; 24.4.87 (US 87–42 404)/9.11.88]
59. Corain B, Jerabek K (1996) Macro- and microporous synthetic organic supports in indus-trial catalysis. Chim. Ind. (Milan) 78:563, 567
60. JA 71 32 573 (Araco KK; Kyowa Leather Cloth Co. Ltd.; 11.11.93/23.5.93)=JA 93–305 832; JA 61 84 950 (Kyowa Leather Cloth Co. Ltd. 9.12.91/5.7.94)=JA 91–350 261
61. US 5 114 438 (PPG Ind. Inc.; DE Adams, GA Brons, DD Leatherman, JJ McGinley; 29.10.90/ 19.5.92)=US 90–605 283
62. F Chu, Kimura Y (1996) Structure and gas permeability of miocroporous films prepared by biaxial drawing of beta-form polypropylene. Polymer, 37:573; Zhu W, Zhang X, Zhao C, Wu W, Xu M (1996) A novel polypropylene microporous film. Polym for Dav Technol Sussex, 7, No. 9, pp 743–748

63. Mehnert L, Bernstein U (1996) Mikroporöse Strukturen und ihre Anwendung für Schutz-bekleidung. Melliand, No 12, pp 870–872

64. Huang S-L, Ruaan R-Ch, Lai J-Y (1997) Gasdurchlässigkeit von PU- Membranen auf Basis Kupferion enthaltender, hydroxyterminierten Polybutadiene. J Membr Sci 71

65. WO 96 37 668 (Akzo Nobel N.V.; HJM van de Ven, E Maderek, JCW Spijkers; 22.5.95/28.11.96)=DE Appl. 19 518 686; WO 96 37 665 (Akzo Nobel N.V.; HJM van de Ven, E Maderek, JCW Spijkers; 22.5.95/28.11.96)=DE Appl. 19 518 684

66. EP 289 859 (PPG Ind. Inc.; J Young, DD Leatherman; 24.4.87/9.11.88)=US 87–42 404

67. DOS 19 518 624 (Akzo Nobel N.V.; H Roettger, F Wechs; 24.5.95/21.11.96)

68. JA 84–125978 (Toray; 30.12.80/20.7.84)=JA 82–234 783

69. Horiuchi K, Yamaguchi T, Murata T, Tanioka A (1996) Computational simulation for the pro-cess of phosphate elimination by hemodialysis using hollow fibers. Sen i Gakkaishi 52: 566

70. Anon. (1997) Lyocellfasern in der Nähwirktechnologie, Technische Textilien, 40 (April), p 99; Anon. (1997) Courtaulds Lyocell – new developments. Chem Fibers Int 47 (April), p 127; Ortlepp G, Beckmann E, Mieck K-P (1997) Fibrillated lyocell filament yarn – a basis for new yarn structures. Chem Fibers Int 47 (April), p 129; Mieck K-P, Nicolai M, Nechwatal A (1997) Zum Veredelungsverhalten von Lyocell-Geweben. Melliand, No 5, pp 336, 338; Fritsche W (1997) Verarbeitungsversuche mit Lyocell-Fasern nach dem Nähdruckverfah-ren. Melliand No 5,p 320

71. Wehlmann U (1997) Reinigen von Abwasser aus der Textilveredlung mit Membranver-fahren. Melliand No 4,p 249

72. Franken T Hydrophilic membranes in biomedical applications. Membrane Tech, No 82, p 6

73. JA 08 311 779 (Achilles Corp.; S Kazuhiro, O Katsumi; 15.5.95/26.11.96)=JA 95–139 993

74. Ulbricht M, Papra A Polyacrylonitrile enzyme ultrafiltration membranes by adsorption, crosslinking and covalent binding. Enzyme and Microb Technol 20(1):61

75. Marques MJA, Silva MEC (1996) Properties of nonwovens in the healthcare industry. Tecnitex 96, Expo 2000, Torino, pp 119–124

76. WO 96 40 072 (Alkermes Controlled Therapeutics; OL Johnston, MM Ganmukhi, H Bern-stein, H Auer, MA Khan; US Prior. 7.6.95/19.12.96)=US 95–477 725; WO 96 40 073 (Alkermes Controlled Therapeutics; SE Zale, PA Burke, H Bernstein, A Brickner; US Prior. 7.6.95/19.12.96)=US 95–478 502

77. US 5 605 750 (Eastman Kodak Co.; ChE Romano, DE Bugner, WT Ferrar; 29.12.95/25.2.97)=US 95–580 698

78. Anon. (1997) Collagen improves wound dressing properties. Medical Textiles, February, p 6

79. Stelkens A, Mackenbach H (1995) Überblick über Textilien in Medizin und Hygiene – Anfor-derunsgprofile. Aachener Textiltagung, 29–30 November; in (1996) DWI Reports 117:65

80. Schwertfeger A, Hoffmann G, Offermann P (1995) Entwicklung von Vandalismus-Schutz-Textilien für den Bereich öffentlicher Verkehrsmittel. Aachener Textiltagung, 29–30 Novem-ber in (1996) DWI Reports 117:227

81. JA 09 300 504 (Kao Corp.; S Sato; 16.5.96/25.11.97)=JA 96–121 612

82. JA 93 1 863 (Achilles Corp.; K Sugaya, K Oosawa; 10.7.95/4.2.97)=JA 95–196 983

83. Kuroyanagi Y (1997) Artificial skin composed of cultured cells and matrix. Nessho 23(1):9; (1997) Chem Abstr 127:9003 h

84. DOS 19 536 033 (Oxyphen GmbH; HB Lueck; 28.9.95/10.4.97)

85. US 5 415 924 (Aquatic Design Inc.; DJ Herlihy; 3.5.95/20.5.97)=US 95–433 567

86. WO 97 15 378 (WL Gore & Associates GmbH; A Bauer, R Leckenwalder; 27.10.95/1.5.97)=DE Appl. 19 540 141

87. Anon. (1997) Lyocell-based fibre replaced alginate in wound dressing. High Performance Textiles 4:2

88. Carroll TR (1995) New advances in high performance composites for the protective clothing market. J Coated Fabr 24:313

89. Gurian M (1995) An evaluation of the effectiveness of anti-microbial finishes and additives to healthcare interior textiles. J Coated Fabr 25:13

90. Pause B (1995) Developement of heat and cold insulating membrane structures with phase change material. J Coated Fabr 25 (July), p 59

91. Anon. (1997) Biocompatibility promotes membrane dialysis function. Medical Textiles May, p 6; reports about the US 5 505 890 of Akzo

92. Bao YH, Xu DN (1997) Hydrophilic polyurethane groups and their application on roller – compacted concrete dams. Polym Concr Proc 2nd East Asia Symp, London, pp 179–188
93. Ukpabi PO (1998) Polyurethane membranes for surgically grown applications (Infectious Fluids) Diss Abstr Int B 58(58), 2104; UMI Order No DA9 727 883
94. JA 92 62 278 (Japan Synthetic Rubber Co. Ltd.; A Morikawa, K Shiho, N Kawahashi, Y Yamakawa, K Kuroda; 29.3.96/7.10.97)=JA 96–103 262
95. Doi K, Matsuda T (1997) Significance of porosity and compliance of microporous polyurethane based microarterial vessel on neoarterial wall regeneration. J Biomed Mater Res 37:4
96. US 5 674 523 (New Dimensions in Medicine Inc.; JV Cartmell, WR Sturtevant, ML Wolf; 1.9.95/7.10.97)=US 95–523 009
97. US 3 330 791 (Reeves Brothers; CE Matev, GC Wert; 16.12.63/11.7.67)
98. US 5 688 855 (SKY Polymers Inc.; VA Stoy, GA Gontarz; 1.5.95/18.11.97)=US 95–434 573
99. EP 808 859 (Sika AG; vorm. Kapsar Winkler & Co.; K Bosch, U Stadelmann-Sidler, TA Burge; 20.5.96/26.11.97)
100. Poole-Warren LA, Martin DJ, Schindhelm K, Meijs GF (1997) Polymeric biomaterials. Mater Forum 21:241; Nina NMK, Woodhouse KA, Cooper SL (1997) Polyurethanes in biomedical applications. CRC Boca Raton
101. JA 09 308 523 (Daichi Lace Mfg. Co.; T Tachibana; 23.5.96/2.12.97)=JA 96–153 128
102. EP 812 948 (Toray Ind. Inc.; J Tabata, T Kamaya, M Hirata, K Saito, K Hori; 14.2.97/17.12.97)=JA 97–30 106
103. PCT 97 46 267 (Gore Enterprise Holdings Inc.; AD Cook, PD Drumheller; 11.12.97/30.5.97)=US 97–865 800; PCT 97 46 590 (Gore Enterprise Holdings Inc.; D Drumheller; 11.12.97/27.5.97)=US 97–863 263
104. T Hattori, T Yashima (eds) (1993) Studies in surface science and catalysis, Vol. 83: Zeolites and microporous crystals, Proceedings of the International Symposium on Zeolites and Microporous Crystals, Nagoya, 22–25 August, Amsterdam, 1994
105. Andrade JD (1985) Contact angle analysis of biomedical polymers: from air to water to electrolytes. In: Chiellini E, Giusti P, Miglaresi C, Nicolais L (eds) Polymers in Medicine II. New York, London, p 38
106. Tanzi MC, Albonico P, Barozi C, Bolognesi A, Fumero R, Tieghi G (1985) Heparinizable segmented polyurethanes for cardio-vascular application. In: Chiellini E, Giusti P, Miglaresi C, Nicolais L (eds) Polymers in Medicine II. New York, London, pp 91–115; Cooper SL, Lelah ML, Grasel TG (1985) Characterization of polyurethanes for blood-contacting applications. In: Chiellini E, Giusti P, Miglaresi C, Nicolais L (eds) Polymers in Medicine II. New York, London, pp 199–215. Piskin E (1985) Characterization of membranes for artificial organs. In: Chiellini E, Giusti P, Miglaresi C, Nicolais L (eds) Polymers in Medicine II. New York, London, pp 247–252
107. Lingnau G (1998) Reines Silber unter sieben Hüllen. Frankfurter Allgemeine Zeitung, 31 March, p 7
108. PCT WO 98 08 884 (Tyndale Plains – Hunter Ltd.; MH Reich, K Nelson, J Kusma; 26.8.97/5.3.98; US Prior. 26.8.96)=US 97–40 094
109. PCT WO 98 01 166 (Innovative Technologies Ltd.; T Grocott et al.; 4.7.96/15.1.98; GB Prior. 4.7.96)=GB 96–14 034
110. DOS 19 732 994 (Scapa Group PLC; P Wroblewski; F Doran; 9.1.96/12.2.98)=US 96–694 791
111. US 5 707 526 (Kraus-Menachem Israel; M Kraus, J Yocab; 8.8.95/13.1.98)=US 95–512 446
112. JA 10 24 098 (Nippon Sherwood KK; T Kikuchi; 12.7.96/27.1.98)=JA 96–182 927

28
Ecology

For textiles, see: e.g. Peter M, Rouette HK (1989) Grundlagen der Textilveredlung, 13th edn. Frankfurt, pp 854–902

1. (1993) Konsum und Umwelt No 3
2. Oertel G (1993) In: Braun/Becker, Kunststoff-Handbuch, Bd 7. Polyurethane. München, Wien

3. „… certainly true scepticisms keeps its value in a world where the beginning and the end is unknown and its center always is in a steady movement", Burckhardt J Weltgeschichtliche Betrachtungen, Suttgart 1949, p 10

4. Kimmerle G, Paulike J, Prager FH (1992) Rauchgastoxizität von Kunststoff-Verbrennungs- und Verschwelungsprodukten. Kunststoffe 82:1175

5. v. Meysenbug C-M (1978) Die Beständigkeit von Kunststoffen und Gummi. Kunst 68, No. 4, p 251

6. Fabris HJ (1978) Die thermische und oxidative Stabilität von Urethanen. Advan Urethane Sci 6:173

7. (1989) SPD: Technische Textilien belasten die Umwelt. Frankfurter Allg Zeit, 3 April

8. DOS 4 234 335 (BASF Schwarzheide; 12.10.92/14.4.94)

9. Bohnhoff A, Goetz Ch, Klingenberger H (1996) Thermische Verwertung von textilen Bodenbelägen. Melliand, No 2,p 120

10. Wallace GK Recycling nonwovens. Tappi Journal 79, p 215

11. Kettemann B-U, Melchiorre M, Münzmay T, Raßhofer W (1995) Recycling von verunreinig-tem PUR. Kunststoffe 85:1947

12. Riedel B, Seyfath E, Taeger E (1996) KFZ-Schalldämmstoffe durch chemisches Recycling der PUR-Weichschäume aus Autositzen. Techn Textilien 39:30

13. JA 08 002 517 (Shiseido Co. Ltd.; Showa Highpolymer Japan; A Torii, T Watanabe; 17.6.94/9.1.96)=JA 94–136 073

14. DOS 4 236 446 (Herfeld GmbH & Co KG; Hornschuch AG; M Derksen, W Kammerer, B Peters, R Schulze-Kadelbach, B Kammerer; 28.10.92/5.5.94)=WO 94 09 959=EP 620 776

15. Anon. (1996) Keine erhöhte Gefahr bei mit Fluopolymerisaten ausgerüsteten Textilien. Melliand, No 5, p 325

16. Kimura Y, Yamaoka T, Ueda T, Kim S (1996) Abbaumechanismus von Azogruppen ent-haltenden Polyurethanen durch die Flora im Verdauungstrakt. Adv Biomater Biomed Eng Drug Deliv Syst. Springer, Tokyo

17. WO 9316009 [Environmental Technologies Europa Ltd.; Info-Ges. Informatik Management & Consulting; H Fuchs, F Markert, W Knaupe; 11.2.92 (DOS 4203866)+3.6.92 (DOS 4 218 216) + 24.8.82 (DOS 4 227 996)/19.8.93)]=EP 625 962

18. Platzek T (1996) Wie groß ist die gesundheitsschädliche Gefährdung durch Textilien wirk-lich?. Melliand, No 11, p 774; Moll RA (1994) Produktsicherheit – eine kostenlose Produkt-sicherheit?. Melliand No 5, p 392

19. Sewekow U, Weber A (1994) Sicherheitsdatenblätter – ein Beitrag der chemischen Industrie für den sicheren Umgang mit Chemikalien. Melliand No 7–8, p 656

20. Grigat E, Salewski K, Timmermann R, Koch R (1997) Biologisch abbaubar in 60 Tagen. Kunststoffe 87:63; DOS 19 615 348 (Bayer; Biotec Biologische Naturverpackungen GmbH; E Grigat et al.; 15.4.96/23.10.97); mixtures of a polyester-amide with starch for disposable bowls resistant to hot water and biodegradable: PCT 97 48 764 (Biotec Biologische Natur-verpackungen GmbH and Bayer; J Loerks et al.; 29.1.96/24.12.97)

21. Hemmpel W-H (1991) Prüfung der Verrottbarkeit von Web- und Maschenstoffen aus Cellu-lose im Eingrabtest. Melliand 768

22. Anon. (1997) Preventing biological degradation of textiles. Int Dyer, February, p 31

23. Sewekow U, Westerkamp A (1997) Probleme bei Analysen nach textilen Ökostandards und der Bedarfsgegenständeverordnung. Melliand, No 1–2, p 55

24. Friedle R, Rieker J (1996) Untersuchungen zur Prüfung der Mutagenität von Textilien. Melliand, No 12, p 865

25. A selection of publications about the discussion of hazardous chemicals in textiles, en-vironmental impact of textiles etc. is given. Klaschka F (1994) Textilien und die menschliche Haut, Fakten und Fiktionen – eine Situationsbeschreibung aus dermatologischer Sicht. Melliand 193; Platzek T (1996) Wie groß ist die gesundheitliche Wirkung durch Textilien wirklich?. Melliand, October; Hemmpel W-H (1994) Textile Ökologie aus dem Blickwinkel des Konsumenten. Melliand 654; Mecheels J (1994) Textilökologie heute. Melliand 642; Baunhofer R (1994) Textil und Gesundheit – eine Einführung. Textilveredlung 29:78; Hart-mann WD (1994) Pro und Kontra Öko-Label aus human-ökologischer Sicht. Textilvered-lung 29:80 (in this publication eco-labels with an evaluation of their aims are discussed);

Herzog W (1994) Ist das EG-Umweltzeichen für Textilien realisierbar?. Textilveredlung 29:86; Kraticek P (1994) Ökotextilien und Gesundheit. Textilveredlung 29:87; Elsner P (1994) Unverträglichkeiten am Hautorgan durch Textilien. Textilveredlung 29:78

26. Anon. (1996) Möglichkeiten für verbrauchte Textilien. Wäscherei und Reinigungspraxis 12:24, 31
27. Hebestreit G, Petzold I (1996) Kollagen enthaltende Polyurethanfilme. In: Heitz E, Fleming H, Sand W (eds) Microb Influenced Corros Mater, Berlin, pp 403–408; (1997) Chem Abstr 300 833
28. Birgenson B, Sterner O, Zimerson E (1988) Chemie und Gesundheit. Weinheim
29. Ried M (1989) Chemie im Kleiderschrank. Reinbeck
30. Matsumoto A (1997) New recycling method of polyurethanes. Kogyo Zairo 45(13):66; see Chem Abstr 128:49051 t
31. William WF, Gastinger AM, Spanier CE, Buckel RJ, Bailey RE (1998) Sorption and microbial degradation of toluenediamine and methylenedianiline in soil under aerobic and anaerobic conditions. Environ Sci Technol 32(5):598
32. Bastian C, Strobbe G (1997) Polyurethane recycling – a joint industry initiative. ISOPA, Brussels, Polyurethanes World Congress Proc
33. DOS 19633891 (BASF; T Jeschke, A Kriesmann, U Bruns, W Scholz, M-C Luederwald, H Peuker; 22.8.96/26.2.98)
34. Anon. (1998) DSM and Allied signal form joint venture for recycling carpets in U.S. JTN, February, p 119

Part 6

**Trade Names, Marketing History,
Summary of Patent Applications**

CHAPTER 29

Comparison of Different Articles

A comparison of the physical properties of leather substitutes using different manufacturing methods shows that these properties are similar regardless of the method used to obtain them: If the articles are manufactured by coagulation, selective evaporation or by stretching of a homopolymer, articles may be produced fulfilling all the necessary demands.

Sometimes surprising results are also obtained. It was shown that a microporous article had a higher degree of water repellency than a hydrophilic article (15 [26]).

29.1
List of Brand Names

For comparison of the properties of different articles, institutes have published the results of their research [1–3]. In table 29-1 different products are named according to their trademark, suplyer and the year of introduction to the market.

The trademarks named belong mainly to chemical companies. Converters of chemical raw materials, i. e. manufacturing companies for ready-made clothing, also sell products under their own trademarks. A list, for example, is available.

Table 29.1. List of Brand Names

Trademark	Company	Type of material	Year of publication or introduction to the market
1509 D	Tenneco Chem		1970
A 246	Reuter		1966
Kablon	Achilles Kk		1977
Aditex	Adidas	Hydrophilic material	
Alsace	Toyobo		1984 [4]
Alcantara = Ecsaine	Iganto/Eni/ Toray Ind. Inc.	Suede-like	[25]
Adiprene	DuPont	System for coating	
Airy	Teijin	Microporous	2.3 [5]
Amara	Kuraray		1979 [8]
Amaretta	Kuraray		3 [4], 2.3 [2]
Arnavon	Arnav		1963
Aquacel	see Hydrocel		

Table 29.1 (continued)

Trademark	Company	Type of material	Year of publication or introduction to the market
Asahigard	Asahi	Oleophobic agent on FC base	
Astrino	Kuraray	Suede-like	1977
Atenas	Bando Kagaku		1975
Aztran	Goodrich	Leather substitute	1971 [9]
Balan	Balamundi	Leather substitute	
Baycast	Bayer	Reactive PUR coating (see Levacast)	(20.2 [3])
Baygard	Bayer	Oleophobic agent on FC base	
Bayperm	Bayer	PUR-foil	
Barex	Technoplast	Leather substitute	1971 [9]
Bar-Lon	Burlington & Barlo Inc.		1966
Baycast	Bayer	Former name of a reactive PUR coating system	1970
Beledano	Kanebo	ultrafine microfiber	[25, 26]
Belima	Kanebo	ultrafine microfiber	[25]
Belleseime	Kanebo		1981 [8]
Belleza	Kanebo	Coated nonwoven	
Belseta	Kanebo	Material consisting of PET-PA fibers with peach skin effect	[25]
Bonsuede	Aigle		1978
Caban	Reuter		1977
Canbrelle	ICI		1971
Cangoran	Freudenberg		1977
Carletta	Kuraray		1984 [3]
Carfene	ICI		1965
Caron	Hoechst AG		1971
Ceecitta	Freudenberg		1976
Ceef	Freudenberg	Nonwoven with a PVC coating	1973
Ceefala	Freudenberg		1973
Ceeluna	Freudenberg		1976
Ceenova	Freudenberg		1976
Ceepala	Freudenberg		1973
Ceepor	Freudenberg		1975
Ceetina	Freudenberg		1976
Ceeval	Freudenberg		1969
Ceevel	Freudenberg	Nonwoven with a PVC coating	1972
Cetita	Freudenberg		1975
Chervaine	Asahi Chem. Ind.		1976
Clarino/Clarino L	Kurashiki Rayon		1970
Comfortex	Raffi & Swanson		1993 [11]
Corfam	Du Pont		
Cordley	Teijin		1971 2.3 [5]
Coverstar	Blücher GmbH	Hydrophilic PUR	[21]
Crisvon	Dai Nippon Ink (DIC)	Polyurethane	

Table 29.1 (continued)

Trademark	Company	Type of material	Year of publication or introduction to the market
Cybor	Gore	Testing apparatus	(9 [13])
Daltoflex	ICI	Polyurethanes for coating	
Dinkam	Kanebo Textiles		1984 [6]
Demizaz	Toray Ind. Inc.	Water vapor permeable membrane	[27]
Dermizax	Toray	Waterproof, breathable, nonporous membrane for protective clothing	[35,41]
Desmoderm	Bayer	PURs for coagulation	
Desmodur	Bayer	Isocyanates	
Desmopan	Bayer	Thermoplastic PUR	
Desmolin	Bayer	Coating products	
Desmophen	Bayer	Polyol	
Dewopor	Degussa		
Domy 1155	Daiichi Lace		1976
Dorlastan	Bayer	Elastomeric fibers	
Dunova	Bayer	Porous fibers	
Dypor	Dynamit Nobel		1969
Ecofeel	Toray und Wacoal	Coated polyester, polyamide and/ or PUR material	[30]
Ecsaine	Toray		1971 [8]
Eikas	Nippon Cloth Co.		1966
Ekraled PUR	Forschungsinstitut für Leder- und Kunstleder- technologie, Freiberg	Poromeric leather substitute	1974 [32]
Elastollan	Elastogran PU-Chemie (BASF)	PURs for coating	
Empor	3M Co.		1973
Elbian, Elbeyan	Nippon Leather		1967
Entrant	Toray	Microporous coating	1984 [1, 27]
Enkora	Enka-Glanzstoff		
Epok	Vecku Hikaku	Suede-like coating for textiles	1967
Estane	Goodrich	Thermoplastic polyurethane	
Everesh	Unitika	Thin microporous PUR coatings	
Exeltech	Unitika	Glutamate	
Exsheep	Asahi Chemical	Filament yarn for suede-like touch	[39]
F 76	Kuraray		1977
Favor Folie V 163	Dai Nippon Ink-	Suede-like foil	1971
	Stockhausen		
Fieldsensor	Toray	Composed material for apparels	[35]
Frequenta	Freudenberg	PVC-coated split leather	1974
Gacella	Benecke	Material with a suede-like surface structure consist- ing of a coated textile	1979

Table 29.1 (continued)

Trademark	Company	Type of material	Year of publication or introduction to the market
Galon	Everflex Inc.	Coated nonwoven	1970
Gambiten		Artificial PVC leather	[32]
Glore	Mitsubishi	Man-made suede	(2.3 [5])
Grabotter	Graboplast Co.	Microporous PUR	[37]
Gore-Tex	Gore	Microporous membrane consisting of PTFE	
Helia "Dorado"	Freudenberg	Artificial PVC leather	1967
Helsapor	Helsa Werke	Microporous coated textile	1986 [12]
Hilake	Teijin		1978
Hitelac	Toray Co.		1964
Hi-Telac	Toyo Rayon	Coated nonwoven	
Hitachi Saxon	Hitachi Chem Co.		1969
Hydes	Toyo Cloth		1974
Hydrocel	Courtaulds	Wound dressing	(10 [87])
Hygrolette	Göppinger Kaliko	Coated nonwoven	1969
Hypol	WR Grace		[16]
Impraperm	Bayer	PURs for water vapor permeable coatings	
Impranil	Bayer	PURs for textile coating	
Imprafix	Bayer	Isocyanate crosslinker for textile coating	
Janeck	Daiichi Kasei		1974
Jeffamin	Huntsman Chem. Co., Salt Lake City, Utah USA	Diamines of a higher molecular weight	
Jentra	Genset Corp.	Coated nonwoven	1970 [33]
Kanebo Patra	Kanebo Ltd.		1971
Keroy	Benecke		
Kurareno	Kurashiki Rayon		1965
Kurarino	Kurashiki Rayon	Nonwoven with a laminated foil	1966
L 26	Kuraray		1972
Lamous	Asahi		1979 (2.3 [5]), [8]
Lacrona	Hornschuch	With PUR-coated textile	1971
Laif-Nubara	Hornschuch		1978
Laif-Elysée	Hornschuch		1978
Levacast	Bayer	Reactive PUR coating of leather	
Likron	Kalle	Foil on PUR base	1970
Ludolette	Göppinger Kaliko	Coated nonwoven	1967
Lycra	DuPont	Elastomeric fiber	
Lyocell	Courtaulds, Lenzing, Unitika [38]	Spec. cellulosic fiber	[17]
Malon	Malon Korea Corp.		1978
Meage	Teijin		(2.3 [5])
Mikropor	Mikropor*	Process to create pores	[14]
Mirror Frontier	Dai Nippon Printing	Foil laminated on a fabric	1969

Table 29.1 (continued)

Trademark	Company	Type of material	Year of publication or introduction to the market
Momba Cuir	Mombers		1966
Mondur	Bayer Corp USA	Isocyanate	
Morimer	Georgia-Borded Fibers Inc.		1964
Munro	Munro Ltd.	Hydrophilic polyether co-block amide	[37]
Neolite	Goodyear		1965
Neoprene	Du Pont	Polychloroprene latex	
Nordicette 65	Ehrenberg		1972
Nordicette 85	Ehrenberg		1972
Nordicette SL	Ehrenberg		1972
Nordiclog	Ehrenberg		1972
Novenda	Freudenberg		
Nylon	DuPont	Polyamide	
OBTL 3000	Grace		
Oraletto	Rückner	Coated nonwoven	
Orasol	Ciba	Dyestuffs soluble in organic solvents	
Ortix	ICI		1967
P 108, P 109	Reuter		1965
Palpa	Unitika		1984 [7]
Palpas	Dai Nippon Inc.	Foil	
Pattina	Du Pont	Coated textile	1968
Patora	Toyo Gomu	Coated nonwoven	1965
Patra	Toyo Rubber Co.		
PebaTex	Elf Atochem	Hydrophilic polyether co-block amide	[18, 31, 37]
PebaMed	Elf Atochem	Water vapor permeable film consisting of polyether-polyamide for medical use	
Penta Star	Hoechst AG		1978
Perlon		Polyamide	
Perfect Leather	Hikaku Kogyo		1966
Permair	Porvair	Microporous foil	
Permatex	Broadley	Water vapor permeable lining	[24]
Permuthane	Stahl	PUR dispersions	[19]
Piccina	Kanebo	Polyester fabric consisting of ultrafine fibers, peach skin effect	[27]
Piazza	Freudenberg		1974
Platilon	Elf Atochem	Water vapor permeable PUR films for wound dressings	
Polaris	Le Thillot	Artificial PVC leather	1967
Pol-Corfam	Polymex		1974
Porelle	Porvair	Microporous PUR foil	(5 [15])
Poron	Porous Plastics	Lining	1969
Porvair	Porvair	Microporous PUR foil	1969

Table 29.1 (continued)

Trademark	Company	Type of material	Year of publication or introduction to the market
Porzan	DuPont		1968
Proof Ace	Unitika	Breathable, waterproof coating	[40]
Quox	Courtaulds	Coated nonwoven	1962
Renwell	Renwell Chem. Comp.		1975
Replex	Toray Ind. Inc.	Water vapor permeable fabric consisting of ultrafine fibers for functional clothing	[27]
Roy	Benecke		
Salpa	Pirelli	Lining	
Scotchgard	3M	FC products for oleo-phobic and water-repellent appliances	
Silastic	Dow Corning	Polysiloxane for coating and appliances	
Silseim	Unitika	Suede-like material	[34]
Skai Molina	Hornschuch	Artificial PVC leather	1977
Skai Canasta	Hornschuch		1977
Skai Pampero	Hornschuch		1977
Skaidur	Hornschuch		1960
Skailen Montara	Hornschuch		1960
Skailen Florino	Hornschuch		1976
Skailen Sutan	Hornschuch		1976
Skailen Gletscher	Hornschuch		1976
Soluna	Mitsubishi Rayon		1984 [5]
Skailan Tonga	Hornschuch		1977
Sobin	Sobin	Nonwoven consisting of collagen fibers	
Sofrina	Kuraray		1980 [8]
Sofspan	Inmont		1976
Splan	Pirelli	Knitted fabric consisting of cellulosic fibers with a PVC coating	1979
Staycool	The Baxenden		(18.2 [38])
Suedemark	Kuraray		(3 [4])
Super Dry	The Baxenden		(18.2 [38])
Supronyl	Hoechst AG	Foil	1969
Sure Sports Cool & Dry	Toyobo	Knitted fabric consisting of polyester for apparel	[28]
Sympatex	Enka	Hydrophilic polyester	[12]
Syscare		Laminate consisting of polyester fibers and a PTFE membrane between 2 textile layers	
Tactel	ICI		
Tanera	Scott Charthan Co.		1971
Terinda	ICI		
Tetratex	Tetratec Corp.	Microporous PTFE	[37]

Table 29.1 (continued)

Trademark	Company	Type of material	Year of publication or introduction to the market
Texapor	Pfaff	Plasmajet process to produce micropores mechanically	
Technosensor	Teijin	Coated water vapor permeable fabric	[28]
Teflon	DuPont	FC products for oleo-phobic and water-repellent appliances	
Tencel	Courtaulds	Spec. cellulosic fibers	[29]
Texapor	Pfaff		[14], (15 [26])
Texon	Texon GmbH		[13]
Thintech	3M	Membrane consisting of polyolefin for lamination	[20]
Toray 233/Ecsaine	Toray Ind. Inc.		1970
Tricel	Courtaulds		[15]
Tuftane	Goodrich		1968
Ucecoat	Union Chimique Belge	PUR for textile coating	[13]
Ultrasuede	Toray		(3 [4])
Varlon SA			
Velvon	ICI		1965
Vilbond	Freudenberg		1970
Viledon	Freudenberg		1970
Walotex	Wolff Walsrode	Water vapor permeable PUR membrane	[23, 37]
Welkey	Teijin		1984 [2]
Witcoflex	The Baxenden		(18.2 [38])
Witcoflex EcoDry	Baxenden	Water vapor permeable leather substitute	[22]
Wolfin-ecri-WS	Degussa		1972
Xylee	Glanzstoff	Nonwoven with a microporous coating	1974 [10]

The trademarks belong mainly to chemical companies. Convertes of chemical raw materials, i.e. manufacturing companies for ready made clothing sell products under their own trademarks as well. A list of those trademarks is e.g. published in [28 (29), p. 232–237].

CHAPTER 30

The Development of Water Vapor Permeable Materials: Patents and Publications

The initiator in the development of microporous materials and the first supplier of a water vapor permeable leather substitute was DuPont (USA). DuPont, encouraged by the success of Nylon®, a synthetic substitute for silk, from the early 1950s started to develop a substitute for the last natural product, leather. Leather, a material which, up to that point, had not yet been reproduced in its characteristics by man.

"... the chemical industry started a frontal attack against one of the last bastions of natural products – the industry of leather and shoes" [1]. Corfam® was, as a result of this development of DuPont, the first water vapor permeable material which attempted to reach or overcome the properties of leather. In an evaluation of the novelties of our time, L. Lessing, an editor of Fortune Magazine, named Corfam „... one of the most important creations of man ... because it ensures the needs of an important consumer product" [2]. This synthetic substitute, for the first time, enabled the industrial manufacture of shoes without the need for experienced staff, because every part of the material had, in contrast to leather, the same properties. The manufacture of shoe uppers consisting of leather has to be carried out by experienced personnel who know well which part of a hide can be cut for a special part of a shoe. Leather differs in its structural properties in the neck, belly, flank etc. Not every part of a hide can be used, for example, for the toe or the side part of a shoe due to its resistance, *dimension stability*, etc.

Corfam was also a challenge for the leather industry: At that time Corfam offered properties leather did not have. Shoes consisting of Corfam could be cleaned by washing with water. The material was offered as *"easy care"* which leather lacked. Corfam offered a higher *abrasion resistance* than genuine leather.

Therefore a lot of intensive development work started in order to improve the properties of leather also. The main work was done in the finishing area. The polyurethane finish of leather began its success [3].

It is known today that Corfam was unsuccessful. In the same journal, Fortune, whose editor years ago praised the product enthusiastically then referred to it as a "100 Mio $ object lesson" [4]. DuPont sold the equipment to Poland where the material was produced and sold for a while under the name *Polcorfam*.

At about the same time ICI cancelled further work on *Aztran®*, Bayer on *Desmoderm® foil*, Glanzstoff on *Xylee®* and many other companies also cancelled work on their products. Only Japanese companies continued to work on their leather substitutes.

DuPont has been referred to as a very innovative company. If an invention does not fit to the product range of the company they discontinue. So A. L. Gore, a former employee of DuPont, started with his associates a $ 1 Billion project based on an invention originally developed at DuPont by marketing Teflon®-based materials as Gore-Tex® [8]. Gore also started a business based on microporous membranes in the fields of medicine, electronics and filtering [9].

The most important contribution of Japanese chemists in the field of water vapor permeable materials was the invention of the ultrafine fibers. Ultrafine fibers (see Chap. 21) are the basis for nonwovens which enables the production of articles with an extreme softness and isotropic behavior. Furthermore, tensile strength and stress-strain properties proved to be excellent. With this technique, nonwovens no longer needed a textile interlayer as was the case with Corfam. The only disadvantage of fine fiber technology is the high production cost. Although certain consumers are paying the relatively high price this product will probably never be a mass product.

Many coating companies have had a market success with the coagulation technique. These companies buy polyurethane solutions or granulates and produce microporous coatings and/or impregnations for the shoe and clothing industry.

It can be said today (1998) that microporous and hydrophilic products have been successful in the area of sport, and many industrial and medical fields (see Chap. 27). These successes were not anticipated when the first patent applications were made.

Finally it can be stated that many inventions came out of laboratories from companies (1) who were unable to recognize the value of an invention at the time the invention was carried out, (2) who were unable to evaluate the marketing, and (3) who did not want to invest the time to successfully market the product.

As an example all important inventions to achieve water vapor permeability were done in Bayer laboratories:

- a patent was applied for coagulation in 1951 (8 [1]),
- for hydrophilic polyurethanes in 1962 (18.2 [10]),
- for selective evaporation in 1966 (9.2 [3]), and then
- for reactive coating in 1966 (10.1 [4]).

None of these inventions were directly used. Years later products deriving from variations of these processes were marketed.

A recent publication says that German companies lack "in risk taking and dynamic spirit which is necessary in emerging fields of technology" [10]. Hopefully the marketing of the technology of the water vapor permeable materials is not proof of this statement.

The literature used in the compilation of this book since the 1970s is dominated by Japanese authors and companies. Many Japanese patent applications exist only in Japan with no foreign equivalents. Therefore they exist only in Japanese and mostly only as an abstract in English. They are, therefore, difficult to understand and evaluate. A direct translation into English reveals that the basis of the invention often is not precisely described. Additionally the patent strategy of Japanese companies seem to differ from that of a European or American one.

Therefore if "patent flooding" or "piracy" is the basis of Japanese patent regulations [5] or just a different philosophy of work the author is unable to judge.

A recent publication describes the Japanese patent strategy as compared to Clausewitz, a Prussian general of the last century, who stated that "superiority in quantity, in tactics and in strategy principally leads to the victory. Therefore to gain a battle the army should be as strong as possible". If a particular company holds many patents describing partial aspects in a field the competition is greatly hindered, even if the basic invention is held by another company. These patents describing only partial aspects of the original invention may help via an arrangement in marketing of articles described by them.

The costs to develop the technologies described in this book are extremely high. The $100 Mio which were spent by DuPont were most likely also spent by other companies working in this field. Bayer, as mentioned, stopped the production of microporous foils but is still a large supplier of polyurethane systems for water vapor permeable materials. Other than Bayer only a few companies have taken the opportunity to market this available technology. Most of these companies were outsiders. These outsiders took the chances offered by this technology and used it. The ability to achieve success with an adopted technology is surely as creative as the invention itself.

"What does invention mean and who is to say that he or she invented this or that? It is sheer nonsense if somebody claims a priority than confesses to be a plagiarist and achieves success with an invention" [7].

References to Part 6

29
Comparison of Different Articles

1. van der Beke R, Dekoninck L (1986) Neuentwicklungen auf dem Gebiet der mikroporösen Beschichtungen. Melliand 11:824
2. Turner R, Worsick B (1981) Encouraging trial of PUCF uppers. Satra Bull vol. 18, June No 28
3. Anon. (1977) Developments in polyurethane coated fabrics. Satra Bull 17:577

29.1
List of Brand Names

1. Anon. (1984) To promote worldwide marketing. Japan Textile News Nov, p 22
2. Anon. (1984) Pursuit of comfort performance. Japan Textile News Nov, p 24
3. Anon. (1984) Development from fiber stage. Japan Textile News Nov, p 26
4. Anon. (1984) Pursuing comfortability in sportwear materials. Japan Textile News Nov, p 28
5. Anon. (1984) Scientific character and feeling. Japan Textile News Nov, p 32
6. Anon. (1984) To meet the state by composite force. Japan Textile News Nov, p 34
7. Anon. (1984) Waterproof/moisture permeable "Palpa" fabric from Unitika. Japan Textile News Nov, p 44
8. Anon. (1981) To establish a new status surpassing natural suede – Japanese man-made suede. Japan Textile News Feb, p 18; Anon. (1997) Super Belleseime from Kanebo. JTN 508
9. Hayashi T (1974) Development of man-made leather, its present and future. Chem. Economy & Eng News 6(2):26
10. Riess W (1974) The manufacture and properties of fibrous composite poromerics. J Coated Fabr July, p 47
11. Krishnan K (1993) Hydrophilic urethanes for textiles. J Coated Fabr July, p 54; Krishnan K (1995) New applications for breathable hydrophilic and non-hydrophilic coatings. J Coated Fabr 26:103
12. Anon. (1986) Neuentwicklungen bei atmungsaktiven Geweben Chemiefasern/Textilind Jan, p 86; Hürten J, Spijkers J (1997) Laminierung von Sympathex Membranen Teil 1. Adhäsion-Kleben & Dichten 4:1; Spijkers JCW (1997) Sympathex technology and product specialities. Textile Technology Int 23, 24, 26, 28; Hürten J, Spijkers J (1997) Laminieren von Sympathex Membranen. Adhäsion-Kleben, 41(1–2): 34, 41(3): 34, 37
13. van der Beke R, Dekoninck L (1986) Neuentwicklungen auf dem Gebiet der mikroporösen Beschichtungen. Melliand 824
14. Detailleur J-P (1994) Textile Bekleidung – wasserfest und atmungsaktiv. 33rd Int Chemiefasertagung, 28–30 September, Dorbirn, AU
15. (1982) Chemiefasern Text.Ind. 32/84 11, p 760
16. DOS 2 925 318 (WR Gore & Ass.; WR. Gore et al.; 22.6.79/17.1.80; US Prior. 20.6.78)=GB 2 024 100=US 4 194 041

17. Eichinger D (1996) Lenzing – Lyocell – an interesting cellulose fiber for the textile industry. Chem Fibers Int (CFI), vol 46, Jan, p 28; Anon. (1995) Courtaulds Lyocell für technische Textilien. Technische Textilien 38:57; Howie I (1995) Competing materials in shoemaking. Leather July, 41; Johnson P (1996) Courtaulds Lyocell – a cellulosic fibre for special papers and nonwovens. Nonwoven Conf, Atlanta (ISBN 0–89 852–658–2); Albrecht W, Reintjes M, Wulfhorst B (1997) Lyocell Fasern. In: Faserstoff-Tabellen. Melliand No 9, 575
18. Anon. (1994) Atmungsaktive Folien für Arbeitskleidung. Technische Textilien 37:91
19. Anon. (1995) Stahl Holland bv, Waalwijk NL. Technische Textilien 38:62
20. Bucheck DJ (1991) Comfort clothes through chemistry. Chemtech. March, p 142
21. Anon. (1996) Atmungsaktive-Hygienebezüge mit PU-Beschichtung. Technische Textilien 39 (Sept): 143
22. Anon. (1997) Baxenden breathable textile coatings. Technische Textilien 40 (April): E 18
23. Anon. (1997) Wolff Walsrode: atmungsaktive Membranen. Technische Textilien 40 (April): 68
24. Anon. (1997) Broadley: atmungsaktive Wetterschutzfutter und Beschichtungen. Technische Textilien 40 (April): 72
25. Anon. (1996) Changes in ultrafine microfibers. JTN Nov, p 35
26. Anon. (1996) "Beledano", Nubuck-type fabric made of ulta-superfine microfiber 'Belima SX' JTN Nov, p 37
27. Anon. (1996) Toray Industries Inc. JTN Aug, p 50
28. Anon. (1996) Toray 'Entrant-G-II-XT' waterproof/breathable fabrics of outstanding low condensation. JTN Oct, p 89
29. Anon. (1997) "Tencel" fashions the world. JTN Mar, p 15
30. Anon. (1997) Fine structured split yarn from Teijin. JTN Apr, p 98
31. Anon. (1997) Atmungsaktive Klimamembrane von ElfAtochem. Leder und Häutemarkt, 30 May, p 12
32. Anon. (1976) Überblick über die Nomenklatur synthetischer Ledersubstitute in der DDR. Leder, Schuhe, Lederwaren 82
33. Holden RF (1976) Die Zurichtung polymerer Polyurethane. J Coated Fabr 180
34. Anon. (1997) Stretch synthetic suede from Unitika. JTN Aug, p 97
35. Anon. (1997) Fieldsensor, entrant G II-XT, Dermizax, replex, light. JTN Aug, p 60
36. Anon. (1997) OP Textilien der nächsten Generation, OP Abdeckungen aus 'Syscare' auf Inter-hospital in Hannover vorgestellt. Wäscherei und Reinigungspraxis 21
37. Painter CJ (1996) Waterproof, breathable fabric laminates: a perspective from film to market place. J Coated Fabr 26(Oct): 107
38. Anon. (1998) JTN Jan, p 37
39. Anon. (1997) New polyester yarn by Asahi Chemical. JTN Dec, p 87
40. Anon. (1997) Proof ace. JTN Feb, p 71
41. Anon. (1997) Dermizax. JTN Feb, p 73

30
The Development of Water Vapor Permeable Materials: Patents and Publications

1. Anon. (1964) Haut aus der Retorte. Der Spiegel, 23 Dec
2. Lessing L (1964) Der Siegeszug von Corfam. published by DuPont
3. Schröer W (1969) BAYDERM-Zurichtung – ein neues Zurichtverfahren. Das Leder 102
4. Anon. (1971) Fortune, Jan
5. Spero DM (1990) Patent protection or piracy – a CEO views Japan. Harvard Business Rev Sept/Oct 58
6. Rahn G (1994) Patentstrategien japanischer Unternehmen. GRUR Int 377
7. v.Goethe JW Maximen und Reflexionen 1146
8. Miller JA (1997) Discovery research re-emerges in DuPont. Research Tech Manag (Jan/Feb) 24
9. Anon. (1997) Rundum wasserdicht. Schuhtechnik No 5/6, p 19
10. Abramson HN (1998) Technology transfer in Germany. Chemtech Jan, p 14

Part 7

**Summary of the Patent Applications
and Practical Examples**

CHAPTER 31

Summary of the Patent Applications

Companies sometimes change their names; therefore, some of the names of identical companies are shown at the end of Table 31-1.

Patent applications in most cases are characterized by the date of priority, i. e. the first application which is normally done in the country where the inventor lives or where the headquarters of a company are situated. The other dates belonging to a patent application, mentioned in the text, concern the date of application in the specific country and the date of publication which is important for a potential infringement of this patent. An infringement of a patent is only possible when the application is published regardless in whatever country.

If in a section a company is named several times under the same citation number with different dates then the specific patent cited is a compilation of several applications with different priority dates. If in one line several companies are named then the patent was applied for by more than one company. The first column shows in which section of this book the specific application may be found in the text with the relevant literature citation, after the hyphen.

Table 31-1. Summary of the Patent Applications

Section	Company	Date
16.2–4	3M	19.3.59
21–71	3M	16.9.65
21–2	3M	16.9.65
10.1–20	3M	15.8.66
10.1–20	3M	28.6.67
18.2–49	3M	26.8.68
18.1–14	3M	26.8.68
7–36	3M	20.10.69
18.1–16	3M	15.3.71
8.6–51	3M	15.10.74
16–12	3M	4.12.84
16–18	3M	8.10.87
7–6	3M	20.10.69
19–10	A.K.U.N.V	17.4.68
7.1–9	AB Tudor	1.9.66
9.2–13	Abaturova NA	6.12.66
8.5–33	Achilles	6.12.80
18.2–60	Achilles Corp	6.10.86

Table 31-1 (continued)

Section	Company	Date
7–59	Achilles Corp	18.5.89
8–23	Achilles Corp	27.11.90
8–22	Achilles Corp	28.11.90
9.3–28	Achilles Corp	29.11.90
18.2–65	Achilles Corp	13.2.92
9.3–26	Achilles Corp	10.2.93
8.2–63	Achilles Corp	2.6.93
19–30	Achilles Corp	27.7.93
8.5–31	Achilles Corp	24.8.94
20.2–18	Achilles Corp	15.6.92
22–53	Achilles Corp	24.12.92
20.2–12	Achilles Corp	2.3.94
20.2–17	Achilles Corp	24.4.94
20.2–12	Achilles Corp	21.7.94
20.2–12	Achilles Corp	7.10.93
23.1–8	Achilles Corp KK	30.10.92
25.5–73	Achilles Corp	15.5.95
25.5–82	Achilles Corp	10.7.95
8.5–25	Achilles KK	6.12.80
8.1–31	Achilles KK	29.12.80
9.3–30	Adidas	17.3.77
21.1–5	ADW; Inst. f. Technologie d. Fasern	14.7.70
11.2–41	Agricola Reg.Trust	5.3.65
11.1–67	AH Hides & Skins Australia Pty	22.11.93
18.3–28	Aichi Hikaku Kogyo KK	10.3.86
18.3–1	Ajinomoto	13.6.66
25.4–6	Ajinomoto	30.9.68
18.3–14	Ajinomoto	11.5.71
18.3–39	Ajinomoto	15.5.87
18.3–4	Ajinomoto Co. Ind	16.4.69
18.3–20	Ajinomoto Inc	4.9.64
18.3–19	Ajinomoto Inc	25.12.67
18.3–10	Ajinomoto Inc	29.12.67
18.3–11	Ajinomoto Inc	25.12.67
18.3–30	Ajinomoto KK	15.8.91
18.2–64	Ajinomoto KK, Ajinomoto Takara Corp KK	11.3.91
18.4–8	Ajinomoto KK, Seiko Kasei KK	8.11.90
18.4–2	Ajinomoto Takara Corp	9.3.94
18–10	Akad. der Wiss. der DDR	15.3.85
10.6–2	Akademie der Wiss	13.9.71
7–39	AKU	18.5.67
18.2–21	AKU	7.2.68
21–68	AKU-Goodrich	8.4.66
25.1–9	Akzo	1.3.74
16–20	Akzo	9.3.78
8.6–12	Akzo GmbH	15.11.71
8.2–84	Akzo GmbH	20.12.79
18–22	Akzo Nobel NV	25.4.96
25.5–65	Akzo Nobel NV	22.5.95
25.5–67	Akzo Nobel NV	24.5.95
11.2–14	Alderfer, SW	30.1.61

Table 31-1 (continued)

Section	Company	Date
21–24	Alegre, AA.	28.2.61
17.1–18	Alegre, AA	28.2.61
8.6–24	Algemene Kunstzijde Unie NV	18.5.67
25.5–76	Alkermes Controlled Therapeutics	7.6.95
9.1–10	Alkor	9.12.44
21–76	Alkor-Werk	10.2.43
17.1–2	Allegre, AA	17.12.53
10.3–9	Allen Ind. Inc.	25.4.60
2.2–4	Allied Chemicals	13.1.70
7–44	Am. Cyanamid Co	1.5.69
9.1–8	Am. Viscose Corp	24.12.53
9.1–8	Am. Viscose Corp	22.12.59
23.1–17	Amerace Corp	16.8.74
13–7	American Viscose Corp	3.4.62
18.2–7	Amicon Corp	26.4.66
16.2–29	Amoco Corp	21.12.90
16.2–29	Amoco Corp	23.8.91
20–2	Andrieu, F	14.3.66
25.5–47	Ansell, CWG	6.12.86
25.5–47	Ansell, CWG	16.1.87
25.5–47	Ansell, CWG	22.1.87
8.6–46	Antonio, JP	2.3.84
25.5–85	Aquatic Design Inc	3.5.95
12–26	Aquitaine-Organico	17.2.70
25.5–60	Araco KK	11.11.93
25.5–60	Kyowa Leather Cloth Co Ltd	9.12.91
18.1–49	Aroma Kagaku Kikai Kogyo KK	2.7.92
21–93	Artos	8.4.76
20.1–37	Asahi	23.2.83
25.5–22	Asahi Chem. Ind	17.8.93
21–85	Asahi Chem. Ind	9.6.78
8–20	Asahi Chem. Ind Co Ltd	14.1.92
21.1–6	Asahi Chem. Ind KK	17.9.73
7–71	Asahi Chem. Ind KK	10.9.85
7.1–11	Asahi Chem. Ind	30.6.73
8.2–64	Asahi Chem. Ind	12.5.93
11.1–43	Asahi Chem. Ind.	2.12.69
2.7–14	Asahi Chem. Ind KK	24.813
19–21	Asahi Chem. Ind KK	22.12.78
8.2–73	Asahi Chem. Ind	14.3.95
8.2–76	Asahi Chem. Ind	20.4.95
23.1–22	Asahi Chem. Ind KK	30.9.76
8.6–53	Asahi Chem. Ind KK	28.7.77
8.6–53	Asahi Chem. Ind KK	27.7.77
11.2–16	Asahi Denka Ind Co Ltd	18.6.63
11.2–15	Asahi Electrochem. Co Ltd	22.1.63
18.2–31	Asahi Glass Co Ltd	26.4.88
23.1–4	Asahi Glass KK	16.3.90
25.3–	Asahi Kasei Kogyo KK	30.3.70

Table 31-1 (continued)

Section	Company	Date
25.5–10	Asahi Kasei Kogyo KK	30.4.93
10.2–15	Asahi Kasei Kogyo KK	26.7.93
22–51	Asahi Kasei Kogyo KK	14.2.94
20.1–40	Ashland Oil Inc.	15.5.70
12–1	Balamundi	7.1.64
9.3–31	Bando Chem. Ind Ltd.	23.8.74
8.2–35	Banto Chotai Rubber	25.1.68
14–5	Barnard; KH	18.4.59
7–45	BASF	30.1.64
17.1–13	BASF	13.5.64
7–42	BASF	22.7.65
11.2–35	BASF	22.11.66
10.5–2	BASF	4.8.67
10.5–4	BASF	30.7.68
10.5–9	BASF	21.9.68
10.5–6	BASF	31.10.68
10.5–5	BASF	2.11.68
10.4–1	BASF	2.11.68
9.3–4	BASF	19.11.68
18.1–40	BASF	29.11.68
10.4–7	BASF	31.12.68
18–18	BASF	30.12.69
20.2–8	BASF	19.9.70
18–19	BASF	25.3.71
10.5–10	BASF	20.1.72
17.1–1	BASF	30.5.72
18–16	BASF	23.11.72
18.1–20	BASF	15.5.73
18.1–45	BASF	1.2.74
18.2–26	BASF	25.5.94
28–8	BASF Schwarzheide	12.10.92
16.2–13	Baudou, A	4.7.66
19.1–29	Baudou, AJG	16.8.65
8–1	Bayer	22.5.51
2.2–3	Bayer	25.4.59
18.2–8	Bayer	5.1.62
18.2–10	Bayer	3.3.62
25.1–4	Bayer	3.3.62
18.2–16	Bayer	6.4.62
18.2–9	Bayer	6.4.62
8.2–55	Bayer	26.10.62
25.5–4	Bayer	26.10.62
8.2–53	Bayer	5.12.62
17.1–8	Bayer	5.2.63
8.2–54	Bayer	28.2.63
25.5–5	Bayer	19.9.63
21–38	Bayer	31.10.63
9.1–3	Bayer	4.6.64
10.3–3	Bayer	27.8.64
10.3–3	Bayer	8.7.65
9.1–5	Bayer	5.11.65

Table 31-1 (continued)

Section	Company	Date
9.2–3	Bayer	3.1.66
8.2–52	Bayer	22.6.66
10.1–4	Bayer	26.10.66
10.1–3	Bayer	26.10.66
10.1–3	Bayer	11.5.67
9.2–14	Bayer	22.11.66
8.2–39	Bayer	13.12.66
8.3–8	Bayer	13.12.66
15–4	Bayer	5.1.67
15–2	Bayer	5.1.67
8.5–13	Bayer	10.1.67
12–8	Bayer	28.4.67
12–6	Bayer	28.4.67
12–5	Bayer	28.4.67
10.1–6	Bayer	12.5.67
12–7	Bayer	16.6.67
10.2–9	Bayer	9.8.67
10.2–9	Bayer	7.12.67
8.6–9	Bayer	9.8.67
9.2–5	Bayer	9.8.67
10–17	Bayer	6.9.67
9.2–7	Bayer	12.10.67
9.3–9	Bayer	25.10.67
10.2–1	Bayer	7.12.67
8.3–25	Bayer	8.3.68
10.1–10	Bayer	2.4.68
8.2–51	Bayer	30.4.68
22–17	Bayer	3.5.68
15–16	Bayer	1.8.68
11.1–28	Bayer	22.8.68
16.2–7	Bayer	21.2.69
16.2–28	Bayer	21.5.69
25.3–4	Bayer	17.10.69
8.4–6	Bayer	26.5.70
10.3–14	Bayer	11.7.70
10.2–10	Bayer	11.7.70
17.1–4	Bayer	11.7.70
25.4–9	Bayer	10.9.70
21–90	Bayer	8.2.71
10.2–12	Bayer	14.5.71
12–13	Bayer	31.5.72
25.4–15	Bayer	6.9.72
12–2	Bayer	23.3.73
25.5–1	Bayer	23.3.73
10.2–4	Bayer	6.6.73
8.2–56	Bayer	7.9.73
17–3	Bayer	27.9.73
12–11	Bayer	10.10.73
18–12	Bayer	7.11.73
18.1–12	Bayer	7.11.73

Table 31-1 (continued)

Section	Company	Date
17.1–17	Bayer	11.4.74
22–31	Bayer	28.5.74
10.1–22	Bayer	22.4.76
7.1–26	Bayer	16.2.77
2.2–18	Bayer	1.4.78
2.2–19	Bayer	6.2.80
2.2–18	Bayer	13.4.83
25.5–12	Bayer	16.11.84
9.3–29	Bayer	2.3.85
9.2–23	Bayer	2.3.85
9.2–23	Bayer	19.6.85
9.3–29	Bayer	26.3.86
18.2–43	Bayer	29.10.87
9.3–29	Bayer	22.10.88
25.5–19	Bayer	19.12.92
19–31	Bayer	9.3.95
28–20	Bayer; Biotec Biologische Naturverpackungen GmbH	15.4.96
25.3–6	Bayer	7.1.69
18.1–2	Benecke	20.4.54
13–17	Benecke	25.3.80
20–6	Bertolaia, G	27.1.61
8–24	Besana, R	5.4.91
15–23	Besnier-Flotex SA	20.11.72
9.3–10	Bick, HC	21.11.51
8.3–33	Blücher, H von	14.1.82
11.2–60	Blücher, H von	17.8.81
18.2–40	Blücher, H von	7.5.88
18.2–5	Blücher, H von	1.9.90
7.2–20	Bocciardo et C SpA	27.5.70
18.2–67	Bocciardo, P	9.12.91
12–14	Bofors	8.12.71
20–13	Bonded Faber Fabric Ltd	28.5.65
21–19	Bonded Fibre Fabric Ltd	30.11.60
11–7	Borg-Warner Corp	13.2.64
25.5–37	Borries, H von	10.8.94
11.2–42	Bridgestone Tire Co Ltd	11.3.65
8.2–91	British Millerain	13.7.71
21–18	British Nylon Spinners	30.9.60
21–53	British Nylon Spinners Ltd	5.2.63
17.1–3	British USM Co Ltd	22.11.56
17.1–20	British United Shoe Mach Corp	1.3.61
12–29	Buchmann, RC	17.5.67
25.5–56	Budinger, W	5.2.93
11.1–30	Bukflex Processes Ltd	28.1.65
25.5–41	Bundesamt für Wehrtechnik und Beschaffung; Forschungsinstitut für Leder und Kunstleder	8.11.84
21–39	Bunse u. Biach	24.10.63
8.3–34	Burlington Ind Inc	3.9.86
8.3–34	Burlington Ind Inc	9.9.86

Table 31-1 (continued)

Section	Company	Date
11.1–29	Burlington Ind Inc	23.9.69
20.1–24	C.T.C	27.11.68
11.2–8	Cacella, AF.	3.11.67
11.1–46	Celanese	18.4.66
16.2–14	Celanese	15.8.66
13–1	Celanese Corp	14.1.71
16.2–23	Celanese Corp	30.5.73
16–3	Celanese Corp	1.6.79
16.2–16	Celanese Corp	26.7.65
16.2–16	Celanese Corp	1.10.65
16–20	Celanese Corp	13.11.69
16–20	Celanese Corp	28.10.70
16.2–8	Celanese Corp	13.11.69
16.2–8	Celanese Corp	28.10.70
8.3–29	Cesk.zav.gum. a plast	17.3.66
8.4–22	Ceskoslov Zavody Gumarenke & Plastikarske	25.5.66
8.3–20	Ceskoslov Zavody Gumarenske & Plastikarske	13.4.67
8.2–28	Ceskoslov Zavody Gumarenski & Plastikavske	13.4.66
9.2–15	Ceskoslov Zavody Gumarenski & Plastikavske	18.7.66
16.2–25	Chakoyan, AS	22.10.62
24–9	Chaussures Andre SA	17. 3.70
24–9	Chaussures Andre SA	15.5.70
24–9	Chaussures Andre SA	23.7.70
24–9	Chaussures Andre SA	19.1.71
24–9	Chaussures Andre SA	19.11.70
10.4–4	Chem.Fabrik Kalk GmbH	27.3.62
2.2–6	Chem.Werke Worms-Weinheim; P Spindler Werke KG	3.9.59
9.3–1	Chemgene Corp	3.2.65
20.1–38	Chemiaro	1.9.77
25.2–	Chemie-Anlagenbau Bischofsheim GmbH	3.11.73
23.1–3	China Textile Inst	14.2.94
8.5–29	China Textile Res Inst	26.11.93
23.1–5	Chinese Textile Industry Research Center	7.2.94
22–22	Ciba Geigy	4.3.70
23.1–6	Ciba Geigy AG	19.1.94
22–23	Civardi, FP	21.12.70
20.1–9	Collagen Corp	25.11.66
20.1–8	Collagen Corp	13.8.69
20–11	Collagen Corp	13.5.68
17–5	Compo Ind Inc	22.5.78
10.1–1	Continental	14.12.64
16.2–3	Continental	17.5.66
10.1–5	Continental	11.6.66
10–26	Continental	2.11.70
10.4–2	Continental Gummi Werke AG	21.11.64
10.4–2	Continental Gummi Werke AG	25.11.64
10.4–2	Continental Gummi Werke AG	5.12.64
10.4–2	Continental Gummi Werke AG	22.11.65

Table 31-1 (continued)

Section	Company	Date
10.4–10	Continental Gummi Werke AG	4.11.65
10.4–9	Continental Gummi Werke AG	20.4.66
10.3–12	Continental Gummi Werke	22.12.69
25.2–5	Continental Tapes Inc	29.6.70
10.4–11	Continnental Gummi Werke AG	15.12.64
8.2–74	Cordley Chem	7.9.95
10.4–5	Courtaulds Ltd	18.7.63
10.4–5	Courtaulds Ltd	14.1.64
7.2–26	Courtaulds Ltd	18.7.63
7.2–26	Courtaulds Ltd	14.1.63
16.1–15	Courtaulds Ltd	3.12.63
16.1–15	Courtaulds Ltd	18.12.63
16.2–6	Courtaulds Ltd	3.12.63
16.2–6	Courtaulds Ltd	18.12.63
7.2–19	Courtaulds Ltd	19.5.65
18.2–44	CPC Intern Inc	14.2.72
18.2–44	CPC Intern Inc	14.4.72
18.2–44	CPC Intern Inc	26.6.72
8.2–69	Dai Nippon Ink & Chem KK	17.6.92
18.1–52	Dai Nippon Ink & Chem. Inc	30.11.93
25.4–27	Dai Nippon Ink & Chem.KK	13.11.92
8.5–23	Dai Nippon Ink & Chem.Co	14.8.70
25.5–9	Daicel Chem. Ind. Ltd	4.3.93
18.2–66	Daicel Huels KK	23.4.92
18.2–66	Daicel Huels KK	24.4.92
8.6–18	Daiichi Chem. Ind	29.8.70
7–70	Dai-Ichi Kogyo Seiyaku Co Ltd	7.12.84
18.2–42	Daiichi Kogyo Seiyaku Co Ltd	17.2.86
18.2–42	Daiichi Kogyo Seiyaku Co Ltd	26.2.86
8.1–11	Daiichi Lace Co	20.3.70
22–13	Daiichi Lace Co Ltd.	29.10.69
15–22	Daiichi Lace Co Ltd	19.8.70
18.3–17	Daiichi Lace Co Ltd	20.11.72
8.5–22	Daiichi Lace KK	31.1.67
8.2–18	Daiichi Lace KK	29.9.69
22–32	Daiichi Lace KK	6.10.72
8.1–30	Daiichi Lace KK	7.12.79
27–4	Daiichi Lace KK	27.10.86
19–32	Daiichi Lace KK	1.3.90
8.1–23	Daiichi Lace KK	3.9.91
7–73	Daiichi Lace KK	26.9.91
16.2–32	Daiichi Lace KK; Hosokawa Micron KK	26.9.91
16.2–33	Daiichi Lace KK; Hosokawa Micron KK	26.9.91
8.6–54	Dainichi Seika Kogy KK	3.6.77
25–4	Dainichi Seiko Kogyo KK	7.9.73
8.2–71	Dainichiseika Color & Chem. Mfg	6.9.91
9.3–37	Dainichiseika Color & Chem. Mfg Co + Ltd; Ukima Colour & Chem.	2.8.83

Table 31-1 (continued)

Section	Company	Date
25.5–32	Dainichiseika Color & Chem. Mfg	19.6.94
18.2–36	Dainichiseika Color Chem	16.8.80
18.2–33	Dainichiseika Color Chem	17.9.84
18.2–45	Dainichiseika Color Chem	5.10.84
18.2–34	Dainichiseika Color Chem	5.10.84
9.3–38	Dainichiseika Color Chem	26.11.87
22–49	Dainichiseika Color Chem	28.12.92
8.9–10	Dainichiseika Color Chem	12.7.95
18.2–70	Dainichiseika Color Chem. Co	17.9.84
18.3–9	Dainichiseika Colour & Chem. Mfg. Co Ltd	14.5.69
11.1–35	Dainichiseika Colour & Chem. Mfg. Co Ltd	15.10.71
2.2–9	Dainippon Ink	10.9.73
18.2–15	Dainippon Ink & Chem. Inc	30.11.93
25.1–10	Dainippon Ink & Chem. KK	18.3.92
25.1–10	Dainippon Ink & Chem. KK	13.11.90
25.1–10	Dainippon Ink & Chem. KK	12.12.91
25.1–10	Dainippon Ink & Chem. KK	13.11.90
25.1–11	Dainippon Ink & Chem. KK	24.8.83
25.5–11	Dainippon Ink & Chem. KK	22.11.90
25.4–32	Dainippon Ink & Chem. KK	18.3.92
18.2–59	Dainippon Ink and Chemicals Inc	19.5.86
8.3–27	Dainippon Ink Chem. KK	19.8.80
18.2–37	Dainippon Ink Chem. KK	21.8.81
22–10	Dainippon Ink. & Chem	8.12.72
11.1–52	Dainippon Ink. & Chem. Co	3.8.71
18.1–27	Dainippon Ink & Chem. Inc	17.3.69
18.1–27	Dainippon Ink & Chem. Inc	7.4.69
15–10	Daisei Kako Co Ltd	27.6.63
22–33	Daito Chem. Ind KK	14.12.72
25.5–54	Daiwabo Create Co Ltd	28.1.91
7–65	De Bell and Richardson Inc	30.11.71
15–5	Deering Milliken Research Corp	30.4.71
21–29	Degussa	15.7.61
14–9	Degussa	21.11.62
17–4	Degussa	6.5.63
19–7	Degussa	22.5.71
11–1	Degussa	31.10.53
2.1–1	Deutsche Celluloid Fabrik	4.6.37
8.2–34	Deutsches Lederinstitut	15.5.68
8.4–7	Deutsches Lederinstitut	11.12.70
25.4–1	Deutsches Lederinstitut	16.3.71
8.4–31	Deutsches Lederinstitut	21.3.72
25.5–5	Diya Gum KK	2.7.85
25.5–39	Doerken AG	20.9.95
10.5 –1	Dow	22.1.63
10.1–21	Dow Chem. Co	7.7.93
10.2–16	Dow Chem. Co	7.6.93
2.2–28	DSM NV	5.6.96
2.2–28	Japan Synthetic Rubber Rubber Co Ltd	5.6.96
2.2–28	Japan Fine Coating Co Ltd	5.6.96

Table 31-1 (continued)

Section	Company	Date
11.1–9	Dunlop	17.10.62
11.1–9	Dunlop	9.9.63
15–11	Dunlop	9.5.64
17.1–14	Dunlop	22.7.64
10–12	Dunlop	6.4.66
11.1–47	Dunlop	16.7.66
11.2–33	Dunlop	11.1.67
11.2–33	Dunlop	21.3.67
11.2–32	Dunlop	11.1.67
11.1–17	Dunlop	21.3.67
11.2–7	Dunlop	5.4.67
11.2–7	Dunlop	4.11.67
11.2–57	Dunlop	21.4.67
11.1–6	Dunlop	21.4.67
11.2–55	Dunlop	4.11.67
11.2–55	Dunlop	2.4.68
11.2–54	Dunlop	4.11.67
11.2–54	Dunlop	2.4.68
11.1–4	Dunlop	4.11.67
11.1–4	Dunlop	2.4.68
11.2–52	Dunlop	10.1.68
11.2–52	Dunlop	28.3.68
11.1–16	Dunlop	10.1.68
11.1–16	Dunlop	28.3.68
11.2–27	Dunlop	2.4.68
11.2–28	Dunlop	16.4.68
11.2–25	Dunlop	20.4.68
11.1–7	Dunlop	20.4.68
11.2–26	Dunlop	29.6.68
21–61	Dunlop	10.7.68
11.1–3	Dunlop	1.5.70
10.5–8	Dunlop AG	1.8.67
11.2–36	Dunlop Co Ltd	6.4.66
24–6	Dunlop Co Ltd	13.10.66
11.2–53	Dunlop Co Ltd	20.4.68
24–3	Dunlop Rubber Co	6.11.63
7–21	DuPont	29.7.52
16–21	DuPont	4.11.52
16.2–5	DuPont	4.11.52
10–2	DuPont	30.6.53
12–4	DuPont	30.6.53
25.5–6	DuPont	30.6.53
12–4	Dupont	21.5.54
8–36	DuPont	1.9.54
8.2–1-	DuPont	1.10.54
4–2	DuPont	28.1.57
8.3–1	DuPont	28.1.57
8.4–2	DuPont	28.1.57

Table 31-1 (continued)

Section	Company	Date
4–4	DuPont	28.1.58
8.2–12	DuPont	3.7.58
9.1–1	DuPont	22.1.59
8.2–12	DuPont	3.2.59
4–2	DuPont	17.3.59
8.2–2	DuPont	24.8.59
4–2	DuPont	17.11.59
8.2–6	DuPont	17.11.59
9.2–11	DuPont	7.2.60
21.1–27	DuPont	9.5.60
21–20	DuPont	1.12.60
8.3–3	DuPont	21.2.61
25.4–16	DuPont	6.4.61
7.2–33	DuPont	31.10.61
8.4–14	DuPont	.31.10.61
8.4–12	DuPont	5.1.62
10.1–23	DuPont	1.3.62
8.3–2	DuPont	13.8.62
17.1–19	DuPont	27.5.63
22–70	DuPont	30.12.63
25.4–17	DuPont	30.12.63
8.2–5	DuPont	6.3.64
8.3–5	DuPont	6.3.64
8.2–26	DuPont	27.3.64
8.2–25	DuPont	27.3.64
8.2–24	DuPont	27.3.64
8.4–15	DuPont	31.8.64
10.3–4	DuPont	6.11.64
21–80	DuPont	8.3.65
21.1–26	DuPont	3.11.65
8.2–38	DuPont	22.12.65
8.3–6	DuPont	22.12.65
22–11	DuPont	12.1.66
22–11	DuPont	29.12.66
21.1–2	DuPont	29.9.66
8.4–32	DuPont	19.12.66
8.4–3	DuPont	1.2.67
23.2–3	DuPont	9.5.67
11.1–32	DuPont	6.11.67
22–71	DuPont	29.11.67
18–2	DuPont	18.6.68
19–3	DuPont	18.6.68
19–1	DuPont	18.6.68
8.3–30	DuPont	22.10.68
21–81	DuPont	16.6.69
8.3–18	DuPont	23.6.69
21–14	DuPont	23.6.69
17.1–21	DuPont	21.5.70
8.4–28	DuPont	1.9.71
21–94	DuPont	7.9.78
22–83	DuPont	29.3.96

Table 31-1 (continued)

Section	Company	Date
12–40	DuPont + Procter & Gamble	29.5.96
12–40	DuPont + Procter & Gamble	6.11.96
11–8	Dynamit Nobel AG	8.3.66
8.6–55	Dynic Corp	7.7.77
18.1–19	Eastman	20.11.62
18–13	Eastman	20.11.62
25.5–77	Eastman Kodak Co	29.12.95
25.5–28	Ebara Inflilco KK	10.8.76
18.1–39	Ehrnberg	3.9.69
11.2–12	El-Baz Nouchy, C	24.4.69
10.6–3	Electrical Storage Battery Co	6.10.59
13–8	Electro-Chem. Ltd	27.12.63
15–15	Elie Adjiman	22.12.66
25.5–45	Enviromental Technologies Europa Ltd	11.2.92
25.5–45	Enviromental Technologies Europa Ltd	3.6.92
25.5–45	Enviromental Technologies Europa Ltd	24.8.92
28–17	Enviromental Technologies Europa Ltd	11.2.92
28–17	Info-Ges. Informatik Management & Consulting	3.6.92
28–17	Enviromental Technologies Europa Ltd	24.8.82
21–92	Es Cipoipari Kutato Ungar	26.4.82
21–92	Bor-Mu Bor Budapest	4.3.80
7–51	ESB-Reeves Corp	27.6.62
16.2–9	Et. A. Chromarat et Cie	17.11.69
15–19	Et. La Chaignaud	23.12.69
7–63	Evans Prod. Co	13.2.79
25.5–40	Exxon Chem. Pat. Inc	20.9.94
9.2–1	Feldmühle	13.9.63
24–11	Ferguson, Shies Ltd	14.1.61
7–50	Fiber Ind Inc	3.4.67
10–18	Fiber Ind Inc	8.6.67
8.8–7	Fil Mat. Leather	5.8.66
22–66	Film Mat. Synth. Leather	13.6.75
22–34	Film Mater Synth	10.3.72
18.4–4	Film Material Res Inst: Mosc. Light Ind Techn Inst	30.4.91
8.5–30	Film Materials Synt Leather Res Inst	4.6.91
18.2–25	Film Materials Synth Leather Res.Inst	4.6.91
17.1–9	Forschungsinstitut für Textiltechnologie	30.9.68
18.1–51	Forschungsinstitut für Textiltechnologie	6.11.69
7.1–25	Forschungsinstitut für die Leder- & Kunstledertechnologie	19.11.74
7.1–21	Forschungsinstitut für die Leder- & Kunstledertechnologie	5.11.79
25.5–53	Forschungsinstitut für Leder & Kunstleder	22.12.87
8.5–34	Forschungsinstitut für Leder & Kunstleder	4.4.79
8.1–19	Forschungsinstitut für Leder- & Kunstledertechnologie	8.1.73
8.5–35	Forschungsinstitut für Leder- & Kunstledertechnologie	17.3.78
20–20	Forschungsinstitut Leder & Kunstleder; Lederfaserwerk Siebenlehn	9.5.88
20–22	Forschungsinstitut Leder & Kunstleder; Lederfaserwerk Siebenlehn	9.5.88

Table 31-1 (continued)

Section	Company	Date
20–21	Forschungsinstitut Leder & Kunstleder; Lederfaserwerk Siebenlehn	9.5.88
7.1–22	Fourier, St	29.10.79
18.2–41	Frauenhofer-Ges	8.2.89
7–7	Freudenberg	1.10.48
12–30	Freudenberg	18.6.60
21–25	Freudenberg	30.1.61
21–26	Freudenberg	23.5.61
21–77	Freudenberg	12.2.62
21–30	Freudenberg	16.5.62
21–31	Freudenberg	27.10.62
21–40	Freudenberg	21.5.63
7–34	Freudenberg	4.10.63
9.1–13	Freudenberg	25.10.63
9.1–18	Freudenberg	30.10.63
11.1–10	Freudenberg	1.11.63
21–41	Freudenberg	29.11.63
21–54	Freudenberg	24.10.64
8.4–10	Freudenberg	9.3.65
8.4–11	Freudenberg	8.4.65
8.4–16	Freudenberg	24.4.65
8.5–16	Freudenberg	24.4.65
8.5–14	Freudenberg	24.4.65
8.5–15	Freudenberg	2.9.65
11.1–1	Freudenberg	1.11.65
21–67	Freudenberg	16.2.66
13–6	Freudenberg	7.4.66
12–22	Freudenberg	26.10.66
11.1–27	Freudenberg	27.10.66
17.1–11	Freudenberg	26.9.67
15–17	Freudenberg	28.2.68
10.2–11	Freudenberg	22.8.70
21–12	Freudenberg	15.9.71
17.1–16	Freudenberg	22.9.71
9.1–20	Freudenberg	6.4.78
17–6	Freudenberg	10.7.81
7.1–8	Fuji Shashin Film KK	6.10.67
8.1–28	Fuji Spinning	11.6.76
20.1–27	Fuji Spinning Co Ltd	20.9.67
20.1–6	Fuji Spinning Co Ltd	29.4.65
9.2–27	Fujikura Rubber Works	1.6.78
7–67	Fujikura Rubber Works KK	18.3.82
11.1–33	Fujikura Rubber Works Ltd	29.7.63
7.2–27	Fujikura Rubber Works Ltd	21.11.63
7–68	Fujikura Rubber Works Ltd	30.3.82
7.2–8	Fujikura Rubber Works Ltd	20.7.62
18–4	Fujimori Kogyo Co	19.4.93
18–4	Fujimori Kogyo Co	7.7.93
15–27	Fukuoka, K	6.4.67
11.1–44	Gebr. Holzapfel	9.1.68
11.1–54	Gen. Latex and Chem. Corp	6.1.71

Table 31-1 (continued)

Section	Company	Date
7–18	General Electric Co	20.5.68
11.2–18	General Foam Corp	21.10.63
10–14	General Tire	20.1.64
10–13	General Tire	20.1.64
10–16	General Tire	27.8.69
10–18	General Tire	25.6.65
8.3–7	General Tire	14.11.69
8.4–5	General Tire	11.12.69
11.1–31	Genset Corp	2.2.65
11.2–43	Genset Corp	14.9.65
11.2–37	Genset Corp	5.7.66
11.2–29	Genset Corp	5.4.68
8.2–30	Genset Corp	6.6.68
8.2–40	Genset Corp	4.12.68
8.2–57	Geoscience Instr Corp	8.3.73
11.1–20	Glander, W	18.11.69
21–42	Glanzstoff	24.7.63
21–70	Glanzstoff	14.11.66
21–63	Glanzstoff	19.10.67
21–59	Glanzstoff	26.3.69
25.2–4	Glanzstoff	13.7.68
21–78	Glanzstoff AG	14.12.65
8.4–21	Glanzstoff AG	18.6.66
22–8	Glanzstoff AG	23.12.67
8.9–8	Glanzstoff AG	23.1.68
8.4–25	Glanzstoff AG	17.12.68
8.3–19	Glanzstoff AG	16.7.69
21–43	Glanzstoff	17.9.63
18.2–17	Goldschmidt AG	5.2.93
7–17	Goodrich	4.8.54
10.3–1	Goodrich	29.12.54
7–8	Goodrich	23.2.55
10.3–5	Goodrich	30.4.57
14–7	Goodrich	18.12.57
11.1–40	Goodrich	19.12.62
8.4–9	Goodrich	31.10.63
8.5–26	Goodrich	18.5.65
8.2–37	Goodrich	18.8.65
8.4–23	Goodrich	18.8.65
8.5–20	Goodrich	16.8.66
8.2–27	Goodrich	3.10.66
25.1–2	Goodrich	30.8.67
17.1–10	Goodrich	29.4.68
8.1–13	Goodrich	28.3.69
8.5–21	Goodrich	24.11.69
9.-19	Goodrich	24.7.70
10–23	Goodrich Co	30.4.57
10–15	Goodyear Tire & Rubber Co	3.11.65
11.2–4	Goodyear Tire	17.11.69
7–9	Göppinger Kaliko	8.2.54
7–10	Göppinger Kaliko	4.10.54
7–9	Göppinger Kaliko	16.11.54

Table 31-1 (continued)

Section	Company	Date
7–19	Göppinger Kaliko	4.10.55
10–11	Göppinger Kaliko	18.4.57
10–10	Göppinger Kaliko	17.9.58
18.1–3	Göppinger Kaliko	20.2.60
9.3–15	Göppinger Kaliko	22.2.60
7–29	Göppinger Kaliko	12.12.63
7–20	Göppinger Kaliko	21.12.63
9.3–7	Göppinger Kaliko	7.11.66
7–37	Göppinger Kaliko	22.7.68
19–5	Göppinger Kaliko	13.8.70
7–72	Göppinger Kaliko und Kunstleder-Werke GmbH	16.11.64
18.1–21	Gorce, M	9.12.66
16–1	Gore	21.3.62
16–2	Gore	3.10.69
16–16	Gore	22.8.77
18–3	Gore	5.5.94
16–8	Gore & Ass	3.7.73
29.1–16	Gore & Ass	20.6.78
16–7	Gore & Ass	29.6.78
25.5–6	Gore & Ass	22.4.80
16–6	Gore & Ass	19.11.82
25.5–7	Gore & Ass	6.7.88
16–10	Gore & Ass	27.7.88
25.5–	Gore & Ass	17.1.89
16–13	Gore & Ass	16.3.93
25.5–24	Gore & Ass	9.8.93
23.1–27	Gore & Ass Inc	14.1.97
18.2–76	Gore & Ass Inc.	17.4.96
16–15	Gore & Ass Inc	12.7.91
25.5–86	Gore & Ass GmbH	27.10.95
16–	Gore &Ass	19.11.82
16–9	Gore &Ass	29.6.88
18.1–47	Gore and Ass Inc	24.6.94
25.5–44	Gore Hybrid Technologies Inc	23.12.94
7.2–21	Grace	6.7.62
18.2–46	Grace & Co	3.5.72
18.2–61	Grace & Co	9.2.76
18.2–48	Grace & Co	9.2.76
18.2–47	Grace & Co	9.2.76
25.5–15	Grace & Co	1.4.93
7–16	Grace Co	19.5.66
21–69	Grace Co	27.4.65
17.1–15	Grace	9.10.64
13–10	Griffine	21.6.63
14–8	Gummi-Werke Richterswil	13.12.57
9.3–2	Gurit AG	28.12.66
25.4–18	Hanny Chem. Co Ltd	7.1.72
25.4–13	Haobas, F	17.3.66
21–37	Hercules Powder Co	10.4.62
28–14	Herfeld GmbH & Co KG; Hornschuch AG	28.10.92
22–57	High Cedar Enterprises Co Ltd	27.11.91
16–14	Himont Inc	3.8.93

Table 31-1 (continued)

Section	Company	Date
18.1–25	Hiroshima Kasei Co Ltd	19.12.63
20.1–11	Hisao, S	24.8.67
18.3–7	Hitachi Chemical	10.4.67
25.4–3	Hitachi Chemical	10.4.67
8.6–2	Hitachi Kasei KK	27.10.65
20.2–15	Hodogaya Chem. Co Ltd	10.6.94
22–7	Hodogaya Chem. Co Ltd	26.7.72
22–39	Hodogaya Chem. Ind KK	5.2.73
8.2–79	Hoechst AG	19.12.77
11.2–51	Holzapfel & Co KG	21.6.72
18.1–13	Holzstoff SA	23.1.74
20.1–20	Honey Chem. Co	4.8.69
22–24	Honey Chem. Co Ltd	20.2.70
8.8–3	Honey Chem. Co	24.2.71
18.3–22	Honey Chem. Co Ltd	1.9.69
8.5–32	Honey Chemical Industry	4.2.77
18.3–12	Honey Chemicals	6.11.68
18.3–13	Honey Chemicals	23.1.69
18.3–13	Honey Chemicals	22.3.69
18.3–44	Honey Chemicals	21.12.76
18.3–44	Honey Chemicals	4.2.77
18.1–34	Honey Kasei KK	20.2.70
18.2–24	Honey Kasei KK	20.2.70
11.2–2	Honey Otafokuwata	25.12.65
2.2–8	Hooker Chem	2.4.70
19–16	Hooker Chem. & Plastics Corp	3.7.72
11.1–38	Hooker Chem. Corp	18.12.67
11.1–38	Hooker Chem. Corp	20.11.68
8–25	Hornschuch	30.10.81
8.2–87	Hornschuch AG	3.10.81
18.2–28	Hosokawa Micron KK	11.10.91
18.2–6	Hosokawa Micron KK; Daiichi Lace KK	1.3.90
11–3	Hutchinson Co Nat. du Caoutchouc	24.3.71
11–3	Hutchinson Co Nat. du Caoutchouc	27.12.71
2.1–3	I.G. Farben	19.1.39
9.3–22	IBM	30.12.63
8.2–80	Ichikawa Keori KK	11.11.77
8.4–30	Ichikawa Woolen Textile Co	3.1.67
11.1–22	ICI	5.7.57
10–6	ICI	9.3.60
21–27	ICI	11.7.61
21–27	ICI	20.6.62
21–44	ICI	7.5.62
21–32	ICI	7.5.62
21–49	ICI	5.2.63
21–45	ICI	5.2.63
21–48	ICI	1.3.63

Table 31-1 (continued)

Section	Company	Date
21–47	ICI	4.3.63
21–47	ICI	15.7.63
21–46	ICI	29.11.63
21–56	ICI	19.8.64
11.2–44	ICI	3.2.65
11.1–24	ICI	10.8.65
11.1–65	ICI	12.7.68
21–75	ICI	31.8.72
20.1–13	ICI Ltd	30.5.49
11.1–2	ICI Ltd	17.9.68
13–5	ICI	26.11.65
18.2–68	Idemitsu Petrochem Co	30.11.90
11.2–63	Immont	19.9.77
9.217	Immont Corp	16.5.66
24–14	Immont Corp	23.5.79
22–42	Immont Corp	3.9.71
20–14	Ind. Fibre e Cartoni Speciale	20.1.70
8.9–4	Inmont	2.12.68
11.2–62	Inmont	30.1.75
23.2–2	Inmont Corp	9.12.68
25.1–5	Inmont Corp	25.4.69
8.9–1	Inmont Corp	22.7.69
8.9–1	Inmont Corp	20.10.69
11.2–48	Inmont Corp	4.12.70
25.4–2	Inmont Corp	13.1.72
12–31	Inmont Corp	23.5.79
12–31	Inmont Corp	23.5.79
23–4	Inmont Corp	9.6.69
23–4	Inmont Corp	16.6.69
4–9	Inmont Corp	19.9.73
8.9–9	Inmont Corp	19.10.74
7.1–2	Institut für Textiltechnologie	18.10.66
8.2–78	Interbrinderea de Piele Sintetitca Bucure	12.8.76
7–28	Interchemical Corp	15.3.63
9.2–25	Interchemical Corp	16.6.65
9.–18	Interchemical Corp	8.10.65
11.1–23	Interchemical Corp	10.2.58
18–11	Ions Exchange and Chemical Corp	25.10.55
7–22	IRPC	6.2.73
10.6–1	IRPC Corp	6.2.73
15–13	Ishizuka, K	7.5.63
7–53	Ishizuka, T	27.12.62
7–56	Ishizuka, T	2.5.63
22–52	Itakura Shoji KK	13.10.92
7.2–46	Ivan Chem. Tech Inst	17.4.78
18.1–24	Izeki, T	5.2.62
7–49	Jacog, EJ	7.7.70
25.4–25	Japan Energy Corp	28.4.94
25.5–13	Japan Gore-Tex Inc	29.8.92
16–11	Japan Gore-Tex KK	1.3.83

Table 31-1 (continued)

Section	Company	Date
16–4	Japan Gore-Tex. Inc	11.12.85
16–4	Japan Gore-Tex. Inc	7.2.86
18.4–6	Japan Steel Works Ltd	15.10.90
25.5–94	Japan Synthetic Rubber Co Ltd	29.3.96
8.6–29	Japan Vilene	16.9.70
21–73	Japan Vilene Co Ltd	24.9.70
12–9	Johnson & Son Inc	14.1.57
14–10	Jung & Simmons	22.4.64
21–62	K Narodni podnik	29.7.68
8.1–9	Kabushiki Kaisha Kobunshi Oyo Kenkyusho	27.1 71
20.1–39	Kako Seisakusho	11.3.96
8.6–30	Kakuda Kagaku KK	13.11.68
8.8–1	Kalle	4.4.64
8.2–8	Kalle	31.7.64
21–79	Kalle	2.11.66
12–12	Kalle	22.12.66
11.2–39	Kalle	30.12.66
21–64	Kalle	30.11.67
8.2–9	Kalle	30.11.67
8.1–7	Kalle	21.12.67
22–18	Kalle	14.3.68
8.1–10	Kalle	29.1.69
12–19	Kalle	3.4.69
8.1–4	Kalle	11.8.69
22–3	Kalle	14.8.69
11.1–34	Kalle	15.10.69
8.1–12	Kalle	24.10.69
12–20	Kalle	22.12.69
21–10	Kalle	22.9.70
21–72	Kalle	30.10.70
7.2–18	Kanebo	8.8.61
8.6–50	Kanebo	26.7.68
8.1–17	Kanebo	7.3.70
8.6–22	Kanebo	6.7.70
25.4–10	Kanebo	28.12.72
8.6–56-	Kanebo	12.7.77
8.9–11	Kanebo	18.11.77
21.1–9	Kanebo	27.2.78
22–75	Kanebo	2.3.78
22–75	Kanebo	30.1.78
15–25	Kanebo	31.10.78
8–26	Kanebo	16.12.78
18.2–74	Kanebo	19.12.78
18.3–25	Kanebo	17.9.82
8.5–43	Kanebo	25.2.85
8.5–45	Kanebo	15.5.85
8.6–58	Kanebo	30.9.87
8.4–39	Kanebo	24.12.87
8.1–26	Kanebo	24.12.87
8–27	Kanebo	22.6.88
21.1–30	Kanebo	17.2.92

Table 31-1 (continued)

Section	Company	Date
23.1–9	Kanebo	25.12.92
25.5–51	Kanebo	10.2.95
19–36	Kanebo	25.5.95
8,1–32	Kanebo	26.2.85
8.6–8	Kanegafuchi Boseki KK	17.5.67
8.5–11	Kanegafuchi Boseki KK	18.11.67
8.6–23	Kanegafuchi Boseki KK	19.7.69
8.6–14	Kanegafuchi Boseki KK	13.4.70
8.6–3	Kanegafuchi Spinning	13.11.65
8.6–21	Kanegafuchi Spinning	1.12.66
8.5–27	Kanegafuchi Spinning Co	8.7.66
7.2–23	Kanegafuchi Spinning Co Ltd	10.12.62
7.2–9	Kanegafuchi Spinning Co Ltd	10.12.62
18–21	Kanegafushi Boseki KK	4.12.67
18.2–23	Kanegafushi Boseki KK	4.12.68
18.2–22	Kanegafushi Boseki KK	28.7.69
18.2–22	Kanegafushi Boseki KK	15.11.68
18–20	Kanegafushi Boseki	27.7.68
18–20	Kanegafushi Boseki	15.11.68
12–39	Kao Corp	26.1.96
12–41	Kao Corp	24.4.96
25.2–6	Kao Soap	17.6.69
8.4–35	Kao Soap Co	12.6.70
7.2–10	Katakura Ind Co Ltd	25.10.62
7.2–22	Katakura Ind Co Ltd	25.10.62
7.2–40	Katakura Kogyo Co Ltd	12.2.63
7.2–40	Katakura Ind Co Ltd	2.10.62
15–24	Kaufmann, P	19.11.73
11.1–21	Kawaguchi Rubber Ind Co	30.9.71
22–29	Kawaguchi Rubber Ind Co	7.4.72
16.2–10	Kawaguchi Rubber Ind Co Ltd	11.11.69
11.1–11	Kawaguchi, R	2.8.63
11.2–22	Kay-Metzeler Ltd	14.1.69
7–32	Kemper, RC	25.5.64
9.1–11	Kendall Co	23.5.60
9.1–2	Kendall Co	20.7.64
9.1–19	Keppeler, M	21.7.94
16.2–26	Kimoto, Y	19.12.62
20–10	Kimura, R	20.12.63
10.1–14	Kinyosha Co Ltd	15.6.70
24–18	Kleber Textil	6.7.94
7–23	Kobe Jushi Co Ltd	15.7.60
11.2–46	Koepp & Co	7.7.64
16.2–24	Kogoku Chem. Ind KK	1.10.73
7.2–1	Kogyo Gijutsu-in	4.9.65
8.7–2	Kohkoku Chem. Ind	5.12.70
8.7–1	Kohkoku Chem. Ind	30.12.70
2.2–11	Kohkoku Kagaku Kogyo KK	11.12.63
22–19	Kokoku Chem. Ind	5.12.68

Table 31-1 (continued)

Section	Company	Date
8.1–27	Kokoku Chem. Ind	10.5.84
8.1–27	Kokoku Chem. Ind	23.5.84
8.5–24	Kokoku Chem. Ind Co	3.6.69
8.2–92	Kokoku Chem. Ind Co	27.1.70
7.2–38	Kokoku Chem. Ind Co Ltd	10.11.60
8.5–38	Kokoku Chem. Ind KK	3.6.83
8–34	Kokoku Chem. Ind KK	30.1.84
8.2–77	Kokoku Chem. Ind KK	28.12.84
8.1–33	Kokoku Chem. Ind KK	29.3.85
23.1–24	Kokoku Chem. Ind KK	1.7.86
23.1–24	Kokoku Chem. Ind KK	22.9.86
8.1–24	Kokoku Chem. Ind KK	23.12.88
7–58	Kokoku Chem. Ind KK	20.1.89
11.1–25	Kokoku Chem. Ind Co Ltd	16.7.65
11.2–56	Kokoku Chem. Ind Co Ltd	3.9.65
11.1–12	Kokoku Chem. Ind Co Ltd	3.9.63
15–21	Kokoku Chem. Ind Co Ltd	16.5.69
11.2–20	Kokoku Chem. Ind Co Ltd	9.11.63
11.2–20	Kokoku Chem. Ind Co Ltd	11.11.63
11.1–14	Kokoku Chem. Ind Co Ltd	15.8.64
15–20	Kokoku Chem. Ind Co Ltd	15.9.69
11.2–19	Kokoku Kagaku Kogyo Co Ltd	27.8.63
8.3–31	Komatsu Seiren KK	3.12.81
2.1–2	Kötitzer Ledertuch- & Wachstuchwerke	8.7.37
7–10	Kötitzer Ledertuch- & Wachstuchwerke	23.3.54
20.1–3	Kotov, MP	26.10.64
19–13	Koyo Kogyo Co Ltd	27.11.62
15–9	Kufner Textilwerke KG	22.3.71
8.6–62	Kunimine Kogyo	13.6.79
8.2–85	Kunimine Kogyo KK	13.6.79
7.2–11	Kunstzijdespinnerij Nyma NV	28.4.61
13–13	Kuraray	27.2.13 12.6.70
8.2–49	Kuraray	3.2.65
8.6–38	Kuraray	29.7.65
8.5–5	Kuraray	25.11.65
8.5–6	Kuraray	4.3.66
8.3–12	Kuraray	26.1.67
8.7–3	Kuraray	17.2.67
19–9	Kuraray	4.10.67
7–38	Kuraray	8.2.68
8.6–42	Kuraray	17.7.68
8.6–27	Kuraray	13.12.68
8.1–15	Kuraray	9.4.69
25.1–3	Kuraray	9.4.69
25.1–7	Kuraray	11.4.69
7.2–29	Kuraray	26.4.69
23.1–18	Kuraray	26.4.69
25.3–3	Kuraray	16.5.69
8.9–2	Kuraray	7.6.69
7.2–34	Kuraray	29.9.69

Table 31-1 (continued)

Section	Company	Date
22–16	Kuraray	29.9.69
25.1–6	Kuraray	11.11.69
22–72	Kuraray	25.12.69
22–26	Kuraray	24.3.70
18.2–13	Kuraray	27.3.70
22–5	Kuraray	13.5.70
2.2–5	Kuraray	23.6.70
16.1–10	Kuraray	6.8.70
8.6–45	Kuraray	15.10.70
21–82	Kuraray	7.12.70
25.1–8	Kuraray	22.12.70
25.1–8	Kuraray	21.12.70
21–74	Kuraray	2.9.71
22–4	Kuraray	10.3.72
21.1–18	Kuraray	22.3.72
19–8	Kuraray	11.4.72
11–13	Kuraray	10.5.72
21.1–22	Kuraray	6.6.72
21.1–22	Kuraray	10.6.72
8.5–8	Kuraray	22.6.72
8.5–10	Kuraray	30.8.72
8.5–46	Kuraray	7.10.72
22–48	Kuraray	17.2.73
8.5–12	Kuraray	31.3.73
22–36	Kuraray	20.4.73
22–35	Kuraray	25.4.73
18.2–71	Kuraray	7.5.73
24–16	Kuraray	7.5.73
22–37	Kuraray	17.5.73
23.2–4	Kuraray	19.7.73
21.1–7	Kuraray	21.8.73
23–3	Kuraray	2.10.73
8.2–58	Kuraray	25.12.74
22–69	Kuraray	18.5.76
22–74	Kuraray	14.6.76
22–76	Kuraray	5.9.77
21.1–14	Kuraray	28.11.77
8.2–81	Kuraray	9.12.77
8.2–83	Kuraray	7.2.78
8.2–82	Kuraray	24.3.78
7.1–19	Kuraray	24.3.78
22–47	Kuraray	21.6.78
22–46	Kuraray	17.1.79
19–22	Kuraray	17.1.79
7–64	Kuraray	10.7.79
25.4–33	Kuraray	3.8.79
22–61	Kuraray	13.2.80
8.6–60	Kuraray	21.7.81
8.5–36	Kuraray	16.11.81
25.4–21	Kuraray	30.9.82

Table 31-1 (continued)

Section	Company	Date
25.4–20	Kuraray	30.11.82
25.3–8	Kuraray	13.12.83
8.5–42	Kuraray	28.12.83
8.5–44	Kuraray	19.7.84
8.5–41	Kuraray	11.2.85
19–44	Kuraray	8.3.85
8.5–40	Kuraray	23.7.87
8.2–59	Kuraray	28.9.87
8.5–39	Kuraray	16.3.88
8–21–	Kuraray	16.11.90
25.4–28	Kuraray	1.7.91
25.4–28	Kuraray	4.9.91
25.4–29	Kuraray	18.10.91
25.4–30	Kuraray	6.12.91
23.1–23	Kuraray	19.5.92
23.1–7	Kuraray	6.12.92
23.1–7	Kuraray	6.11.92
23.1–7	Kuraray	30.10.92
19–33	Kuraray	22.1.93
21–88	Kuraray	15.2.93
23.1–10	Kuraray	22.3.93
19–29	Kuraray	3.6.93
23.1–12	Kuraray	4.6.93
19–27	Kuraray	29.10.93
21.1–10	Kuraray	9.11.93
8.2–66	Kuraray	1.12.93
8.2–66	Kuraray	24.11.93
25.4–23	Kuraray	1.12.93
17.1–24	Kuraray	29.9.95
22–82	Kuraray	11.3.96
22–82	Kuraray	19.3.96
2.2–27	Kuraray	14.3.96
22–15	Kuraray	29.3.96
24–8	Kuraray	30.9.68
8.6–25	Kuraray	28.12.68
25.3–2	Kuraray	16.5.69
22–8	Kuraray	30.4.70
22–14	Kuraray	1.3.96
22–79	Kuraray	6.3.96
18.1–36	Kurare Plastics Co Ltd	20.4.64
10.3–8	Kurashiki Boseki KK	12.11.65
21–52	Kurashiki Rayon	19.3.63
21.1–19	Kurashiki Rayon	19.3.63
7.2–43	Kurashiki Rayon	16.4.63
14–11	Kurashiki Rayon	15.7.63
21–51	Kurashiki Rayon	20.9.63
8.3–4	Kurashiki Rayon	15.11.63
8.3–4	Kurashiki Rayon	11.12.63
8.3–4	Kurashiki Rayon	29.1.64

Table 31-1 (continued)

Section	Company	Date
21.1–8	Kurashiki Rayon	12.12.63
21–57	Kurashiki Rayon	4.7.64
22–12	Kurashiki Rayon	7.9.64
21.1–17	Kurashiki Rayon	25.9.64
8.5–1	Kurashiki Rayon	14.11.64
8.2–29	Kurashiki Rayon	3.12.64
8.3–26	Kurashiki Rayon	4.12.64
8.5–2	Kurashiki Rayon	3.2.65
8.5–2	Kurashiki Rayon	2.8.65
8.2–32	Kurashiki Rayon	15.10.65
8.5–4	Kurashiki Rayon	25.11.65
8.5–3	Kurashiki Rayon	25.11.65
8.1–1	Kurashiki Rayon	12.3.66
8.1–1	Kurashiki Rayon	22.3.66
11.2–3	Kurashiki Rayon	18.5.66
11.2–38	Kurashiki Rayon	24.5.66
8.5–7	Kurashiki Rayon	27.5.66
8.5–7	Kurashiki Rayon	9.6.66
18.1–8	Kurashiki Rayon	13.10.66
8.2–11	Kurashiki Rayon	10.2.67
25.4–19	Kurashiki Rayon	10.2.67
22–9	Kurashiki Rayon	30.6.67
8.9–7	Kurashiki Rayon	26.10.67
8.9–5	Kurashiki Rayon	4.11.67
8.9–3	Kurashiki Rayon	21.11.67
8.9–3	Kurashiki Rayon	23.1.68
8.9–6	Kurashiki Rayon	29.12.67
8.4–38	Kurashiki Rayon	17.1.68
8.5–9	Kurashiki Rayon	17.1.68
8.1–18	Kurashiki Rayon	5.2.68
8.1–3	Kurashiki Rayon	8.5.68
8.1–3	Kurashiki Rayon	16.5.68
8.1–3	Kurashiki Rayon	8.6.68
8.1–3	Kurashiki Rayon	1.8.68
8.1–3	Kurashiki Rayon	27.5.68
25.4–12	Kurashiki Rayon	16.12.68
25.4–12	Kurashiki Rayon	19.12.69
25.4–12	Kurashiki Rayon	15.3.69
8.6–40	Kurashiki Rayon	2.4.69
8.6–40	Kurashiki Rayon	24.4.69
8.6–40	Kurashiki Rayon	26.4.69
10.3–10	Kurashiki Spinning	31.3.65
17–2	Kurashiki Spinning	13.8.73
18.4–9	Kurashiki Spinning Corp; Shohikagaku Kenkyusho KK	19.2.96
19–19	Kurashiki Spinning KK	15.8.73
7–31	Kureha Kagaku Kogyo KK	8.5.70

Table 31-1 (continued)

Section	Company	Date
16.2–1	Kureha Kagaku Kogyo KK	23.7.71
16.2–1	Kureha Kagaku Kogyo KK	18.11.71
7–47	Kureha Kagaku Kogyo KK	22.6.72
16.2–22	Kureha Kagaku Kogyo KK	5.10.71
7.2–44	Kureha Seni Co. Ltd.	27.1.65
16.1–2	Kyodo Kasei Kogyo Co Ltd	29.12.62
20.1–32	Kyoeisha Kagaku KK	31.7.92
7–75	Kyoeisha Kagaku KK	24.12.92
16.2–34	Kyoeisha Kagaku KK; Osaka Prefecture, Tomen KK	24.12.92
8.4–1	Kyowa	7.9.63
18.3–6	Kyowa Hakko Kfgyo Co Ltd	20.4.70
25.4–5	Kyowa Hakko Kogyo Co Ltd	22.11.67
25.4–4	Kyowa Hakko Kogyo Co Ltd	20.4.70
18.3–8	Kyowa Hakko Kogyo Co Ltd	22.11.67
18.3–3	Kyowa Hakko Kogyo Co Ltd	2.9.68
18.3–5	Kyowa Hakko Kogyo Co Ltd	16.4.70
18.4–1	Kyowa Hakko Kogyo Co Ltd	20.5.70
18.3–2	Kyowa Hakko Kogyo Co Ltd	20.10.67
18.3–2	Kyowa Hakko Kogyo Co Ltd	4.6.68
18.3–21	Kyowa Leather Cloth	19.2.66
22–40	Kyowa Leather Cloth	18.12.67
8.6–26	Kyowa Leather Cloth	23.5.68
8.4–24	Kyowa Leather Cloth	7.7.69
8.1–16	Kyowa Leather Cloth	9.10.69
18.3–16	Kyowa Leather Cloth	14.12.70
8.4–36	Kyowa Leather Cloth	15.12.70
8.4–34	Kyowa Leather Cloth	15.12.70
7–62	Kyowa Leather Cloth	31.8.74
8.3–36	Kyowa Leather Cloth.	7.11.72
11.2–11	Kyowa Leather Co Ltd	9.4.64
7.2–32	Kyowa Leather Co Ltd.	27.10.60
7.2–12	Kyowa Leather Co Ltd.	27.10.60
8.1–29	Kyowa Rubber Ind KK	8.10.77
8–29	Kyowa Rubber Ind KK	8.10.77
8–29	Kyowa Rubber Ind KK	11.10.77
20.2–3	Lankro Chem. Ltd	12.3.68
20.2–3	Lankro Chem. Ltd	9.12.68
18.1–7	Lantor Ltd	9.12.59
21–33	Lantor Ltd	24.1.62
10.3–11	Lantor Ltd	9.3.65
22–20	Latsuraya Fine Goods KK	23.5.68
25.5–57	Ledergerber, WJ	22.2.91
12–21	Licencia Talalmanyokat Ertekesitö Vallalat	16.12.67
20.1–14	Linorusta	2.6.72
18.1–22	Little Inc	21.6.60
11.1–49	Little Inc	23.8.65
13–18	Lonseal	15.4.78
8–28	Lonseal Corp	20.3.77
20–1	Lorant, I	9.10.68
15–28	Lorica SpA	19.6.87

Table 31-1 (continued)

Section	Company	Date
22–59	Lorica SpA	19.3.91
12–25	Ludlow Corp	21.9.64
12–25	Ludlow Corp	25.6.65
11.1–26	Marles-Kuhlmann-Wyandotte	9.4.65
11.1–26	Marles-Kuhlmann-Wyandotte	19.8.65
20.1–35	Maruhatchi Muramatsu KK; Nitta Gelatine KK	1.3.91
11–9	Matsumoto Yushi Seiyaku	16.4.79
25.5–43	Matsushita Reiki KK; Mitsubishi Jukogyo KK; Komatsu Seiren KK	13.12.94
8.2–70	Med.Polymer Res Inst	9.6.92
25.5–9	Membrane Techn	23.7.86
21–65	Metallgesellschaft AG	13.12.67
7.2–47	Millipore Corp	25.5.77
18.3–27	Mitsubishi Chem. Ind KK	31.1.83
18.3–29	Mitsubishi Chem. Ind KK	23.1.86
8.1–21	Mitsubishi Kagaku KK	19.1.94
25.5–36	Mitsubishi Kagaku KK	24.5.94
9.3–6	Mitsubishi Kasei Corp	24.10.91
21–87	Mitsubishi Paper Mills	28.11.72
8–18	Mitsubishi Paper Mills Ltd	11.10.72
20–15	Mitsubishi Petrochem KK	13.8.73
11.1–64	Mitsubishi Petrochem KK	13.8.73
21–66	Mitsubishi Petrochemical	18.8.67
16.1–11	Mitsubishi Plastics Ind Ltd	30.6.70
8.8–5	Mitsubishi Rayon	8.3.66
8.2–20	Mitsubishi Rayon	2.6.66
8.1–2	Mitsubishi Rayon	14.6.66
8.4–26	Mitsubishi Rayon	22.12.66
25.2–1	Mitsubishi Rayon	23.10.67
25.2–3	Mitsubishi Rayon	31.10.67
8.4–18	Mitsubishi Rayon	2.11.67
7.1–5	Mitsubishi Rayon	17.9.68
8.1–14	Mitsubishi Rayon	12.3.70
7.1–10	Mitsubishi Rayon	3.6.70
11.1–60	Mitsubishi Rayon	11.6.73
23.2–9	Mitsubishi Rayon	13.11.76
8.6–57	Mitsubishi Rayon	14.12.77
7–57	Mitsubishi Rayon	3.6.80
19–23	Mitsubishi Rayon	6.7.81
8.3–32	Mitsubishi Rayon	30.9.82
8.5–37	Mitsubishi Rayon	17.1.83
21–86	Mitsubishi Rayon	11.5.83
11.1–13	Mizuno, Y	31.8.67
15–12	Mizuno, Y	9.10.65
21–17	Mölnycke Väfveriaktiebolag	10.3.60
7.2–3	Mombers, NV Lederfabriek L	26.4.63
7.2–3	Mombers, NV Lederfabriek L	22.6.73
20.2–6	Mombers, Lederfabriek LPJ	10.3.69
7.2–28	Mombers, NV Lederfabriek L	14.7.67
21–34	Monsanto	4.4.62
21–60	Monsanto Co	24.3.69

Table 31-1 (continued)

Section	Company	Date
16.2–12	Montecatini-Edison	7.11.67
12–15	Motte-Bossut SA	7.10.60
8.2–41	Murphy, WT	16.2.71
7.2–25	Nagao, K	9.3.62
23–6	Nairn-Williamson Ltd	13.11.69
2.2–10	Nairu-Williamson Ltd	5.8.69
11.1–55	Nakamura, M	17.6.71
23.2–7	Nan Ya Plastics Corp	30.7.93
9.3–20	Nankai Gum Co Ltd	12.2.62
8.1–25	Nankai Gum KK	4.4.88
11.1–37	Naphthachimie	28.1.71
18.1–28	Nat. Patent Devel. Corp	15.10.68
18.1–9	National Polychemicals	20.10.65
15–3	Neumann, W	5.8.68
25.5–96	New Dimensions in Medicine Inc	1.9.95
20.1–25	Nichiko Co Ltd	13.5.68
20.1–22	Nihon Hikaku KK	12.11.63
20.1–10	Nihon Leather Kogyo Kabussiki Kaisha	25.11.66
7.2–36	Nihon Matai Co Ltd	6.6.70
7.2–24	Nikko Kasei Kogyo Co Ltd	12.4.62
7.2–13	Nikko Kasei Kogyo Co Ltd	12.4.62
9.3–5	Nikko Physiochem. Shik	26.2.73
9.3–3	Nikko Physiochem. Shik	26.2.73
9.3–24	Nikko Shikiryo Kogyo KK	4.11.72
18.1–6	Nino	8.5.64
11.2–23	Nippon Art Paper Mfg. Co Ltd	13.8.69
18.1–18	Nippon Burakah Kogyo KK; Nippon Miractoran KK and Toray Ind	7.5.80
8.4–29	Nippon Cloth Ind	7.5.66
8.6–37	Nippon Cloth Ind	30.9.72
22–73	Nippon Cloth Ind	8.9.72
21–16	Nippon Cloth Ind Co Ltd	23.2.71
8.6–11	Nippon Cloth Ind	27.12.68
7.2–2	Nippon Cloth Ind Co Ltd	26.7.69
16.2–21	Nippon Cloth Ind Co Ltd	31.7.70
16.2–20	Nippon Cloth Kogyo Co Ltd	26.8.70
21–35	Nippon Felt Kogyo Co Ltd	12.1.62
7.2–30	Nippon Ikoru Kagaku Kogyo KK	10.2.69
11.1–45	Nippon Kakoh Seishi KK	21.2.69
14–6	Nippon Leather Industry	17.9.65
9.3–19	Nippon Mikusani Kogyo KK	21.8.61
7–55	Nippon Orimono Kako Co Ltd	29.12.63
18.1–30	Nippon Rubber Co	16.8.63
18.1–10	Nisshin Vinyl Kogyo KK	16.9.66
18.2–63	Nitta Gelatine KK	24.1.92
16.1–13	Nitto Denko Corp	16.12.93
2.2–7	Noberasco, M	12.3.65
20.2–3	Norwood Ind Inc	20.8.68
17–9	Norwood Ind Inc	14.7.82
11.2–61	Nylco Corp	19.3.84
16.1–12	Oakwood Ind Inc	15.7.82

Table 31-1 (continued)

Section	Company	Date
15–29	Okabe Kinzoku Kogyo	19.3.81
18.2–4	Omnium de Prospective Industrielle	19.7.69
20.1–36	Otsuga Kagaku Yakuhin KK	28.6.91
25.5–84	Oxyphen GmbH	28.9.95
15–8	Palladium	30.1.52
15–8	Palladium	21.4.52
19–14	Pandel-Bradford Inc	17.8.67
11.2–1	Pandel-Bradford Inc	20.1.71
2.2–15	Pannenbecker, H	11.9.71
20.2–3	Pechiney-Ugine-Kuhlmann	5.4.72
13–3	Peiler	3.8.55
13–3	Peiler	10.12.55
11.1–50	Peltex	10.9.65
11.1–50	Peltex	24.11.65
9.3–32	Peltex	13.9.65
18.2–3	Pennel & Flipo	22.7.68
11.1–53	Pennel & Flipo	2.3.71
11.2–45	Peters, TV	6.7.65
20.2–3	Phelan-Faust Paint Mfg Co	2.1.64
12–27	Philip Morris Inc	10.8.70
21–28	Phrix-Werke AG	5.7.61
24–2	Pirelli	6.3.62
10.5–3	Pittsburgh Plate Glass	1.2.66
24–5	Plastic Coating Ltd	13.3.68
15–7	Plotnikov, IV	7.12.58
9.3–21	Plotnikov, IV	15.1.62
2.2–12	Plymouth Rubber Co Inc	3.8.68
8.2–75	Polimersintez Combi	22.1.87
9.2–12	Polymer Corp	15.2.61
7.2–35	Polymer Corp US Prior.	2.4.64
25.2–8	Polymersintez Res Prodn Assoc	22.2.91
10.2–13	Polysar Ltd	16.2.73
7–27	Porous Plastics	25.11.63
7–5	Porous Plastics	25.11.63
7–5	Porous Plastics	21.10.64
7–15	Porous Plastics	1.4.65
7–48	Porous Plastics	15.6.65
7–43	Porous Plastics	15.6.65
8.6–5	Porous Plastics	7.3.66
8.6–4	Porous Plastics	7.3.66
8.6–4	Porous Plastics	11.1.67
23–1	Porous Plastics	11.5.67
7–40	Porous Plastics	31.7.67
8.6–7	Porvair	11.1.67
8.6–6	Porvair	11.1.67
8–20	Porvair	1.5.68
8.6–13	Porvair	9.12.69
8.6–13	Porvair	28.4.70

Table 31-1 (continued)

Section	Company	Date
23–5	Porvair	12.3.70
8.6–15	Porvair	11.12.70
8.6–16	Porvair	2.6.71
8.6–17	Porvair	25.7.73
8.2–90	Porvair	20.3.74
8.6–61	Porvair	1.12.81
24–4	Porvair	18.6.69
23–2	Porvair	12.3.70
7–41	Porvais	30.3.66
16.2–2	Potters Ind Inc	15.4.71
25.5–66	PPG Ind Inc	24.4.87
25.5–58	PPG Ind Inc	24.4.87
25.5–61	PPG Ind Inc	29.10.90
9.3–13	PPG Ind Inc	22.6.70
24–7	Premier Footwear (Fleetwood) Ltd	20.7.68
7–11	Pritchett and Goldand EPS Comp Ltd	17.2.58
7–11	Pritchett and Goldand EPS Comp Ltd	26.1.59
9.3–34	R. T. Vanderbilt Co Inc	6.12.62
25.5–97	Reeves Brothers	16.12.63
18.1–18	Reeves Brothers	16.12.63
7–24	Reeves Corp	17.11.60
20–8	Reizin, RE	26.11.62
25.5–21	Rexham Ind	7.4.93
7–25	Rhodiaceta	5.7.63
8.8–6	Rhodiaceta	23.7.63
12–1	Rhodiaceta	20.5.64
8.2–16	Rhodiaceta	20.5.65
8.2–4	Rhodiaceta, Soc	27.5.64
8.2–17	Rhodiaceta, Soc	17.9.65
8.2–17	Rhodiaceta, Soc	28.12.65
8.4–33	Rhodiaceta, Soc	28.12.65
12–23	Rhone Poulenc	1.7.66
11.2–49	Richardson Co	18.2.70
20–9	Rishin, AE	27.4.62
12–33	RM Ind Products Inc	11.7.84
12–16	Rogers Corp	5.9.62
12–17	Rogers Corp	20.12.62
11.1–48	Rohm and Haas Co	28.6.66
20.2–9	Rohm and Haas Co	12.3.71
22–30	Rohm and Haas Co	2.4.73
11.1–62	Rohm and Haas Co	3.7.74
11.1–18	Röhm and Haas Co	9.7.73
21–21	Rohm and Haas Co	12.5.60
15–18	Roser Neudorf GmbH	5.8.68
11.1–41	Rubber Ind Co Ltd	24.12.63
25.5–98	SKY Polymers Inc	1.5.95
25.5–25	Saint Gobain Vitrage; Asahi Glass Co Ltd	21.10.93
19–34	Saito Yoshimitsu	27.6.94
24–10	Saltel, LC	15.10.70

Table 31-1 (continued)

Section	Company	Date
22–21	Sandoz AG	13.2.64
22–21	Sandoz AG	2.3.64
22–21	Sandoz AG	11.12.64
20.2–14	Sandoz Ltd; Sandoz Patent GmbH	16.9.92
8–35	Sanyo Chem. Ind Ltd	7.1.84
25.4–24	Sanyo Chem. Ind Ltd	15.12.93
25.4–26	Sanyo Chem. Ind Ltd	28.4.94
8.2–3	Sanyo Chem. Ind	3.12.70
18.2–12	Sanyo Chem. Ind Co Ltd	21.6.65
25.1–1	Sanyo Chem. Ind Co	21.12.66
25.4–31	Sanyo Chemical	31.3.78
7–12	Sapilevskii, PF	5.1.63
12–35	Sarna Patent und Lizenz AG	19.4.94
20.1–4	Sato, H	20.2.64
20.1–29	Sato, H	15.7.67
20.1–28	Sato, H	20.9.67
23.2–5	Satra	1.7.68
23.2–5	Satra	10.12.68
13–2	Sauterer	20.8.52
13–2	Sauterer	1.3.53
20.2–7	SAVA	5.5.69
20.2–13	Schaefer, Ph	29.3.94
24–17	Schaefer, Ph	21.3.95
12–32	Schaetti & Co	28.3.83
20.1–31	Schaller, H & G	21.7.69
14–4	Schmidt, HJ	1.12.56
14–3	Schoeler, A	22.6.56
16.2–17	Schumann, W	29.4.65
11.2–30	Scott Paper Co	20.5.68
8–30	Sehren KK	23.1.79
8.3–35	Sehren KK	12.9.86
18.3–24	Seiko Kasei KK	23.8.82
8.8–8	Seiko Kasei KK	9.1.96
22–58	Seiren Co Ltd	14.5.92
16.2–8	Seiren Co Ltd	13.12.95
11.1–58	Sekisu Chem. KK	30.5.66
11.2–47	Sekisui Chem. Ind Co Ltd	3.12.64
11.2–40	Semperit	15.9.66
25.5–29	Seton Healthcare Group PLC	8.6.94
17.1–12	Shell	20.12.66
7–33	Shibata Gomu Kogyo KK	4.10.63
11.1–39	Shibata Gomu Kogyo KK	24.5.61
18.3–15	Shibata, K	12.8.64
7–4	Shiga Akiresu	16.4.73
19–18	Shiga Akiresu KK	16.4.75
7.2–4	Shigeo	28.12.60
25.5–20	Shimizu, Y	6.8.93
18.2–50	Shirley Institute	19.4.84
18.2–50	Shirley Institute	18.4.85

Table 31-1 (continued)

Section	Company	Date
18.2–35	Shirley Institute	23.6.81
28–13	Shiseido Co. Ltd; Showa Highpolymer Japan	17.6.94
16.1–3	Shoshichi	3.8.62
18.4–5	Showa Denko KK	8.8.90
22–54	Showa Denko KK	8.2.93
18.1–46	Showa Denko KK	20.5.93
18.1–48	Showa Denko KK	2.9.93
18.1–11	Showa Rubber Co Ltd	31.5.67
13–11	Siemens-Schuckert-Werke AG	3.3.64
25.5–99	Sika AG; form. Kapsar Winkler & Co	20.5.96
16.2–18	Skin-Yamato Gomugaku Mfg Co Ltd	13.5.64
18.1–5	Skirodova, KM	18.5.59
18.1–17	Soc. des Produits Tiffine	18.7.63
8.2–62	Soko Seiren KK	24.6.93
7–1	Spirit, J et al.	10.9.56
17.1–23	Stahl Chem. Ind BV	4.6.84
9.2–20	Stamicarbon BV.	30.5.74
8.2–36	Statni Vyzkumny Ustav Eozedelny	6.5.68
8.3–17	Statni Vyzkumny Ustav Kozedny	5.3.70
22–63	Stauffer	24.1.79
22–27	Steinhardt, R	29.1.71
7–2	Stern, IA	19.12.59
7–13	Stern, IA	18.5.61
7–54	Stern, IA	31.8.62
7–3	Stern, IA	1.2.61
11–6	Stern, IA	18.5.61
8.6–19	Stockhausen	5.3.70
19–11	Stockhausen	26.2.71
17–7	Stockhausen GmbH	4.10.86
7.2–48	Styled; Soc Resp Limit	8.12.82
8.4–17	Suehiro Seni Kogyo KK	27.12.66
25.5–52	Sumitomo Chem. Co Ltd	26.6.92
25.5–52	Sumitomo Chem. Co Ltd	7.4.92
11.1–42	Sumitomo Chem. Ind Co Ltd	22.8.63
7.1–20	Sumitomo Chemical	26.7.79
16.1–9	Sumitomo Electric Ind	4.11.64
16.1–8	Sumitomo Electric Ind	16.12.64
18.4–3	Sumitomo Seika Chem. Co Ltd	12.6.92
2.2–2	Sun Star Chem. Ind Co Ltd	22.7.68
18.2–62	Surface Coatings Inc	16.7.92
18.2–62	Surface Coatings Inc	11.1.93
8–31	Suzutora Seisen KK.	26.3.79
20.1–12	Svit	13.11.65
8.8–4	Svit	13.7.66
8.2–10	Svit, NP	14.7.65
18.1–15	SVUK	6.5.68
20.1–26	SVUK	16.8.68
7.2–41	Taira Okuda	5.7.67
12–18	Takasago Toryo Co Ltd	11.6.62
16.2–19	Takase Co	20.8.70

Table 31-1 (continued)

Section	Company	Date
25.5–26	Takeda Badische Urethane Ind Ltd	5.7.93
25.4–8	Tanaka, J	6.9.71
22–50	Taoka Kagaku Kogyo KK	21.1.94
20–17	Tatsuyama, H	3.4.63
7.1–6	Teijin	6.4.63
16.2–27	Teijin	4.7.63
9.1–14	Teijin	27.11.63
8.6–33	Teijin	21.8.64
25.3–5	Teijin	2.11.64
9.1–16	Teijin	24.2.65
9.3–8	Teijin	7.11.67
9.3–16	Teijin	14.10.68
9.3–17	Teijin	31.1.69
9.3–11	Teijin	31.1.69
9.3–18	Teijin	2.4.69
18.2–51	Teijin	24.4.69
9.3–12	Teijin	25.12.69
19–4	Teijin	9.7.70
8.4–37	Teijin	22.7.70
9.3–25	Teijin	17.4.73
18.2–73	Teijin	23.3.77
8.2–72	Teijin	8.5.78
22–64	Teijin	8.5.78
22–60	Teijin	8.5.78
22–77	Teijin	2.6.78
22–77	Teijin	8.5.78
22–62	Teijin	2.6.78
18.2–72	Teijin	7.2.80
18.2–69	Teijin	7.2.80
8.2–86	Teijin	14.5.80
19–24	Teijin	13.6.80
9.2–24	Teijin	31.3.83
9.3–35	Teijin	4.7.83
9.3–36	Teijin	18.7.83
25.5–11	Teijin	9.8.85
25.5–10	Teijin	6.3.86
25.5–10	Teijin	20.6.89
21.1–28	Teijin	18.6.92
21.1–28	Teijin	9.11.92
25.5–16	Teijin	19.8.92
9.2–26	Teijin	10.3.93
19–28	Teijin	2.11.93
19–38	Teijin	7.9.95
7.1–12	Teijin Cordley Ltd	13.7.73
18.2–52	Teijin Cordley Ltd	3.10.74
18.2–52	Teijin Cordley Ltd	28.9.84
17.1–22	Teijin Cordley Ltd	24.12.91
8.6–47	Teijin Cordley Ltd	13.2.93
8.6–47	Teijin Cordley Ltd	16.2.93

Table 31-1 (continued)

Section	Company	Date
8.1–20	Teijin Cordley Ltd	10.3.93
19–26	Teijin Cordley Ltd	10.3.93
20–7	Teikoku Kasai Kogyo Co Ltd	10.2.61
8.2–31	Tenneco	6.6.68
11.2–24	Tenneco	22.1.69
11–4	Tenneco	6.3.72
19–17	Texon Inc	8.2.73
14–13	Textilausrüstungsgesellschaft Schroers & Co	10.3.72
10–21	Textron Inc	1.9.67
20.2–4	Th. Böhme KG	7.8.70
7–19	The Chloride Electrical Storage Comp	6.7.53
7–19	The Chloride Electrical Storage Comp	7.7.53
25.5–27	The Polymer Technology Group	27.6.95
18.2–11	Thiokol Chem. Corp	18.5.64
18.2–53	Thoratec Lab Corp	21.5.84
12–24	Thorne	10.5.65
13–12	Tiefenbacher & Co	2.3.68
12–2	Tiefenbacher & Co	2.3.68
23.1–20	Toa Gosei Chem Ind Ltd	20.12.91
19–42	Toa Nenryo Kogyo KK	1.11.79
22–56	Toda Kogyo Group Corp	22.11.91
22–28	Tokai Plastics Kogyo KK	19.2.71
16–19	Tokuyama Soda KK	30.7.87
16–19	Tokuyama Soda KK	12.6.86
11.2–5	Tokyo Toyo Rubber	5.9.63
18.1–26	Tomikawa, A	19.2.63
18.2–27	Tonen Kagaku KK	18.8.94
22–41	Toppan Printing KK	10.9.73
16.2–30	Toppan Printing KK	25.12.84
13–15	Toppan Printing KK	25.12.84
8.2–19	Toray	21.11.66
8.6–34	Toray	10.8.67
8.6–34	Toray	17.4.68
11.2–34	Toray	29.8.67
21.1–1	Toray	10.1.68
8.6–39	Toray	20.1.68
8.6–32	Toray	29.3.68
8.4–8	Toray	17.4.68
21.1–20	Toray	26.4.68
8.5–17	Toray	3.6.68
8.6–35	Toray	10.7.68
8.5–19	Toray	9.10.68
2.2–13	Toray	4.11.68
8.3–10	Toray	12.12.68
8.2–42	Toray	29.1.69
15–14	Toray	25.2.69
8.3–16	Toray	19.3.69
8.3–9	Toray	22.5.69
8.6–36	Toray	31.5.69
7.1–7	Toray	31.5.69

Table 31-1 (continued)

Section	Company	Date
8.1–6	Toray	31.5.69
8.3–15	Toray	12.6.69
8.1–5	Toray	3.7.69
8.6–41	Toray	14.7.69
8.5–18	Toray	6.8.69
8.1–8	Toray	21.8.69
8.2–33	Toray	27.11.69
8.6–28	Toray	27.11.69
21.1–23	Toray	29.1.70
21.1–4	Toray	2.2.70
8.6–43	Toray	5.2.70
8.6–44	Toray	21.2.70
19–6	Toray	21.2.70
7–66	Toray	26.2.70
21.1–24	Toray	28.2.70
8.2–93	Toray	13.3.70
12–28	Toray	25.6.70
21.1–15	Toray	15.7.70
8.6–31	Toray	14.8.70
19–15	Toray	7.11.70
8.2–21	Toray	19.2.71
8.2–44	Toray	24.10.72
8.3–37	Toray	22.12.72
8.3–37	Toray	22.2.73
8.5–47	Toray	11.9.74
19–37	Toray	13.2.78
25.3–9	Toray	13.2.78
8.5–28	Toray	6.12.78
8.6–59	Toray	11.12.79
8.6–49	Toray	11.12.79
18.1–19	Toray	7.5.80
19–25	Toray	9.7.80
8.5–28	Toray	6.12.80
25.5–68	Toray	30.12.80
18.2–54	Toray	23.3.81
18.2–54	Toray	26.2.85
19–43	Toray	15.4.81
23.1–25	Toray	22.2.82
21–95	Toray	31.3.82
21–95	Toray	6.5.82
14–14	Toray	19.11.82
18.2–39	Toray	16.6.84
7.1–24	Toray	23.7.84
7.1–24	Toray	25.7.84
16.2–31	Toray	7.1.85
21–96	Toray	21.6.85
8.6–48	Toray	11.10.85
8.6–48	Tray I	4.5.86
8.6–48	Toray	19.5.86

Table 31-1 (continued)

Section	Company	Date
18.3–40	Toray	31.10.85
18.3–40	Toray	4.3.86
18.3–40	Toray	19.5.86
8.2–67	Toray	21.5.93
18.4–10	Toray	24.5.96
23.1–2	Toray; Daikyo Chem; Daiichi Lace KK	30.6.93
23.1–21	Toray	14.12.94
25.5–50	Toray	20.10.95
13–19	Toray	28.12.95
18.2–75	Toray Industries Inc. & Daiichi Lace Mfg Co Ltd	14.2.96
10.3–6	Toyo	21.5.64
7.2–39	Toyo Cloth	28.6.60
7.2–14	Toyo Cloth	28.6.60
7.2–16	Toyo Cloth	1.2.62
8.3–21	Toyo Cloth	17.7.62
8.3–21	Toyo Cloth	23.10.6
8.3–21	Toyo Cloth	18.5.63
7.2–15	Toyo Cloth	17.7.62
7.2–15	Toyo Cloth	23.10.62
7.2–15	Toyo Cloth	18.5.63
7.2–4	Toyo Cloth	16.10.62
7.2–4	Toyo Rayon	16.10.62
7.2–4	Toyo Cloth	16.10.62
17.1–7	Toyo Cloth	2.11.62
8.6–52	Toyo Cloth	5.11.63
8.8–2	Toyo Cloth	13.11.63
8.3–14	Toyo Cloth	2.7.68
18.3–18	Toyo Cloth	27.9.68
8.4–27	Toyo Cloth	23.10.72
11.1–57	Toyo Cloth	21.6.73
11.1–61	Toyo Cloth	22.6.73
11.1–69	Toyo Cloth	17.2.76
22–67	Toyo Cloth	20.4.77
19–39	Toyo Cloth	13.6.77
25.2–9	Toyo Cloth	20.9.77
7–61	Toyo Cloth	9.10.80
7–69	Toyo Cloth	24.1.83
18.3–26	Toyo Cloth	22.2.83
18.4–7	Toyo Cloth	25.10.90
16.1–14	Toyo Cloth	2.9.93
7.2–7	Toyo Cloth	4.9.62
8.2–89	Toyo Cloth	7.3.81
11.2–6	Toyo Gomu Kagaku Kogyo KK	30.12.65
8.3–22	Toyo Gomu KK	21.3.64
13–9	Toyo Gomu Kogyo KK	27.12.63
7.2–5	Toyo Gomu KK	21.3.64
19–40	Toyo Ink Mfg Co	10.6.77
18–8	Toyo Ink Mfg Co.	3.10.94
8.9–12	Toyo Prod KK	27.1.78

Table 31-1 (continued)

Section	Company	Date
21.1–3	Toyo Rayon	19.3.66
8.4–4	Toyo Rayon	13.10,67
21.1–21	Toyo Rayon Co Ltd	5.11.68
8.2–47	Toyo Rayon KK	19.9.66
8.2–47	Toyo Rayon KK	4.3.67
11.1–51	Toyo Rubber Chem.	26.5.70
8.3–24	Toyo Rubber Chem. Ind	5.12.67
11.2–31	Toyo Rubber Chem. Ind Co Ltd	6.5.68
11.2–21	Toyo Rubber Co Ltd.	8.8.63
11.2–21	Toyo Rubber Co Ltd.	17.10.63
8.2–15	Toyo Rubber Ind	18.2.63
8.2–13	Toyo Rubber Ind	11.11.63
8.6–10	Toyo Rubber Ind	18.12.63
8.2–14	Toyo Rubber Ind	27.12.63
8.6–1	Toyo Rubber Ind	13.5.64
8.2–43	Toyo Rubber Ind	6.7.65
8.2–46	Toyo Rubber Ind	10.1.66
8.2–7	Toyo Rubber Ind	20.5.66
8.2–45	Toyo Rubber Ind	22.7.66
8.4–19	Toyo Rubber Ind	5.8.66
8.3–23	Toyo Rubber Ind	12.11.66
8.2–48	Toyo Rubber Ind	18.7.67
8.3–11	Toyo Rubber Ind	9.8.67
8.3–13	Toyo Rubber Ind	10.8.67
8.4–20	Toyo Rubber Ind	26.1.68
10.6–4	Toyo Rubber Ind	2.8.61
2.2–16	Toyo Rubber Ind	27.12.63
9.1–12	Toyo Rubber Ind	13.7.64
8.2–60	Toyo Rubber Ind	5.8.66
25.4–14	Toyo Rubber Ind	25.8.67
18.2–55	Toyo Rubber Ind	19.6.82
18.2–56	Toyo Rubber Ind	22.6.84
18.2–56	Toyo Rubber Ind	21.5.86
9.3–33	Toyo Rubber Ind	22.5.70
10.3–2	Toyo Rubber Kogyo	11.2.63
8.2–22	Toyo Rubber Kogyo	15.3.65
7.2–6	Toyo Rubber	6.7.65
7–52	Toyo Spinning	14.1.61
10–8	Toyo Spinning	28.8.62
7.1–13	Toyo Spinning	29.10.62
8.4–13	Toyo Spinning	22.10.64
25.2–2	Toyo Spinning	3.4.65
8.2–50	Toyo Spinning	9.4.65
16.2–15	Toyo Spinning	28.10.66
16.2–11	Toyo Spinning	8.10.68
16.1–6	Toyo Spinning	8.10.68
16.1–1	Toyo Spinning	17.7.69
25.4–11	Toyo Spinning	19.11.70
7–30	Toyo Spinning	2.9.63

Table 31-1 (continued)

Section	Company	Date
8.3–28	Toyobo KK	27.3.78
7.2–17	Toyota Central Res and Dev	15.2.72
7.2–37	Toyota Chem. Res Dev Lab Inc	6.12.71
8–19	Toyota Jidosha KK	12.12.91
20–19	Tray Textiles Inc Kohoku Chem. Ind Co Ltd	7.5.80
18.1–37	Tschech Akademie	5.11.71
8.2–23	Tsunoda Kagaku	7.12.72
25.4–7	Tsunoda Kagaku	7.12.72
11.1–63	Tsuyaei Kogyo Co KK	22.9.73
20.2–5	UCB	21.9.70
10–27	Ugine Kuhlmann	5.4.72
10.3–13	Ugine-Kuhlmann	30.10.69
7–46	Uhde GmbH	22.10.71
11.2–17	Union Carbide	26.7.63
11.2–58	Union Carbide	13.12.63
11.1–8	Union Carbide	13.12.63
18.1–32	Union Carbide	20.11.64
7–35	Union Carbide	27.6.69
10.2–14	Union Carbide	18.6.74;
16.1–5	Uniroyal	21.6.63
16.1–5	Uniroyal	3.9.65
9.1–7	Uniroyal	31.5.66
9.1–6	Uniroyal	31.5.66
9.1–9	Uniroyal	18.5.67
16.1–7	Uniroyal	8.11.68
11–12	Uniroyal	26.2.69
19–12	United Merchants and Manufacturers	9.10.61
20.1–1	United Shoe Mach Corp	14.10.59
20.1–2	United Shoe Mach Corp	21.11.61
20.1–5	United Shoe Mach Corp	3.5.60
20.1–5	United Shoe Mach Corp	27.3.61
7.2–31	Unitika	27.9.71
7.2–45	Unitika	22.6.73
11.1–59	Unitika	13.7.73
19–41	Unitika	17.4.78
22–78	Unitika	13.2.79
18.3–41	Unitika	23.1.84
18.3–41	Unitika	17.2.84
18.3–41	Unitika	31.8.84
18.3–41	Unitika	7.9.84
18.3–41	Unitika1	3.9.84
18.2–18	Unitika	26.1.84
18.3–23	Unitika	13.2.84
18.3–45	Unitika	12.6.84
18.3–45	Unitika	25.12.86
8.4–40	Unitika	21.1.87
7.1–23	Unitika	11.5.87

Table 31-1 (continued)

Section	Company	Date
18.3–42	Unitika	27.10.87
18.3–42	Unitika	29.2.88
18.3–43	Unitika	18.7.88
23.1–1	Unitika	22.3.93
22–55	Unitika	22.4.93
8.2–61	Unitika	10.8.93
9.3–27	Unitika	22.11.93
8.2–65	Unitika	15.12.93
8.1–22	Unitika	31.1.94
8.1–22	Unitika	31.1.94
25.5–42	Unitika	6.12.94
8.2–94	Unitika	13.10.95
25.5–33	Unitika	16.1.96
21.1–16	Unitika	29.2.96
10.4–8	Universiteit Twente	8.2.93
14–1	US Rubber	10.9.54;
7–26	US Rubber	4.6.58
7–19	US Rubber	4.6.58
18.1–4	US Rubber	10.7.59
11.2–13	US Rubber	8.3.60
10.3–7	US Rubber	20.4.62
16.1–4	US Rubber	11.6.63
10–20	US Rubber	6.8.63
20–16	US Secr of the Army	8.11.62
20.1–7	USM	21.12.55
15–6	USM	7.8.58
15–6	USM	1.3.61
20.1–21	USM	25.9.61
20.1–15	USM	31.1.62
20.1–	USM	14.7.64
9.3–23	USM	18.12.64
9.3–14	USM	18.12.64
10.1–2	USM	21.10.65
20.1–30	USM	2.8.66
10.1–7	USM	17.4.67
10.1–8	USM	19.5.67
10.1–9	USM	14.11.67
10.1–11	USM	17.5.68
10.1–13	USM	16.10.69
10.1–12	USM	12.12.69
10.1–15	USM	8.4.70
10.1–16	USM	16.6.72
10.1–16	USM	7.8.72
10.1–16	USM	9.5.73
10.1–17	USM	21.6.74
10.4–3	VEB Chemiefaserkombinat Schwarza	3.11.72
18.1–38	VEB Chemiefaserkombinat Schwarza	10.7.74
9.1–4	VEB Kunstblume Sebnitz	27.12.63

Table 31-1 (continued)

Section	Company	Date
19–35	VEB Vogtländische Kunstlederfabrik	26.2.85
11.1–68	VEB Vogtländische Kunstlederfabrik Tannenbergsthal	17.6.87
7–60	VEB Vowetex	28.11.83
21–91	VEB Vowetex Plauen	12.3.86
21–9	Vepa AG	6.2.69
21–8	Vepa AG	31.7.69
21–11	Vepa AG	5.10.70
11.1–15	Vogtländ.Kunstlederfabrik	16.9.65
23.1–11	Volg Poly	9.1.92
14–2	Votteler's Nachf GmbH	10.8.54
12–37	Vsecjusnyj nautschno-issledovatelskij; Institut sintetischeskich smol VI-25	21.12.76
20.2–19	Vyzkumny uytav kozedelny	29.8.75
11–10	WL Gore & Ass Ind	6.4.84
7–14	Walkerlan Ltd	11.4.62
18–14	Westo GmbH	4.1.64
18–14	Westo GmbH	21.10.64
10–22	Wharton Ind Inc	9.5.68
20.2–3	Wharton Ind Inc	9.5.68
20.2–3	Wharton Ind Inc	10.4.69
10–5	Will, G	30.7.58
10–7	Will, G	21.3.61
10.4–13	Will, G	2.11.61
10.4–13	Will, G	18.11.65
10.4–13	Will, G	2.1.65
10.4–13	Will, G	6.6.66
10.4–13	Will, G	14.11.63
10.4–12	Will, G	11.1.66
10.5–7	Will, G	12.11.68
10–4	Will, G	4.5.62
18.2–1	Wolff Walsrode	19.11.93
8.9–13	WTZ des VEB Kombinats Wolle und Seide	12.12.79
20–12	WTZ Techn.Text	6.2.67
4–5	Wünschmann, M	30.10.13
7.1–4	Wünschmann, M	30.10.13
25.5–7	Wyandotte	2.3.64
25.5–8	Wyandotte	27. 4.64
12–3	Wyandotte	27.4.64
10–2	Wyandotte Chem. Corp	27.4.64
8.2–88	Yamada, T	30.7.87
14–12	Yamanashi Kasei Kogyo Co	1.8.70
23.2–8	Yamazen Sangyo KK	8.6.91
21–55	Yanagimachi, S	14.3.64
12–10	Yoshihige, M	18.10.73
22–68	Yoshimura Abura	12.11.76
9.2–9	Yuasa Battery	22.7.63
9.2–16	Yuasa Battery	25.3.66
9.2–10	Yuasa Battery	29.3.66
9.1–17	Yuasa Battery	30.6.66

Table 31-1 (continued)

Section	Company	Date
9.2–21	Yuasa Battery	29.12.70
7.2–42	Zaionckovskii, AD	2.2.61
18.1–35	Zaionckovskij, AD	25.4.61
18–15	Zaionckovskij. AD	25.4.61
18.1–12	Zantsev, VK	15.1.92
21–23	Zwoboda, O	8.3.60
21–22	Zwoboda, O	14.3.60
11–5	Zwoboda, O	14.10.60
21–36	Zwoboda, O	15.2.62
21–83	Zwoboda, O	14.7.62
18.1–23	Zwoboda; O	14.6.60

Bayer = Farbenfabriken Bayer
British United Shoe Machinery = USM
Dai Nippon Ink = DIC = Dynic
Enka-Glanzstoff = AKU = AKZO = Allgemeen Kunstzijdenspinnerij
Francolor = Ugine Kuhlmann
Hoechst = Farbwerke Hoechst
Kurashiki Rayon = Kuraray
Porvair = Porous Plastics
PPG = Pittsburgh Plate and Glass Co
SVUK = Vyzkumny ustav kozedelny
Toyo Rayon = Toray
Vepa=Sandoz.

CHAPTER 32

Some Practical Examples

In the examples given in Table 32-1, formulations are given which can be used for application directly onto a textile (direct coating) or in one or several layers onto a releasing agent such as a release paper (indirect process) and finally the textile is laminated into the polyurethane mass. The products selected are products of the LS or SP division of Bayer AG, D 51368 Leverkusen, Germany.

Example of a Coagulation (courtesy of H. Mergard). A polyurethane is prepared from polybutylene glycol adipate, MDI and butane-1,4-diol. The polyurethane should have a Shore A hardness of around 85. The polyurethane is dissolved in dimethylformamide in a concentration of 20%. Then 1% silica and 10% of ground cellulose fibers are stirred into this solution. After 48 h storage at room temperature the solution is diluted with DMF to a concentration of the PU of 10%. The diluted solution is applied onto a napped and sheared polyester cotton fabric with a doctor's knife in a thickness of 1 mm. Then the fabric is dipped into a DMF/water mixture (75:25). The fabric is then washed for 3 h in fresh water, squeezed between rollers to remove part of the water included and dried at 80°C. The resulting microporous coated fabric has a water vapor permeabilty of ca. 8–10 mg/hcm^2.

Table 32-1. Practical Examples (by courtesy of W. Baelz)

	Article	Release paper	Substrate	Parts	Quantity of polyurethane (solid) g m⁻²	Temperature
One component direct coating	Raincoat	®Stripcoat VNS glatt, matt (S.D. Warren)	Knitted polyamide		30 g/m²	90-120-140 °C
1st coat						
Impranil ENB-03 solution				1000		
Desmoderm matting agent				100		
Xeroderm L70-01 dissolved in toluene				5		
Bayderm white CB-TO				45		
Bayderm red B-TO				79		
Bayderm red-violet B-TO				15		
				6		
2nd coat						
Impranil EWN 13 solution N Toluene				1000 100		
Coating with polyurethane dispersions	Shoe upper	®Highlight 801 Pearl (Wiggins Teape)			25 g/m²	80-100-150 °C
1st coat (top coat)						
Impranil DLV dispersion				600		
Impranil DLV dispersion				400		
EUDERM darkbrown C				113		
EUDERM caramel D-C				30		
EUDERM bordo D-C				8		
Rheolate 205 (5% in water)				60		
2nd coat (adhesive coat)					25 g/m²	80-100-150 °C
Impranil DLV dispersion				600		
Imprafix CIN solution				15		
EUDERM darkbrown C				75		
EUDERM caramel D-C				20		
EUDERM bordo D-C				5		
Rheolate 205 (5% in water)				60		

Table 32-1 (continued)

Article	Release paper	Substrate	Parts	Quantity of polyurethane (solid) g m^{-2}	Temperature
Two component coating in a direct working process[a]					
Anorak	Direct coating	Polyamide			
1st coat (base coat)				20 g/m²	100-120-120°C
Impranil C solution			1000		
Impranil TH solution			50		
Imprafix BE solution			50		
2nd coat				25 g/m²	100-120-120°C
Impranil C solution			100		
Desmoderm matting agent MC			80		
Imprafix TRL solution			80		
Imprafix BE solution			50		
3rd + 4th coat				20 g/m²	90-120-150°C
Impranil ELH solution			1000		
Desmoderm matting agent MC			100		
Xeroderm L 70-01 dissolved in toluene			10		
			90		
High solid system	Shoe upper	®Transkote Patent OA (S.D. Warren) woven fabric bonded by coagulation			
1st coat				10 g/m²	70-90-110°C
Desoderm Finish HX			250		
Isopropanol			375		
Toluene			375		
2nd coat				180 g/m²	160-160-160°C
Impranil HS-90			1000		
Impranil HS-C			90		
Levacast fluid SN			10		
Ruß Elftex 415 (Degussa)			40		

Table 32-1 (continued)

Article	Release paper	Substrate	Parts	Quantity of polyurethane (solid) g m^{-2}	Temperature
3rd coat Impranil EWN 13 solution Imprafix TH solution Preventol AB dissolved methyethyl ketone			1000 15 5 25		
36 g/m^2	80-120-140 °C				
Water vapor permeable coating with a hydrophylic polyurethane[b,c]		Polyamide ca. 80 g/m^{-2}			
Raincoat	Direct coating				
1st and 2nd coat Impraperm LH-03 solution Isopropanol Toluene Imprafix SV solution Imprafix SK solution			1000 75 75 50 7	15 g/m^2	70-90-90-110 °C
3rd coat Impraperm AD-01 solution Impraperm AD-03 solution Desmoderm matting agent MC Xeroderm L 70-01 dissolved in toluene			400 600 120 10 90	20 g/m^2	90-130-150 °C

[a] Pretreatment – 20 g l^{-1} Baygard AFF, drying max. 100 °C.
[b] Pretreatment – 8 g l^{-1} Baygard AFF, drying max. 100 °C.
[c] Treatment of finished material – 30 g l^{-1} Baygard AFF, 30 g l^{-1} Isobutanol, 30 g l^{-1} Isopropanol; drying and condensation 120/160 °C.

Subject Index